D1274600

The Life of the
Honourable Robert Boyle F.R.S.

Plate I. Robert Boyle, *aet.* 37. *Drawing by William Faithorne.*

The Life of the Honourable ROBERT BOYLE F.R.S.

R. E. W. MADDISON Ph.D.
Librarian, Royal Astronomical Society

TAYLOR & FRANCIS LTD.
LONDON
1969

Published 1969 by Taylor & Francis Ltd., Cannon House, Macklin Street, London, W.C.2.

Printed and bound in Great Britain by Taylor & Francis Ltd., Cannon House Macklin Street, London, W.C.2.

© 1969 R. E. W. Maddison

All rights Reserved. No part of this publication may be reproduced, stored in a retrieval system, or transmitted in any form or by any means, electronic, mechanical, by photocopying, recording or otherwise, without the prior permission of the Copyright owner.

SBN 85066 028 9

Distributed in the United States of America and its territories by Barnes & Noble Inc. New York.

85361

To

Adélaïde

85361

Preface

Thomas Birch published *The Life Of the Honourable Robert Boyle* in 1744.* On the fly-leaf of his own copy, which is in the British Museum Library, he has written:

> 'Tho. Birch
> I began this Life of Mr.
> Boyle June 9th 1743, & finish'd
> it on wednesday 27th July following.'

This extraordinary achievement would have been impossible had he not had the valuable assistance of the Reverend Henry Miles,† who carried out all the laborious preparatory work for the collected edition of Boyle's works, which appeared under the name of Thomas Birch as editor. Miles himself very modestly wrote to Birch: " I am taking all the pains I can to furnish you with Stones and Timber, you must be the architect, & I shall think my Self happy if I can ease you of the trouble of collecting materials, in any measure, knowing you will have work eno' on y[r] hands to form them meet for the Edifice, and to put them toge[r],"‡ In his preface, Birch said " Some very considerable additions are made in this edition, which were never before published . . . These additions the public owes to the reverend and learned Mr. *Henry Miles* of *Tooting* in *Surrey*, and F.R.S. who is possessed of those manuscripts of Mr. *Boyle*, which were put into his hands, with leave to make use of them for the public good, by the late worthy Mr. *Thomas Smith*, apothecary in the *Strand*, who lived seventeen years with Mr. *Boyle*, and was with him at his death; and which have been lately increased by a part of the original collection, that had been communicated to Dr. *William Wotton* . . . and were restored by his

---

* 8vo. London. The *Life* is prefixed to *The Works of the Honourable Robert Boyle*. 5 vols. Folio, London, 1744; 6 vols. 4to, London, 1772.

† M. B. Hall. 'Henry Miles, F.R.S. (1698–1763) and Thomas Birch, F.R.S. (1705–66).' *N & R.* (1963) **18,** 39.

‡ B.M., Add. MSS. 4314, fol. 64. Letter dated 16 August 1742.

son in law the reverend *William Clarke* . . . I must not conclude this preface without returning my sincerest acknowledgments to Mr. *Miles* abovementioned, to whose great labour, judgment, and sagacity, the conduct and improvements of this edition are chiefly to be ascribed . . ." Nevertheless, the extent of the contribution made by Miles is not apparent except to those who have worked through the mass of surviving Boyle papers.

The *Life* written by Birch has served as the basis of all subsequent accounts of Robert Boyle, wherever they have been published, right down to the present day.* In 1878 Thomas Henry Huxley tried in vain to persuade John Tyndall to write a book of about 200 pages on Robert Boyle and the Royal Society (including Robert Hooke), even though he was willing to wait a couple of years for Tyndall to dispose of his outstanding commitments and to plod through the extensive writings of Boyle and Hooke.† The account of Robert Boyle's life by the present writer is not intended as an entire replacement of the one by Birch, which will always retain its value and utility; but, apart from correcting some minor errors, and modifying some of his statements, which with the passage of time seem to have been wrongly interpreted, it does deal with various aspects of the subject that Birch does not mention. No attempt is made to discuss in detail the content of Boyle's numerous publications, nor the importance of his position, nor the influence of his work on contemporaries and posterity. The sole purpose of the present work is to describe, the author hopes accurately, the events of Robert Boyle's life so as to supplement Birch, and to provide a guide through the manifold aspects of his life and work.

The present work is based on a life chronology of Robert Boyle, compilation of which was started by the author over twenty-five years ago. Although articles by him published in various journals have been freely drawn upon for this work, it is nonetheless more than a mere compilation, for fresh material has been incorporated and the relevant literature has been utilized. A Leverhulme Research Fellowship, though awarded for another purpose, namely, the preparation of an annotated edition of Robert Boyle's correspondence, helped considerably the completion of the present work, and is gratefully acknowledged.

---

* R. E. W. Maddison. 'A Summary of Former Accounts of the Life and Work of Robert Boyle.' *Annals of Science* (1957. Published 1958) **13,** 90.

† W. R. Dawson. *The Huxley Papers* . . . London, 1946, p. 162. Items Nos. 179–183. Correspondence between Huxley and Tyndall, December 1878.

For permission to reproduce material from manuscripts or printed books and to use certain illustrations, grateful thanks are due to

The Right Hon. the Earl of Cork and Orrery
The Visitors of the Ashmolean Museum, Oxford
Bord Fáilte Éireann
The Trustees of the British Museum
The Trustees of the Chatsworth Settlement
The County Archivist, Essex Record Office
The Provost and Fellows of Eton College
The Curator, Museum of the History of Science, Oxford
The President and Council of the Royal Society, London.

For information and assistance in various ways thanks are due to

His Grace the Duke of Devonshire, P.C., M.C.
The Hon. Mrs. Reginald Boyle
Lt. Cdr. the Hon. John W. Boyle, D.S.C.
Sir Gavin de Beer, D.Sc., F.S.A., F.R.S.
Mr. I. Kaye, Librarian, The Royal Society, London, as well as his assistants, past and present, *viz.*, Messrs. R. K. Bluhm, N. H. Robinson, and L. P. Townsend.
Mr. F. R. Maddison, M.A., Curator, Museum of the History of Science, Oxford.
Mr. R. E. Maddison, M.A., A.C.I.S.
M. Michel Rochaix, Director, Station Fédérale d'Essais Agricoles, Lausanne.
Professor G. Nebbia, Istituto di Merceologia, Università degli Studi, Bari.
Mlle C. Santschi, L'archiviste d'État adjoint, Archives d'État, Genève.
Mr. P. Strong, Keeper of the Library and Collections, Eton College.
Mr. T. S. Wragg, M.B.E., T.D., Librarian, Chatsworth.

In conclusion I have to record my indebtedness to my publishers for the great interest they have taken in the production of this book; acknowledgement must also be made of the skill and patience of the compositors and production staff.

R. E. W. M.

Burlington House,
London, 1968.

Sources

Manuscripts

Royal Society

Boyle Letters in 7 volumes
Boyle Papers in 46 volumes
Boyle Notebooks. MSS. 185–200
Henry Oldenburg's Commonplace Book. MS.1
Correspondence. Early Letters

British Museum

Add MSS. 4228, 4229, 4314, 6210, 18023, 19832

Chatsworth

Earl of Cork's Letter Book, 24 April 1634–19 July 1641
Lismore MSS.

National Library of Ireland

Orrery Papers

Printed Books

Contemporary as well as later works are cited at the appropriate point in the text. Many others have been consulted too, but it is not necessary to give them special mention.

Abbreviations and Titles of cited works

Biogr. Brit.	*Biographia Britannica, the Lives of the most eminent persons who have flourished in Great Britain and Ireland.* 6 vols. London, 1747–1766.
B.L.	Royal Society. Boyle Letters.
B.M. Add. MSS.	British Museum. Additional Manuscripts.
B.M. Slo. MSS.	British Museum, Sloane Manuscripts.
Birch *History*	Thomas Birch. *The History of the Royal Society of London for Improving of Natural Knowledge.* 4 vols. London, 1756–1760.
Bittner	L. Bittner and L. Gross. *Repertorium der diplomatischen Vertreter aller Länder.* 2 vols. Berlin, 1936.
Bompiani *Autori*	*Dizionario Letterario Bompiani degli Autori di tutti i tempi e di tutte le letterature.* 3 vols. Milan, 1956–1957.
Bompiani *Opere*	*Dizionario Letterario Bompiani delle Opere e dei Personaggi di tutti i tempi e di tutte le letterature.* 12 vols. Milan, 1947–1950.
Boulton	R. Boulton. *The Theological Works of the Honourable Robert Boyle, Esq; Epitomiz'd.* 3 vols. London, 1715.
B.P.	Royal Society. Boyle Papers.
Calendar of Orrery Papers	*Calendar of Orrery Papers*, edited by Edward MacLysaght, Dublin, 1941.
C.J.	*Journals of the House of Commons.*
C.R.	*Comptes rendus hebdomadaires des Séances de l'Académie des Sciences.*
C.S.P. Colonial	*Calendar of State Papers: Colonial Series (America and West Indies).* London, 1860–.
C.S.P. Domestic	*Calendar of State Papers: Domestic Series.* London, 1856–.
C.S.P. Ireland	*Calendar of State Papers relating to Ireland.* London, 1860–.

D.N.B.	*Dictionary of National Biography.* 22 vols. London, 1908–1909.
Evelyn (Bray)	*Diary and Correspondence of John Evelyn F.R.S.* A new edition. Edited by William Bray. 4 vols. London, 1854.
Evelyn (de Beer)	*The Diary of John Evelyn.* Edited by E. S. de Beer. 6 vols. Oxford, 1955.
Firth & Rait	*Acts and Ordinances of the Interregnum, 1642–1660.* Edited by C. H. Firth and R. S. Rait. 3 vols. London, 1911.
Fulton	John Farquhar Fulton. *A Bibliography of the Honourable Robert Boyle.* Second edition. Oxford, 1961.
II Fulton	*Idem.* 'Second Addenda to a Bibliography of the Honourable Robert Boyle.' *Oxford Bibliographical Society Proceedings and Papers.* N.S. (1947). I. Fasc. 1.
G.E.C.	G.E.C(ockayne). *Complete Peerage.* New edition. 13 vols. London, 1910–1953.
Gunther	Robert Theodore Gunther. *Early Science in Oxford.* 15 vols. Oxford, 1923–1967.
Hooke *Diary*	*The Diary of Robert Hooke, M.A., M.D., F.R.S., 1672–1680.* Edited by Henry W. Robinson and Walter Adams. London, 1935.
Hutchins	John Hutchins. *The History and Antiquities of the County of Dorset.* 4 vols., 3rd edition, London, 1861–1873.
Huygens	*Oeuvres Complètes de Christiaan Huygens* publiées par la Société Hollandaise des Sciences. 23 vols. The Hague, 1888–1950.
I Lismore	*The Lismore Papers.* Edited by the Rev. A. B. Grosart. (First Series.) 5 vols. Privately printed, 1886.
II Lismore	*Ibid.* (Second Series.) 5 vols. Privately printed, 1887–1888.
L.J.	*Journals of the House of Lords.*
Madan	Falconer Madan *Oxford Books.* 3 vols. Oxford, 1895–1931.
Masson	Flora Masson. *Robert Boyle.* London, 1914.
New Peerage	*The New Peerage; or, Ancient and Present State of the Nobility of England, Scotland, and Ireland.* Second edition. 3 vols. London, 1778.
N & Q.	*Notes and Queries.*

Newton. *Correspondence.*	*The Correspondence of Isaac Newton.* Edited by H. W. Turnbull (Vols. I–III); by J. F. Scott (Vol. IV–). Cambridge, 1959– .
N & R.	*Notes and Records of The Royal Society of London.*
N.S. (For dates.)	New Style. Gregorian Calendar.
N.S. (For periodicals.)	New series.
N.Y.	No year.
Orrery Papers	The Countess of Cork and Orrery. *The Orrery Papers.* 2 vols. London, 1903.
O.E.D.	*Oxford English Dictionary.*
O.S.	Old Style. Julian Calendar.
RB.	Used in footnotes to designate Robert Boyle.
R.C.H.M.	*Royal Commission on Historical Manuscripts.*
Townshend	Dorothea Townshend. *The Life and Letters of the Great Earl of Cork.* London, 1904.
Turnbull, *H.D. & C.*	George H. Turnbull. *Hartlib, Dury and Comenius.* London, 1947.
Venn	J. Venn and J. A. Venn. *Alumni Cantabrigienses.* 4 vols. Cambridge, 1922–1927.
Verney	F. P. Verney (Vols. I & II); M. M. Verney (Vols. III & IV). *Memoirs of the Verney Family.* London, 1892–1899.
Works Fol.	*The Works of the Honourable Robert Boyle. To which is prefixed The Life of the Author.* Edited by Thomas Birch. 5 vols. London, 1744.
Works	*Ibid.* 6 vols. London, 1772. [Although described on the title-page as 'A New Edition', it is a reprint of the folio edition of 1744.]
Worthington	*The Diary and Correspondence of Dr. John Worthington.* Edited by James Crossley. 2 vols. Chetham Society. (1848) **13**; (1855) **36**; (1886) **114**.

Illustrations

Plates II to XL follow page 254

The coat of arms on the front cover is that of the Boyle family: *Party per bend embattled, argent and gules*

I

Childhood, Eton, the Grand Tour

(1627–1644)

Robert Boyle's autobiographical sketch, which covers most of this period, has the title ' An Account of Philaretus during his Minority '.* The exact date when it was written is not known, though the Rev. Mr. Miles surmised that the date was soon after Boyle's return from his continental travels. Evidence that it was written during his Stalbridge period (1645–1655) is provided by the ' Account ' itself, in which he mentions the burning of his youthful poetical effusions " the day he came of age ", and says that of his brothers and sisters " five women and four men do yet survive ". The ' Account ' was therefore written between 25 January 164^{7}_{8}, when Boyle came of age, and the early part of 165^{5}_{6}, when his sister, Joan, died. The date may perhaps be even more closely determined. He says, " from that Howre to this; Agues & he haue still been perfect stranger: " but he had an attack in July 1649. We conclude, then that the ' Account ' was written after January 1648 and before July 1649. It is appropriate to reproduce this ' Account ' in Boyle's own words, and to supplement them by comments, explanations, or elaborations, where necessary. His frequent inclusion of alternative words pending a final choice is noteworthy.

The surviving pages of the ' Account ' end with Boyle's arrival at Marseilles when on the way to Geneva for his second stay in that town. If the ' Account ' were indeed ever completed, then the concluding pages have long since disappeared, for they were not available even to Birch in the early part of the eighteenth century; so, the continuation of the story of Robert's minority has to be put together from other sources.

---

* The original manuscript is in the archives of the Royal Society (B.P. **XXXVII**). The version of the ' Account ' given here is reproduced therefrom, and includes some slight omissions from the version printed by Birch in his *Life of the Honourable Robert Boyle* (***Works Fol.* I; *Works.* I.**).

AN ACCOUNT OF PHILARETUS DURING HIS MINORITY*

[WRITTEN BY ROBERT BOYLE]

Not needlesly to confound the Herald with the Historian, & begin a Relation by a Pedigree†, I shall content my selfe to informe yow, that the immediate Parents of our Philaretus were, of the Female Sex, [a]the Lady Catherine Fenton[a],‡ (a Woman that wanted not Buty, & was rich in Vertue;) & on the Father's side, that Richard Boyle Earle of Corke, who by God's blessing on his prosperous Industry, vpon[b] very inconsiderable Beginnings built so plentifull & so eminent a Fortune, § that his prosperity has found many Admirers, but fewe Parallels.

He was borne‖ the 14 Child of his Father (of which 5 women & 4

[a–a] Partly conjectured; paper worn away. [b] Interlined above *from* deleted.

---

* Philaretus was a name frequently used by RB to indicate himself.

† In spite of what RB says it is nevertheless desirable at the outset to provide some account of the Boyle family in order that the relationship between its numerous members during the seventeenth century may be clear. For our purpose it is sufficient to start with Roger Boyle who belonged to that branch of the family, which had been settled in Herefordshire for several generations. See Appendix III with its accompanying notes.

‡ Catherine Fenton was the Earl of Cork's second wife. Further ramifications of the family through this marriage are shown below.

Robert WESTON, Lord Chancellor of Ireland
d. 1573
|
Sir Geoffrey FENTON(2) = Alice WESTON(1) = Hugh BRADY, Bishop of Meath
d. 1608 | *d.* 1631 *d.* 1584
|
Sir William FENTON   Catherine FENTON = Richard BOYLE, Earl of Cork
d. 1630

(*D.N.B: I. Lismore.* **II**, pp. 102–104.)

§ Consult Terence O. Ranger. 'Richard Boyle and the making of an Irish Fortune, 1588–1614.' *Irish Historical Studies*, (1957), **10**, 257.

‖ Robert Boyle was born 25 January 162$\frac{6}{7}$, and the Earl of Cork made the following entry in his diary:

> " My wife, god ever be praized, was about 3 of the clock in thafternoon of this day, the Sign in *Gemini*, libra, Safely delivered of her seaventh son at Lismoor: god bless him. for his name is Robert Boyle."

On the 8th of February following he made the further entry:

> " The viijth my seaventh son was Xtned in my chapple of Lismoor by my chapleyn and kinseman Mr Robert Nayler: and named Robert: His godfathers the Lo Digby Barron of Geshill, and Sir Frances Slingsby, his godmother the countess of castlehaven. The god of heaven make him happy, & bless him with a long lyffe & vertuous: & mak him blessed in having many good & Relegeows children." (*I. Lismore.* **II**, pp. 112, 207).

RB has stated the date of his birth in accordance with Church Style.

men do yet suruiue*) in the Yeare 1626 vpon St. Paul's Conversion-day, being the 25th of January; at a Cuntry-house of his Father's cal'd Lismore,† then one of the noblest Seates & greatest Ornaments of the Prouince of Munster, in which it stood; but now so ruin'd by the sad Fate of warre, that it serues only for an Instance & a Lecture, of the Instability of that Happinesse, that is built upon the incertin[a] Possession of such [b]–fleeting Goods–[b], as it selfe was.

To be such Parent's Son, & not their eldest,[c] was a Happinesse that our Philaretus would mention with greate expressions of Gratitude; his birth so suiting his Inclinations & his Desseins, that had he been permitted an Election, his Choice would scarce haue alter'd God's Assignement. For as on the one side a Lower[d] Birth wud haue too much expoz'd him to the Inconveniences of a meane Discent; which are too notorious to need specifying; soe on the other side, to a Person whose Humor indisposes him to the distracting[e] Hurry of the World; the being borne Heire to a Greate Family, is but a Glittering kind of Slauery; whilst obliging him to a Publick & entangled[f] course of Life to support the Credit of his Family, & tying him from[g] satisfying his dearest Inclinations, it often forces him to build the Aduantages of his House[h] vpon the Ruines of his owne Contentment. A man of meane Extraction [i]–(Tho neuer so aduantag'd by Greate meritts)–[i] is seldome admitted to the Priuacy & the secrets[j] of greate ones promiscuously; & scarce dares pretend to it, for feare of being censur'd sawcy, or an Intruder. And titular Greatnesse is euer an impediment to the[k] Knowledge of many retir'd /[Blank]/ Truths, that cannot be attain'd without[l] Familiarity with meaner Persons, & such other Condiscentions, as fond opinion in greate men disapproues & makes Disgracefull. But now our Philaretus was borne in a Condition, that neither was high enuf to proue a Temptation to Lazinesse; nor low enuf to discourage him from aspiring.

And certinly to a Person that so much affected an Vniversall /[Blank]/ Knowledge, & arbitrary vicissitudes of Quiet & Employments; it

---

[a] Interlined. [b–b] Interlined above *things* deleted. [c] Followed by *our* deleted. [d] Followed by *Ex* deleted. [e] Interlined. [f] *—tangled* interlined above *—gaged* deleted. [g] Followed by *the* deleted. [h] Interlined above *Family* deleted. [i–i] In margin. [j] Altered from *secrecy*. [k] Followed by *at* deleted. [l] Followed by *an Int* deleted.

---

* The five surviving women were his sisters Alice, Joan, Katherine, Dorothy and Mary; the four men were his brothers Richard, Roger, Francis and himself. The 'Account' was therefore written after RB's return from the continent mid–1644 and before 11 March $165\frac{5}{6}$ when Joan died. See also note * on p. 21.

† See Samuel Lewis. *A Topographical Dictionary of Ireland.* 2 vols 4to London, 1842. Vol. **II**, p. 283.

The castle was attacked on various occasions during the troubles in Ireland, and suffered much damage. It descended to the Earls of Cork and Burlington, and eventually to the Duke of Devonshire in whose family possession it remains. It was restored in 1812.

could not be vnwelcome to be born to a Quality, that was a handsom stirrop to Preferrment without (being) an Obligation to court it: & which might at once, both protect his higher Pretensions from the Guilt /[Blank]/ of Ambition, & (secure) his Retirednesse from Contempt.

When once Philaretus was able without Danger to support the Incommoditys of a Remoue, his Father, who had a perfect auersion for their Fondnesse,[a] who vse to breed their Children so nice & tenderly, that a hot Sun[b], or a good Showre of Raine as much [c]–endangers them as–[c] if they were made of Butter or[d] of Sugar; sends him away from home; & commits him to the Care of a Cuntry-nurse; who by early[e] invring him by slow degrees to a Coarse (but cleanly)[f] Dyet, & to the Vsuall Passions /[Blank]/ of the Aire, gaue him so vigorous a Complexion, that both Hardships were made easy to him by Custome, & the Delights of conveniencys & ease, were endear'd to him by their Rarity.

Some few yeares after this, two greate Disasters befell Philaretus; the one was the Decease of his Mother, whose Death would questionless haue excessiuely afflicted him, had but[g] his Age[h] permitted him to know the Value of his Losse. For he wud euer reckon it amongst the Cheefe Misfortunes of his Life, that he did ne're know her that gaue it him: her free & Noble Spirit[i] (which had a handsome Mansion to reside in) added to her kindnesse & sweete carriage to hir owne, making her[j] hugely regretted by her Children, & so lamented by her Husband; that not only he annually dedicated the Day of hir Death to solemne Mourning for it; but burying in her Graue all thoughts of Aftermarriage, he reiected all Motions of any other Match, continuing a constant Widdower till his Death.

The second Misfortune that befell Philaretus, was his Acquaintance with some Children of his owne Age, whose stuttring Habitude he so long Counterfeited that he at last contracted it. Possibly a iust Judgment vpon his Derision, & turning the Effects of God's Anger into the Matter[k] of his Sport.[1] Diuers Experiments, beleeu'd the probablest meanes of Cure, were try'd with as much successlesnesse as Diligence; so Contagious & catching are men's Faults, & so Dangerous is [m]–the familiar–[m] commerce of those (condemnable) Customes, that by being imitated but in ieast, come to be learn'd (& acquir'd) in earnest.

---

[a] Altered from *Fondlynesse*. [b] Interlined above *Day* deleted. [c–c] Interlined. [d] MS has *of*. [e] Interlined. [f] Followed by *& who* deleted. [g] Interlined. [h] Followed by *but* deleted. [i] *humor* is interlined above; no deletion. [j] Followed by *so* deleted. [k] *subiect* is interlined above; no deletion. [1] Followed by *Many* deleted. [m–m] Interlined.

But to show that these Afflictions made him not lesse the Obiect of Heu'n's Care; he much about this time escap't a Danger, from which he ow'd his Deliuerance wholly to Prouidence; being so far from contributing to it himselfe, that he did his Endeuor to oppose it. For waiting on his Father[a] vp to Dubline, there to expect the Returne of his Eldest Brother (then landed out of England with his new Wife, the Earle of Cumberland's Heire;)* as they were to passe ouer a Brooke,[b] at that time suddenly by immoderate showers swell'd to a Torrent; he was left alone in a Coach only with a Footboy to attend him: where a Gentleman of his Father's[c] very well hors't accidentally espying him, in spite of[d] some others & his owne Vnwillingnesse & Resistance, (they not beleeuing his stay Dangerous) carry'd him in his Armes ouer the Rapid Water; which prou'd so much beyond expectation, both swift & Deepe, that[e] horses with their Riders were violently[f] hurry'd downe the Streame, which easily ouerturn'd the vnloaded[g] Coach; the horses; (after by long struggling; they had broke their harnesse) with much adoe, sauing themselues by swimming.†

As soone as his Age made him capable of (admitting) Instruction, his Father (by a Frenchman‡ & by one of his Chaplaines) had him taught both to write a Faire hand, & to speake French & Latin; in which (especially the first) he prou'd no ill proficient; adding to a reasonable Forwardnesse in Study, a more then vsuall Inclination to it.

This Studiousnesse obseru'd in Philaretus, endear'd him very much vnto his Father; who vs'd (highly) to commend him both for it & his

[a] Followed by *vpon his Father* not deleted. [b] Followed by *im* deleted. [c] Followed by *acci* deleted. [d] Followed by *his* deleted. [e] Followed by *many very narrowly scap't Drowning* deleted. [f] Interlined. [g] Interlined above *Empty* deleted.

* Richard Boyle, Viscount Dungarvon, with his wife and retinue arrived in the King's ship *The Ninth Whelp* at Howth on Sunday evening 14 September 1634. The Earl of Cork fetched them in three coaches to his house in Dublin, where he lodged them. (*I. Lismore*. **IV**, p. 45).

† The accident mentioned here must be the one that occurred on the journey from Dublin to Lismore, which is noted by the Earl of Cork in his diary as having happened 20 December 1634. "After my coach was overthrown in the 4 myles water, and my coache horses in danger of drowninge, we all, god be praized, came safely to Lismoor." (*I. Lismore*. **IV**, p. 65). The Four-Mile-Water was a village in the parish of Kilronan, co. Waterford, near the banks of the River Suir, about 6 miles S.W. of Clonmel. (*The National Gazetteer*. 3 vols. London, [n.d.]).

‡ The Earl of Cork mentions 'Mounsier Frances de Carey, my childrens French tutor', 30 June 1630. (*I. Lismore*. **III**, p. 41). At this date RB was barely four years old: we are not told if he received instruction, then or later, from this tutor.

The following year RB was given a personal servant. The Earl's diary records 20 October 1631:

> "My old servant, Robert Bridges, with his wyffe and children, being at dublin, and fallen into great poverty, I sent W[m] Barber with v[li] vnto them for their releeff . . . & took his eldest son, Richard Bridges, from him & appointed him to attend my yongest son, Robert Boyle: So to give his father ease of chardges, & the sons breeding." (*I. Lismore*. **III**, p. 107).

Veracity: of which (latter) he would often giue him this Testimony; that he neuer found him in a Lye in all his Life time. And indeed Lying was a Vice both so contrary to his nature & so inconsistent with his Principles, that as there was scarce any thing he more greedily desir'd then to know the Truth, so was there scarce any thing he more perfectly detested; then not to speake it. Which brings into my Mind a foolish Story I haue heard him jeer'd with, by (his Sister,) my Lady Ranalagh; how she hauing giuen strict[a] order to haue a Fruit-tree preseru'd for his sister in Law, the Lady Dungaruan, then big with Childe; he accidentally comming into the Garden &[b] ignoring the Prohibition, did[c] eat[d] halfe a score of them[e]: for which being chidden by his sister Ranalagh; (for he was yet a Childe:) & being told by way of aggrauation, that he had eaten halfe a dozen Plumbs; Nay truly sister (answers he simply to her) I haue eaten halfe a Score. So perfect an enemy was he to a Ly, that he had rather accuse himselfe of another fault, then be suspected to be guilty of that. This triuiall Passage I haue mention'd now; not that I thinke [f]that in it selfe[f] it deserues a Relation /to be recorded/, but because as the Sun is seene best at his Rising &[g] his Setting, so men's natiue[h] Dispositions are clearlyest perceiu'd, whilst they are Children, & when they are Dying. And certainly these little sudden Accidents are the greatest Discouerers of men's true Humors. For whilst the Inconsiderablenesse /[Blank]/ of the thing affords no temptation to dissemble; & the suddennesse of the Time /[Blank]/ allowes no Leasure to put Disguizes on; men's Dispositions do appeare in their true, genuine, shape: whereas most of those[i] actions that are done before others, are so much done for others; I meane most[j] /solemne/ Actions are so personated; that we may much more probably guesse from thence, what men desire to seeme, then what they are: such Publicke, formall Acts, much[k] rather being aiusted /fitted, acc[ommodated]/to men's Desseins, then flowing from their Inclinations.

Philaretus had now attain'd & /or/ somewhat past the Eighth yeere of his Age, when his Father; (who supply'd what he wanted in Schollership himselfe by being both a Passionate Affecter & eminent Patron of it)[l] ambitious to improue[m] his early Studiousnesse; & considering[n] that greate men's Children's breeding vp at home, tempts them to Nicety, to Pride, & Idlenesse; & contributes much more to giue them a Good opinion of themselues, then to make them deserue it;

[a] Interlined. [b] Interlined. [c] Interlined. [d] Followed by *eve them almost all vp* deleted. [e] Followed by *vp* deleted. [f–f] Interlined. [g] Followed by *wh* deleted. [h] Interlined. [i] Followed by *Publick* interlined above *thi*, all deleted [j] Followed by *Publicke* deleted. [k] Followed by *mor* deleted. [l] Followed by *desi* deleted. [m] Interlined above *cultiuate* deleted. [g] Altered from *considera-tion*.

resolues to send ouer Philaretus in the company of[a] (Mr F. B.) his elder brother; to be bred vp at Eaton Colledge* (neere Windsor) whose Prouost at that time was (one) Sr Henry Wotton,† a Person, that was not only a fine Gentleman himselfe, but very well skill'd in th'Art of making others so: betwixt whom & the Earle of Corke an Ancient Friendship had been constantly cultiuated by Reciprocall Ciuilitys. To him therefore the good old Earle recommends Philaretus: who hauing lay'd a weeke at Youghall for a wind, when he first put[b] to Sea, was by a storme beaten backe againe, (not only a Tast but an Omen (& and Earnest) of his future Fortune:/after Destiny./) But after eight Dayes further stay; vpon the second summons of a promising[c] Gale, they went aboord once more; & (tho the Irish Coasts were then sufficiently infested with Turkish Gallyes) hauing by the way touch't[d] at Ilford-combe, & Myn-head, at last they happily[e] arriu'd at Bristoll.

The Second Section

Philaretus, in the Company of his Brother, after a short stay to repose & refresh themselues at Bristoll; shap't his Journey directly for Eaton-colledge: where a Gentleman‡ of his Father's, sent to conveigh him[f] thither, departing, recommended him to the especiall Care of Sr Henry Wotton: & left with him, partly to instruct & partly to atte[n]d

[a] Followed by *his elder* deleted. [b] Repeated in MS. [c] Interlined above *happy* deleted. [d] *landed* interlined above; no deletion. [e] Restored after deletion with *safelly* interlined above. [f] Altered from *them.*

* The Earl of Cork wrote in his diary 9 September 1635:

"This day (the great god of heaven bless, guyde, and protect them) I sent my two youngest sons Frances and Robert Boyle, with Carew their servant, vnder the chardge of my servant Thomas Badnedge, from Lismoor to yoghall to embarcque for England to be schooled and bredd at Eaton, as my worthie frend Sir Henry wotton, provost of Eaton colledg, should direct and order them, to whome I wrott to that purpose . . . I gaue Badnedg 50[li] ster: in ready money at his departure to defray their chardges and his, with my lettres of creddit to the now L. Maior of London to supply him with any moneis he should demaund not exceeding 300[li] ster: And I gave iij[li] to my sons."

(*I. Lismore.* **IV**, p. 125).

On 24 September 1635

"This day Tho. Badnedge with my two yongest sons Frances and Robert Boyle embarcqued at yoghall for Englande."

(*I. Lismore.* **IV**, p. 127).

On 2 October 1635

"Badnedge delivered my 2 yongest sons Frances & Robert at Eaton colledg vnto the chardge of my worthie frend & cowntrey man Sir Henry wotton, provost of Eaton, And to the tuicon of Mr John Harrison, cheef Schoolmaster there. God bless and prosper them in true Religeon & larning." (*I. Lismore.* **IV**, p. 129).

† Sir Henry Wotton (1568–1639); poet and diplomat; ambassador at Venice; engaged in various diplomatic missions abroad; M.P.; provost of Eton, 1629–1639. (*D.N.B:* L. P. Smith. *The Life and Letters of Sir Henry Wotton.* 2 vols. Oxford, 1907).

‡ Thomas Badnedge, a trusted servant of the Earl of Cork; married Bridget, a widowed daughter of Sir John Leeke; made captain of the town guard at Youghal. (Verney, I. pp. 204–6).

him one R.C.* one that wanted neither vices nor cunning to dissemble

---

* Robert Carew was a dissembler, who long deceived the Earl of Cork as well as Sir Henry Wotton. This last called him 'one of the most loving and zealous servants that I haue euer observed in lyfe'; and on another occasion said, 'For truely theare can not be a more tender attendant about your sweete Children.' (*II. Lismore*. **III**, pp. 228, 262: L. P. Smith. *The Life and Letters of Sir Henry Wotton*. Vol. **II**, pp. 358, 361.) RB through daily contact with the man accurately assessed him, and made his strictures accordingly.

Robert Carew, in a letter dated 29 February 163^{5}_{6} to his father Richard Carew, says that he is enamoured of a maiden, and wants to cast his fortunes in that direction if his father will help. His wages are only £5 a year, even though he tutors his young masters in French and Latin. He asks for some competency from his father so that he can marry. He says that nobleman's service is no use. (*Lismore MSS*. **XVIII**, No. 120). This is undoubtedly the matter touched upon by Sir Henry Wotton in his letter 6 June 1636 to the Earl of Cork, where he writes:

> "Truely (My good Lord) I was shaken with such an amazement at the first percussion thereof, that till a seconde pervsal I was doubtfull whether I had readd right: For we are all heere so well perswaded of young Mr Caries [i.e. Carew] discretion and temper and zeale in his charge and in the whole carriadge of himself as it wilbe harde to stampe vs with any new impression: yet because your Lordships letter was so confident, I bestowed a Daye in a little Inquisitivenesse. And founde indeede that betweene him and a younge Mayed, dawghter to owre vnderbaker and almost (like Fathers), I doe not altogether (I must confesse) [think] vnhandsome, nor so far otherwise as she thinks her self theare had passed long since certayn civil, which she was content to call amorous language. But it is neere half a yeare that he hath not been with her: Tyme enough I dare swere to refrigerat more love then was euer betweene them. So as in that point your Lordship may quiet your thoughts: yet glad I am for the letter you were pleased to write, because it will give me an apt occasion at his returne to warne him how carefull and vigilant he ought to be in preserving his person from scandal, when such a levitie as this (whether serious or sporteful) is flowen ouer so suddaynly to youre Lordship: of whose good opinion it behoveth him to stande in much awe." (*II. Lismore*. **III**, p. 261: L. P. Smith. *The Life and Letters of Sir Henry Wotton*. Vol. **II**, p. 360).

In spite of his expressed desire to escape from nobleman's service, Carew remained in attendance on the two boys. At some time in 1638 he appears to have induced Francis Boyle to be surety for £6, or, in the words of the Earl of Cork, 'he basely gott my son Franck to be bownd for him.' (*I. Lismore*. **V**, p. 61, under date 16 October 1638). More of Carew's misdemeanours came to light after the boys had left Eton, when their schoolmaster John Harrison wrote to Francis Boyle as follows:

> "And hee [R. Carew] that was the occasion both of the multiplying of your expences & of some other excesses whereof the blame hath redouded to mee I beleeve now repenteth himself to late. I could not bee satisfyed till I had found out what became of your silver tankard. And by the enclosed copie of the bill of sale of it you may see which way it went, & where it is, & whence you may have it vppon payment of the vse & of the monie for the time that it runn out."

(*Lismore MSS*. **XIX**, No. 34. Letter dated 15 March [163^{8}_{9}.])
The enclosure was a memorandum signed by Robert Carew 15 March 163^{7}_{8} in connection with the pawning of a silver tankard for £4 at New Windsor. (*Lismore MSS*. **XIX**, No. 35).

Subsequently, he appears to have acted as courier for the Earl of Cork, for we find him at Paris in December 1638 'in poore equipage', when Marcombes, who was in that town with Lewis and Roger Boyle, gave him 25 French livres to bring him home. Marcombes interceded with the Earl to forgive Carew one fault on his journey, on account of which he had suffered much misery. (*II. Lismore*. **III**, p. 283: *Lismore MSS*. **XIX**, No. 78. Letters of Marcombes to Earl of Cork dated $^{18}_{28}$ December 1638).

them. For tho his Primitiue fault was onely a dotage vpon Play, yet the excessiue love of thatt, goes seldome vnattended with[a] a Traine of Criminall /culpable/ retainers: for fondnesse vpon[b] Gaming is the seducingst Lure to Ill Company; & that, the subtlest Pander to the worst Excesses /Debauches./ Wherefore our Philaretus deseruedly reckon'd it both[c] amongst the Greatest & the vnlikelyest Deliuerances[d] he ow'd Prouidence, that he was protected from the Contagion of such Presidents. For tho[u]gh the man wanted not a Competency of Parts, yet peruerted Abilitys make men but like those wandring Fires Filosofers call *Ignes fatui;* whose light[e] serues not to direct but to seduce the[f] credulous Trauellerg[g], & allure him[h] to[i] [j]–followe them in their–[j] Deuiations. And it is very true, that during the[k] Minority of[l] Judgment, Imitation is the Regent in the Soule; &[m] Those, that are least capable of Reason, are most sway'd by Example. A blind man will suffer[n] himselfe to be lead, tho by a Dog /or Child/.

Not long our Philaretus stay'd at schoole,[o] ere its[p] Master (Mr Harrison)* taking notice of some aptnesse & much Willingnesse in him to learne; resolu'd to improue them both by all the gentlest wayes of Encouragement. For he would often dispense from[q] his attendance at schoole at the accustom'd howres; to instruct him priuately & familiarly in his Chamber.† He would often as it were cloy him with fruit & Sweetmeats & those little Daintys that age is greedy of; that

[a] *by* interlined above; no deletion. [b] *of* interlined above; no deletion. [c] Interlined above *not only* deleted. [d] *Benefitts* interlined above; no deletion. [e] Followed by *when the* deleted. [f] Altered from *their*. [g] Interlined above *Followers* deleted. [h] Altered from *them*. [i] Followed by *like* deleted. [j–j] In margin. [k] Followed by *Vnrip* deleted. [l] Followed by *the* deleted. [m] Followed by *whilst P* deleted. [n] Interlined above *let* deleted. [o] Followed by *bef* deleted. [p] Altered from *his*. [q] Followed by *a* deleted.

* John Harrison (?–1642), the Rector or Head Master of Eton College; elected a Fellow in October 1636, when William Norris, previously the Usher, became Head Master. Harrison wrote to the Earl of Cork 19 October 1635 saying that the Provost had placed the two boys in his care, and that ' since theire arrivall here, the place hath seemed to agree wondrous well with theire tempers. (*II. Lismore.* **III**, p. 216).

Francis and Robert were lodged with John Harrison on arrival at Eton, and evidently continued to do so after he became a Fellow, for the Earl of Cork wrote to Sir Henry Wotton expressing his pleasure at the election, and that Harrison had continued in instructing his two sons. (Earl of Cork's Letter Book 24 April 1634 to 19 July 1641, p. 136).

By his will, John Harrison bequeathed his books, sextant and dials to be sold for the benefit of the College library. The books were not sold, and his fine collection of scientific works is still in the library of Eton College. (R. Birley. ' Robert Boyle's Head Master at Eton '. *N. & R.* (1958) **13**, 104. R. Birley. ' Robert Boyle at Eton '. *idem* (1960) **14**, 191).

† Robert Boyle at Eton.

We probably know more about the school days at Eton of RB than of anyone else. Apart from his own reminiscences given in the ' Account ', much additional information is available from other sources. The Audit Books at Eton College contain the names of Boyle *a* and Boyle *i* (*i.e.* major and minor) for the period 1635–8. They were Commensals,

by preuenting the want,[a] he may[b] lessen both the[c] value of them &[d]

[a] Followed by *of them* deleted. [b] Followed by *ta* deleted. [c] *his* interlined above; no deletion. [d] Followed by *his* deleted.

and being sons of noblemen sat at the second table in Hall. Tuition was free, but meals taken in Hall were paid for. Shortly after their arrival at Eton, Francis Boyle wrote to his father to tell of the kindness of Sir Henry Wotton, "and how louingly he did receiue vs, and intertained vs this fust day of our intraunce at his owne table. he hath also lent vs a chambre of his owne, with a bedd furnished afore our owne wilbe furnished: al which I leaue to your Lordships consideration to requit. wee are much bound to the young Lords and specially to the Earle of Peter Borrows sonne, with whom we dine and supp." (*II. Lismore*. **III**, p. 215. Letter dated 17 October 1635). Two days later Robert Carew wrote to say that the boys

> "are very well beloued for their ciuill and transparent carriage towards all sorts, and specially my sweet Mr Robert, who gained the loue of all. S^ir^ Hary Wutton was much taken with him for his discours of Ireland and of his trauels, and he admired that he would obserue or take notice of those things that he discoursed off. he is mighty courteous and loueing towards them, and lent a Chamber furnished vntill wee could furniture soe their owne Chamber: wee inioy it yett, which is a great fauor; and did invite my Masters to his owne table seuerall tymes. . .touching my Masters essence, they dine in the hall with the rest of the boarders, where sitts the Earle of South Hamptons[1] fowre sonns, the Earle of Peterboroughs[2] two sonns, with other knights sonns: they sitt permiscously, noe obserueing of place or qualitie; and at nights they supp in their Chambers, but my Masters in regard our Chamber is not furnished doe supp with my Lord mordaunt the Earle of Peterborows sonne, where they are most kindly intertained; but wee haue their owne Commons brought thither: yett they take it as a great kindnes to be so loueingly vsed: they are very familiar one with another; And my Lord, there is to be obserued the fasting nights, wherevppon the Colledge allowe noe meat Fridays and Saturdayes. wee must vppon those nights haue the cookes meat which is sometymes mighty deare, for he must haue his owne rate, not the Colledges prise; as also for breakefasts for euery day they haue a poore breakfast at two pennys apeice. this will come to money besids their Chambers, accutrements and cloathes, which your Lordship must furnish them withall. . .they are vpp euery morning at a halfe an hour afore 6 and soe to scoole to prayers, and their houres for writing be these from nine to ten in the morning and from fiue to 6 in the euening." (*II. Lismore*. **III**, p. 217).

A subsequent letter from Carew gives further information.

> "Two Irish vagabonds accidentaly coming thorow Eton, and haueing some intelligence of my masters being there, inq[ui]red for their Chamber; where being directed, tould me privately that your Lordship was dangerously sicke and past the judgment of the profundest doctors there. I concealed it from my Masters, but at last they had some intelligence of it, which made them much to lament . . . As for my Masters priuat and publicke devotion, theire ciuil and courteous carriage towards all men accordingly, theire neatnes in aparelling, kembing and washing, and my tender care of theire health and prosperitie in

[1] This must be a mistake for Northampton. Spencer Compton, 2nd Earl of Northampton had six sons, four of whom were at Eton between 1633 and 1636.

James Compton (1622–1681);
Charles Compton (?–1661);
William Compton (1625–1663);
Spencer Compton (?–1636).

(Wasey Sterry. *The Eton College Register, 1441–1698*. Eton, 1943.)

[2] The Earl of Peterborough's sons were

Henry Mordaunt (1623–1697); at Eton 1635–1638; succeeded as 2nd Earl of Peterborough.

John Mordaunt (1626–1675); became 1st Viscount Avalon.

(Wasey Sterry. ***Ibid.***)

the Desire. He would sometimes giue him vnask't Playdayes, & oft[a]

[a] Interlined.

Learning, and all other noble education, I leaue to the approbation of their noble freinds Sir Hary wootton and Mr Harison, whose vigilant care ouer them requires noe lesse then your honorable fauour, and my Masters doe in all matters obey and obseruantly performe their mandats, soe that they giue me more occasion of ioy and felicitie by their actions than cause to complaine: they will neither chopp nor change abooke nor any other thing that is theirs without my consent, who will not see any disaduantage or deceipt vsed against them: they are very studious and diligent to fulfill Gods comandements and your Lordships will and mandats. . .
[Mr Francis] is not soe much giuen to his booke as my most honored and affectionate Mr Robert, who looseth noe houre without a line of his idle time, but one Scooldays he doth compose his exercises as well as them of double his yeares and experience. Thy are vnder ye tuition of ye vsher, in regard thy were placed in the 3d forme; a carefull man he is, yett I thanke god I haue gained their loues soe farr that I cann gett them to doe more then their scoole exercise in the Chamber and am auctorised soe to doe by Mr Harison, who sees that they doe it with much willingnes and facilitie. thy writ euery day most commonly a coppy of the french and Latin, besides their v[er]sions and dictamens, more than any two of their Ranke in the scoole can doe. They practise the french and latine but they affect not the Irish, notwithstanding I shew many reason to bind their minds thereto. Mr Robert Some tymes desirs it and is a litle intred in it: he is growne very fatt and very joviall and pleasantly merry, and of ye rarest memory that euer I knew; he prefers Learning afore all other vertues or pleasures; Mr Prouost does admire him for his excellent genious. he was chosen in a play the 28th of 9ber. He came vppon ye stage. He had but a mute part, but for the gesture of his body, and the order of his pace he did brauely. . .
Sir Hary Wootton hath made choise of a very sufficient man to teach them to play on the vyol and to sing; he doth also vndertake to helpe my Master Roberts defect in pronontiation, which is a principall reason that they should bestow any hours in that facultie, for it is a thing that eleuats the Spirits and may hinder their proceedings in matters of greater moment. Here are two frenchmen. . .and men of good life and conversation, and God being pleased I shall make choise of one with Mr Prouests approbation after Christmas; in the meane time they shall doe as they did hitherto, read a chapter most commonly euery night in a french testament afore they goe to bedd, besides their priuat prayers. I gott thone to read in the English testament and thother in the french and so mutually as may appeare." *II. Lismore*. **III**, p. 223). Letter of [December] 1635).

RB was still affected with the defect of speech he had so wilfully acquired in Ireland. On this matter Sir Henry Wotton told the Earl of Cork:

"I haue commended seriously and with promise of a good rewarde youre spiritay Robin to the Master of oure Coristers heere: who maketh profession (and hath in one or two before, given good proofs thereof) to correct the errors of voices and pronunciation: for which he shall haue fit houres assigned him." (*II. Lismore*. **III**, p. 228: L. P. Smith. *The Life and Letters of Sir Henry Wotton*. Vol. **II**, p. 359).

In a letter received by the Earl of Cork, January $163\frac{5}{6}$, Carew says that his young masters on their arrival at Eton

"were much preiudised, in as much as they could not be accommodated with those things following sooner, at first coming, for the furnishing of their chamber, first for this respect that they were be houlding to the Earle of PeterBorows sonnes for the vse of their chamber and at much extraordinarys besides their commons; for the other were often visited with their neighbouring freinds, for which they made great provisions and it was concluded that my masters should beare halfe when as they had none to trouble them: Secoundly that Mr Harison bestow vpon him such Bals & Tops & other impleme[n]ts of Idlenesse,

furnished them with linnen and a bedd, for which he expects some reward . . . These things were bought by Mr Perkins who is mighty backward in sending of them new suits. . . Sent by Mr Perkins since the 16 of 9^{ber}:

one Feild bedd curtins and valance of blewperpetiana, with lace and frenge and testar defectiue,

one feather bedd, boulster, two pillows, 4 pillow boars and 2 paire of Sheits.

one Rugg, two blanketts; 2 table cloathes and doozen of Napkins diaper,

a tankard, two spunes and a salt silure, with two peuter porengers,

A lookeing glass and 6 towels; andirons, fire shouell and tonges,

A paire of knifes, snuffers and bellows,

A coople of Stooles, two chaires and a carpett suitable to the bed.

a flocke bedd for me, boulster, two course parre of Sheites, a Rugg and 2 blanketts, 4 paire of stockens, 2 hatts with siluer twist, a beddcord and a doozen of trenchers; a close stoole and brush with 2 wooden combes. . .

Those commons which after the rate of fiue and 6^{d} a weeke comes to 43^{li} and 4^{s} sterling besides their fasting nights supper, which are friday and Saturday; and although I keepe an accoũpt of euery seuerall dish I am not able to tell your honor what they cost, for they say it is not the costome to aske afore the pay day. Their breakefast after 4^{d} a day comes to 29^{s} 8^{d} per quarter: their chamber 5^{li} per annum; Their washing 16^{s} a quarter: their Barber which is the colledge chirurgion 5^{s}, for wood and coles Mr Harison laid out 34^{s} 6^{d}; Candles I recived from Mr Harison 16 pound and thinkes my selfe very spareing in not spending more, for this moneth past was an expenciue time for candles by reason of a sport that they vse in the hall euery night and euery recent [comer] must find candles; for my Masters a halfe apound euery night and in play nights a whole pound. Mr Harison bought 3 peuter dishes two saucers and chamber pott, and finds bookes and paper and other necessarys that is vsed in the scoole; but for their tuition the ordinary costome is 10^{s} a peice, yett they leaue it to your honorable bounty. . ." (*II. Lismore.* III, p. 237).

To this most interesting account of conditions at Eton College in the seventeenth century, Carew later added a report on his charges. Writing to the Earl of Cork, 29 February 163$\frac{5}{6}$, he says:

"[Mr Francis] loues learning but his memory much inferiour to Mr Robert's[1] . . . yett all his delight is in hunting and horsemanship, from which desire m^{r} Robert often dissuads him, and exhorts him to learning in his youth, for sayth he there cann be nothing more profitable and honorable; and in such amanner that your Lordship would admire of it or thinke it incredible, what flowing vertues my poore witt cann find in him. He [Robert] is wise, discreet, learned and deuout, and not such deuotion as is accustomed in children, but withall in Sinceritie he honors God, and prefers him in all his actions. He is of a fayre amiable countenance, Growne much in thicknes and talnes, and very healthfull. Praysed be God, dureing his residence here, he had but one fitt of Sicknes, which came by a causualty, not any innate infirmitie. The 10th of January he tooke a conceit against his breakfast as he is alwaies curious of his meat, and would goe fasting to church. It being a could morning the peircing aire interd into his tender stomack and caused a greiuious griping in his belly for the space of two dayes, and Mr Prouost would haue him to take phisicke; but I was against it, and desired respeit till next day, and then I would condescend vnto it, and in the meane space he was well recouered, thanks be to God, and continues, still increaseing in vertues. His delight is in his learning and as well in all other noble facultys, for he takes noe pleasure in playing with boyes nor running abrood, but when I wait vppon him; and to talke latine, he [has] much affected. This quarter he learned to play on musick and sing, and he can giue good account for any thing that was

[1] This sentence thus in MS. Printed version is incorrect.

as he had taken away from others that had vnduely vs'd them. He

---

taught him, but he could not proceed farr, for his teacher had much imployments abrood in the countrey this quarter. My most honorable Lord, what a happines its to me to see how loueingly they live together, that neuer yett two ill words passe betweene them, which is rare to see; and specially when the youngest exceeds the elder in some qualitie, then most commonly there is an emulation betweene them as I see in this Colledge betweene two brothers, and noblemans sonnes. There is such an envy between them alwaies that their seruants cannot liue quitly for them, but the peace of God is with my masters; they liue in such natural fraternitie and amitie that all the colledge of all degrees takes notice of them, and as well for that as for their braue courteous carriage to all men, they [are] highly respected, and very well beloued . . . my master Roberts tombe [=thumb] is sore, wherefore he could not writ. . ." (*II. Lismore*. **III,** pp. 242–4).

Whilst Francis and Robert were at Eton College, their elder brothers, Lewis and Roger, were preparing to make the grand tour under the care of Isaac Marcombes. On their way from Dublin to London they called at Eton, 1 March 163^{5}_{6}, and supped with Francis and Robert. The next day all four brothers went to see Windsor Castle, whence Francis and Robert returned to Eton. (*Lismore MSS*. **XVIII,** No. 123, Letter of Marcombes to Earl of Cork, 16 March 163^{5}_{6}; *ibid*. No. 124, Letter of the same to the same, $^{20}_{30}$ March 163^{5}_{6}.) The Whitsun vacation of 1636 (say 21 May to 18 June) was spent with their sister Lettice, Lady Goring, at Lewes in Sussex. On this matter, Sir Henry Wotton wrote to the Earl of Cork saying:

"I received some dayes since from my Ladie Goring there sister, some few lines expressing a desire to haue them with her at this tyme of owre vacation, when owre Schoole Annually breaketh vp two weekes before Whitsontyde and peeceth agayne a fortnight after; which just and kinde motion was to me an absolute commaunde. And so I sent them to Her at Lewes in Sussex, together with the Captayn of owre Schoole, a well lerned and well tempered Boye: whose Frendes dwell in that Shiere, so as he may serue them bothe for a good Guide and Compaynion. It wilbe a solace for my Ladie, and for them a fine refreshment. And I am glad to tell your Lordship that shee will see Franck in better health and strength then he hath been in either Kingdome before. And Robert will entertayne her with his pretie conceptions, now a greate deale more smoothely than he was wonte. We expect them both agayne vnder Gods fauour on Saterday come seavenight." (*II. Lismore*. **III,** pp. 260–1; L. P. Smith. *The Life and Letters of Sir Henry Wotton*. Vol. **II,** p. 360).

They were with their sister Lettice again during the vacation in 1637, concerning which visit Carew wrote to the Earl of Cork as follows:

"they haue beene this vacation (with the consent of all those that are in authoritie over them) with their deare Sister, where there was nothing wanting, to afford a good and pleasant entertainment if my honourable Lady had not been visited with her continuall guest, griefe and melancholy: which is vncurable while she liues amongst those vnhappy plants, which yelde her nothing but vexation of mynde, yea haue already suckt from her all comfort. my witt is not able to make a full relation of her estate, but my minde is capable of greefe to knowe the truth of it. she prays incessantly that she may inioy your honors presence here, as there. my masters did comfort her as much as they could, but her languishing heart could not receive much comfort, soe that it made them cry often to looke vppon her. they were since 8th of Maj vntill the 7th of this present, being admonished by Sir Henry Wotton their tutor and Schoolmaster . . . to retreat themselves in the countrey at their pleasure; and it was not without some relation to the booke: for most commonly euerj morning they compose a peice of Latine and sometimes they writ vharses: but ye rest of my master Robert's time was employed in reading some pleasant History, wherein he takes noe smale delight, and reaps therebj much profit; for what he reads would sometimes commend others before him to rowze his Emulation, & oftentimes giue him Commendations before others, to engage his

takes such impression in him that it cannot easily gett awaj, and not onlj the Methode, but the same phrase he runnes without booke. . .I received for my Masters 2 scarlet suits without coats, 2 shirts, 4 lacd bands and 5 paire of Cuffs, 2 night Caps, a doozen Napkins, a half doozen towels, a doozen hankchershirs, 2 cloath of siluer dublets. he [? Mr Perkins, the tailor] stayed them here [London] for a day to prouide some other necessaris for them. . ." (*II. Lismore.* **III**, pp. 268–9. Letter dated 9 [June] 1637: Earl of Cork's Letter Book 24 April 1634–19 July 1641, p. 199. Letter of W. Perkins to Earl of Cork dated 29 May 1637).

Whilst the two boys were at Eton College, the Earl of Cork was not unmindful on occasion to send them some token of his affection. When their cousin Boyle Smyth left Dublin to join Lewis and Roger Boyle at Geneva in order to accompany them in Italy (a journey from which he was destined never to return) he carried letters and 'two angels as tokens for Franck & Robin at Eaton.' (*I. Lismore.* **IV**, p. 221.) Later in the same year, 1637, he sent them both '50s in gowld for tokens.' (*I. Lismore.* **V**, p. 34).

At the beginning of August 1638 the Earl of Cork left Youghal to go to Stalbridge, his estate in Dorsetshire, whither his youngest two sons came on the 19th of the month. (*I. Lismore.* **V**, p. 58). At some unrecorded date the two boys returned to Eton; perhaps it was 11 October 1638 (*I. Lismore.* **V**, p. 61) during the Earl's progress towards London. His diary subsequently provides the following entry:

> " This 23 being Frydaie, the 23th of November, I departed London, & came that night to Eaton, from whence I took with me my two yongest sons Frances and Robert, for whose expences there and their man, for three yeares, for diett, and tutaradge, and aparell, I paid 914li. 3s. 9d, and came back to Stalbridg on Monday, being the xxvjth of November, 1638. . ." (*I. Lismore.* **V**, p. 64).

This period provides the earliest surviving letter written by RB: it is dated 18 February 163$\frac{5}{6}$. The number of letters surviving from the correspondence between the Earl of Cork and his two sons at Eton College at this time is extremely small, though we know from the *Lismore Papers* that others were written. An inventory of the Earl's 'Writings, Clothes, etc' at Dublin, dated 5 May 1643, reveals in 'A list of such writinges as are in a trunck covered with red leather . . .' the presence of 'Five files from and concerning Mr Francis and Robert Boyle at Eton.' (*II. Lismore.* **V**, p. 132). Needless to say, it would be most interesting to know what they contained. The possibility that they may come to light is not excluded seeing that the Earl's Letter Book from 25 March 1634 to 19 July 1641, which is mentioned elsewhere in the inventory, reappeared not so many years ago.

Sundry remarks on Eton College will be found in the letters of John Beale, who was a near contemporary of RB there, though a trifle earlier. (*Works Fol.* **V**; *Works.* **VI**, *passim*). At a later date Marcombes passed his strictures on Francis Boyle's lack of profit from his Eton days. He says:

> ". . .for except reeding and writting Inglish he was grounded in nothing of ye wordle . . . neuer any gentleman hath donne lesse profit of his time then he had done when he went out of England; and besides yt if he had been Longer att Eatton he had Learned there to drinke with other deboice scholers, as I haue beene informed by Mr Robert, which is ye finest gentleman of ye wordle. . ."

(*II. Lismore.* **IV**, p. 234).

References to schools, schoolmasters, and discipline at the time will be found in the following: F. Watson. 'Curriculum and Text Books of English Schools in First Half of Seventeenth Century.' *Trans. Bibliogr. Soc.* (1900–2) **VI**, 159–267: *idem. English Grammar Schools to 1660.* Cambridge, 1908. H. M. Smith. 'The Early Life and Education of John Evelyn.' *Oxford Historical and Literary Studies*, Vol. XI, 1920, pp. 45–48, notes 82, 83: Wasey Sterry. *A List of Eton Commensals, 1563–1647.* Eton, 1904: *idem. The Eton College Register, 1441–1698.* Eton, 1943.

Endeauors to deserue them. Not to be tedious, he was carefull to instruct him in such an affable, kind, & gentle way, that he[a] easily preuail'd with him to consider /affect/ Studying not so much as a Duty of Obedience to his Superiors /Parents/ but as the Way to [b]–purchase for himselfe–[b] a most delightsome & invaluable Good. In Effect, he soone created in Philaretus so strong a Passion to acquire Knowledge, that what time he could spare from a Schollar's taskes (which his retentiue Memory made him not find vneasy;) he would vsually employ so greedily in Reading; that his Master would be sometimes necessitated to force him out to Play; on which, & vpon Study, he look't as if their natures were inverted. But that which he related to be the first occasion that made him so passionate a Friend to Reading; was the accidentall Perusall of Quintus Curtius; which first made him in Loue with other then Pedanticke[c] Bookes, & coniur'd vp in [him] that vnsatisfy'd Curiosity[d] of Knowledge, that is yet as greedy, as when it first was rays'd. In gratitude to this Booke I haue heard him hyperbolically[e] say, that not onely he ow'd more to Quintus Curtius then Alexander did; but deriu'd more aduantage from the History of that greate Monark's conquests, then euer he did from the /[Blank]/ Conquests themselues.*

Whilst our youth was thus busy'd about his Studyes, there happened to him an Accident that Silence must not couer. For being one Night gone to bed somewhat early; whilst his Brother was conversing with some Company by the fire side;[f] without giuing[g] them the least warning [h]–or summons to take heed–[h] a greate part of the wall of their chamber, with the Bed, Chaires,[i] Bookes & furniture of the next chamber ouer it, fell downe [j]–at unawares–[j] vpon their Heads. His brother had his band torn about his Neck & his Coate vpon his Backe, & his Chaire crush't & broken vnder him; but by a lusty Youth then accidentally in the roome was snatch't from[k] out[l] the Ruines,[m] by which Philaretus had[n] (in all probability) been remedilesly oppres't; had not his Bed been curtain'd by a watchfull Prouidence; which kept all heauy things from falling on it: but the Dust the Crumbled rubbish raised was so thicke That he might there haue stifled, had not he remembred to wrap his head in the sheet; which seru'd him as a

[a] Followed by *made* deleted. [b–b] Interlined above *attaine* deleted. [c] Followed by *& con* deleted. [d] *Appetite* interlined above; no deletion. [e] In margin. [f] Followed by *at unawares &* deleted. [g] *affording* interlined above; no deletion. [h–h] Interlined. [i] Followed by *&* deleted. [j–j] Interlined. [k] Interlined above *of* deleted. [l] Followed by *of* deleted. [m] Preceded by *rubbish* deleted. [n] Followed by *been* deleted.

* Quintus Curtius Rufus (1st Century A.D.), author of *Historiae Alexandri Magni*.

strainer, through which none but the purer Aire could find a passage. So sudden a Danger & happy an Escape, Philaretus would sometimes mention with expressions both of Gratitude & Wonder. To which he would adde the Relation of diuers other almost contemporary Deliuerances; Of these, one was, that being fallen from his horse, the beast run ouer him, & trod so neare his throate, as within lesse then two inches of it, to make a hole in his band; which he long after reseru'd for a Remembrancer. An other was, that riding thorou a Towne vpon a Nagge of his owne whose starting quality he neuer obseru'd before, his horse vpon a sudden fright [did] rise bolt vpright vpon his hinder feet, & falling rudely backwards with all his Weight against a Wall had infallibly crush't his Rider into Peeces, if [a] by a strange instinct, he had not cast himselfe off at first & [b–]was quit of it[–b] for a slight Bruise. The last was [c–]an Apothecary's mistake.[–c] Philaretus being newly recouer'd of a Fluxe, the doctor had prescrib'd [d–]him a[–d] refreshing Drinke; the fellow that should administer it, instead of it brings him a very strong vomit prepared & intended for another. Philaretus was that morning visited by some of his schoole-fellowes, who; (as we [*sic*] was not Ill belou'd amongst them) presented him with some sweetmeats, which[e] hauing eaten, when afterwards he would haue eaten his Breakfast, his stomacke whither out of Squeamishnesse or Diuination, forc't him to render it againe. To which lucky[f] accident, the Physitian ascrib'd his Escape from the[g] Apothecary's Error; for in the Absence of those that tended him, his[h] Fisick cast him into[i] hideous Torments, the true Cause of which, by [them] neuer dream'd of, remain'd long vnconiectur'd; vntill at last the Effects betray'd it; for after a long struggling, at last the Drinke wrought with such violence, that they fear'd, that[j] his Life would be disgorg'd together with his Potion. This Accident made him long after apprehend more the Fisitians then the Disease & was possibly the occasion that made him afterwards so inquisitiuely apply himselfe to the study of Fisicke, that he may haue the lesse need of them that professe it. But Philaretus wud not ascribe[k] any of these Rescues vnto Chance, but would[l] be still industrious to perceiue[m] the hand of Heu'n in all these Accidents: & indeed he would professe that in the Passages of his Life he had obseru'd so gratious & so peculiar a Conduct of Prouidence, that he should be equally blind & vngratefull, shud he not both Discerne & Acknowledge it.

[a] Followed by *he had* deleted. [b–b] Interlined above *so scap't* deleted. [c–c] In margin. Followed by *This* deleted. [d–d] Interlined above *some* deleted. [e] Followed by *together with his br* deleted. [f] Followed by *acc* deleted. [g] Followed by *violen* deleted. [h] Followed by *hid* deleted. [i] Followed by *greevous* deleted. [j] Interlined. [k] Interlined above *owe*; no deletion. [l] Followed by *st* deleted. [m] Interlined above *Discouer* and *discern* deleted.

Philaretus had[a] now for some 2 yeares been a constant Resident at Eaton; (if yow except[b] a few visits which during the long Vacations, he made his sister my Lady Goring, at Lewes in Sussex;) when about Easter he was sent for vp to London to see his Eldest Brother the Lord Dungaruan;* where being visited with a Tertian Ague, after the Queen's† & other's Doctors Remedyes had been succeslesly essay ['d]; at last he return'd againe to Eaten, to deriue that Health[c] from a Good Aire & Dyet, which Fisick could not giue him. Here to diuert his Melancholy they made him read the stale Adventures[d] *Amadis de Gaule;*‡ & other[e] Fabulous & wandring Storys; which[f] much more preiudic'd him by vnsettling his Thoughts, then they could haue aduantag'd him; had they effected his Recouery; for meeting in him with a restlesse Fancy, then made more susceptible of any Impressions by an vnemploy'd Pensiuenesse; they[g] accustom'd his Thoughts to such a Habitude of Rauing, that he has scarce euer been their quiet Master since, but they[h] would take all occasions to steale[i] away, & go a gadding to Obiects then vnseasonable & impertinent. So great an Vnhappinesse it is, for Persons that[j] are borne with such Busy Thoughts, not to haue congruent Obiects propos'd to them at First. Tis true that long time after Philaretus did in a reasonable[k] measure fixe his Volatile Fancy & Reclaime his Ramage thoughts;[l] by the vse of all those Expedients he thought likelyest to fetter (or at least, to curbe/bridle/)[m] the rouing wildnesse of his wandring /hagard/ Thoughts. Amongst all which the most effectuall Way he found to be, the Extractions of the Square & Cubits Rootes, & specially those more laborious Operations of Algebra, which both accustome &[n]

[a] Altered from *having*. [b] Followed by *some* deleted. [c] Interlined above *Recouery* deleted. [d] Followed by *of th* deleted. [e] Followed by *Rauing Bookes;* deleted. [f] Followed by *did him more ha* deleted. [g] Followed by *taught his* deleted. [h] Interlined. [i] Interlined above *flinch*; no deletion. [j] Followed by *haue* deleted. [k] *consider* interlined above *reasonable;* no deletion. [l] Followed by *but yet not so as perfectly to bridle their rouing wildness* deleted. [m] Followed by *his* deleted. [n] Followed by *th* deleted.

* The date of this visit may be fixed as March 1638. Richard Boyle was living in the parish of St Margaret's, Westminster, during 1637–8. (The Records of St Margaret's Parish, Overseers Accounts, 1637–1641, Vol. 154. Archives Department of City of Westminster Libraries.) He headed the court delegation that greeted the Venetian ambassador in November 1638. (*Calendar of State Papers, Venetian, 1636–1639*. London, 1923. Vol. 24, p. 472. Letter of Giovanni Giustinian dated 19 November 1638. N.S.) Whilst in London RB wrote a dutiful letter to his father. It is dated 14 March [163^{7}_{8}], and is printed in *Biogr. Brit.* **II**, p. 914, note C.

† Sir Theodore Turquet de Mayerne (1573–1655), physician, M.D., Montpellier. Settled in England, and became court physician to James I and Charles I.

‡ Amadis de Gaula, the hero of a famous fifteenth century romance of chivalry, first printed at Saragossa in 1508. The complete cycle of romances in the French translation comprises 24 vols. 12mo, Lyons and Paris, 1557–1615. The earliest parts had been translated into English before RB's time. (Bompiani, *Opere*. **I**, p. 100).

necessitate the Mind to attention, by so entirely exacting the whole Man; that [a]–the least–[a] Distraction, or heedlessnesse, constraines vs to renew our (Taske &) Trouble, & rebegin the Operation. Six weekes had Filaretus been troubl'd with his Ague, when he was free'd from it by an accident, whic[h] is no slender Instance of the force /power/ of Imagination; for the Fisitian hauing sent him a Purge to giue (as he sayd) the fatall Blow to the Disease; our Philaretus had so perfect an auersion to all Fisicke, & had newly essay'd it so vnsuccesfully, that his Complaints induc't the Maydes[b] of the[c] House he lodg'd in, (partly out of complaysance to him, & partly out of a beleefe, that Fisick did but exasperate his Disease;)[d] vnknowne to him to[e] poure out the Potion, & fill the Viall with Syrup of stew'd Prunes; a Liquor so resembling it, that Philaretus[f] (see the force of Fancy) swallow'd it with the same Reluctancy; & found the Tast as loathsome as if't had been the Purge: but being after acquainted with the Cuzenage, whither 'twas that his sicknesse[g] (as hauing already reach't it's Period) would haue expir'd of it selfe, or that his Mirth dispatch't it; I pretend not to determine; but certaine 'tis, that from that Howre to this; Agues & he haue still been perfect stranger. And he had much adoo to refraine from laughter, when going to thanke & reward the Doctor for his Recouery, he found it wholly ascrib'd to the Efficacy of a Potion, he had neuer swallow'd[h] but in Imagination /His fancy only swallow'd./

He had now seru'd well neere halfe an Apprentice-ship at Schoole, when there arriu'd Intelligence of his Father's being landed in England, & gone to Stalbridge, a Place in Dorsetshire then newly purchas'd by him. Thither Philaretus accompanys[i] his sister the Countesse of Kildare, to waite vpon him.* The good old Earle welcom'd him very kindly; for whether it were to the Custome of old People[j] (as Jacob doted[k] most on Beniamin and Joseph) to giue their Eldest Children the largest proportions of their fortunes, but the youngest the greatest shares of their Affections; or to[l] a likenesse[m]

[a–a] *a small* interlined above; no deletion. [b] *Seruants* interlined above; no deletion. [c] Followed by *Place* deleted. [d] Followed by *to* deleted. [e] Followed by *fling* deleted. [f] Followed by *swallow'd it wi* deleted. [g] Followed by *had* deleted. [h] *tooke* interlined above; no deletion. [i] Interlined above *waits vpon* deleted. [j] Interlined above *Men, w* deleted. [k] Followed by *on* deleted. [l] Interlined above *in* deleted. [m] Interlined above *resemblance;* no deletion.

* The Earl of Cork sailed from Youghal, 2 August 1638, and landed at Bristol two days later in the evening. He arrived at, and saw, Stalbridge for the first time on the 6th of that month. Richard Boyle, Viscount Dungarvon, with his wife and daughter were already there. Joan Boyle, Countess of Kildare, together with Francis and Robert Boyle arrived at Stalbridge, 19 August 1638. The family gathering included also Alice Boyle, with her husband, the Earl of Barrymore. (*I. Lismore.* V, pp. 57–58).

obseru'd in Philaretus both to his Fathers Outside /body/ & his mind; or to both an exact & affectionate Obedience his Duty made him pay his Com[m]ands; or, as it seemes most likely, to his neuer hauing liu'd with his Father to an Age that might much tempt him to run in debt & take such other Courses to prouoke his Dislike, as in his elder Children he seuerely disrellish't: To which of these Causes the Effect is to be ascribe[d], it[a] is not my Taske to resolue; but certine it is, that from Philaretus his Birth[b], vntill his Father's Death, he euer continu'd very much his Fauorite. But after some weeke's Enioyment of the summer[c] Diuersions[d] at Stalbridge, when his Father remou'd to London, he left him by the way at Eaton-Colledge,* from whence at his Returne into the West some few months after; he tooke him absolutely away:† after Philaretus had spent[e] in that Schoole (then very much throng'd with young Nobility) not much beneath Foure Yeares; in the last[f] of which he forgot much of that Lattin he had gott: for he was so addicted to more reall[g] Parts of Knowledge, that he hated the study of Bare words, naturally; as something that relish't too much of Pedantry to consort with his Disposition & Desseins: so that by the change of his old Courteous Schoolemaster, for a new rigid Fellow;‡ loosing those encouragements that had formerly subdu'd his Auersion to Verball studys; he quickly quitted his Terence & his Grammar, to read in History their Gallant[h] Acts, that were the Glory of their owne & the Wonder of our Times. And[i] indeed, 'tis a much[j] nobler[k] [Blank] for Ambition, to learne to do[l] things, that may deserue a roome in History; then onely to learne, how congruously[m] to write such actions, in the Gowne-men's Language.

---

[a] Followed by *th* deleted. [b] Followed by *to* deleted. [c] Interlined. [d] Followed by *of* deleted. [e] Followed by *neere 4 yeares* deleted. [f] Followed by *mont* deleted. [g] *solid* interlined above; no deletion. [h] Interlined above *rare noble* deleted. [i] Followed by *tis* deleted. [j] Followed by *fitter subiect* deleted. [k] Followed by *ob* deleted. [l] Followed by *act* deleted. [m] Interlined.

---

* The Earl with his retinue left Stalbridge, Tuesday, 9 October 1638, for London, where he arrived the following Friday. RB and his brother were left at Eton on the way, Thursday, 11 October 1638. (*I. Lismore*. V, p. 61).

† On his way back from London to Stalbridge the Earl of Cork, when passing through Eton, removed his two sons from the College. See note on p.14. Arrangements were made for the return of the linen, bedding, and furniture in the chamber they occupied at Eton. (*I. Lismore*. V, p. 65). It may be mentioned here that the new block of science buildings at Eton College, which was opened in May 1959, has been appropriately named *The Robert Boyle Schools.*

‡ During RB's stay at Eton the Head Masters were:

John Harrison, 1630–1636,
William Norris, 1636–1642;

the Ushers were

William Norris, 1631–1636,
Charles Faldoe, 1637–1642.

RB's reference is presumably directed against the last named. (Wasey Sterry. *The Eton College Register 1441–1698*. Eton, 1943.)

As soone as Philaretus was arriu'd at Stalbridge,* his Father assign'd the care of teachi[n]g him to one Mr W. Douch, then Parson of that Place, & [a-]one of his Chaplaines,[-a] & (to avoid the temptations to Idlenesse, that Home might afford) made him both lodge & dyett where he was taught; tho it were not distant[b] from his Father's house aboue twice musket-shot.† This old Diuine instructing our Youth both with care & Ciuility, soone brought him to renew his first acquaintance with the Roman Tongue, & to improue it so farre that in that Language he could readily enuf expresse himselfe in Prose, & began to be no dull Proficient in the Poeticke straine; which latter he[c] was naturalarry[d] addicted to, resenting a greate deale of Delight in the Conversation of the Muses, which neuerthelesse he euer since that time forbore to cultiuate; not out of any Dislike or Vnderualuing of Poetry, but because[e] in his Trauells hauing by discontinuance forgot much of the Latin Tongue, he afterwards[f] neuer [g-]could find[-g] [occasion] to redeeme his Losses by a serious study of the Ancient Poets; & then for English Verses, he say'd they[h] could not be certaine of lasting applause, the changes of our Language being so greate & nimble /sudden/ that the rarest Poems within few yeares[i] will passe for obsolete; & therefore he vs'd to liken[j] Writers in English verse to Ladys that haue their Pictures drawne[k] with the Clothes now worn; which tho at present neuer so rich & neuer so much in Fashion, within a few yeares hence will make them looke like Antickes. Yet did he at Idle howres write some few verses both in Franch &[l] Latin: & many Copys of amorous, merry, & Deuout ones in English; most of which, vncommunicated, the Day he came of Age he sacrific't to

---

[a-a] Interlined. [b] Interlined. [c] Followed by *much* deleted. [d] Followed by *much* deleted. [e] Followed by *b.h.* deleted. [f] *-wards* altered from *would*; followed by *could* deleted. [g-g] Interlined above *persuade himselfe* deleted. [h] Followed by *Good were* deleted. [i] Followed by *must* deleted. [j] Followed by *the* deleted. [k] Followed by *in* deleted. [l] Followed by *En* deleted.

---

* See note † on p. 19.

† The Earl's entry in his diary under date 2 December 1638 reads:

> "Mr dowch receaved into his hows at Stalbridge, and into his tuicon, my said 2 sons Frances & Robert, and Rydowt their servant at xv[li] the quarter; which is lx[li] per annum".

(*I. Lismore*. V, p. 65). The Rev. William Douch was instituted as Rector of Stalbridge, 1621; outed 'for scandal and delinquency', 1648; died 11 June 1648. (C. H. Mayo. *The Minute Books of the Dorset Standing Committee, 23rd Sept. 1646 to 8th May 1650*. Exeter, 1902.) Subsequent Rectors during RB's lifetime were John Douch, 1648, ejected 1649; Nathaniel Fairclough, intruder; Richard Shute, intruder, 1656, ejected; John Douch, re-installed, 1660; died, 1675; Samuel Rich, instituted, 1675; a petition for his removal for immorality, etc., is in B.P. XL; Thomas Hyde solicited RB to present him to the living, 1689, (*Works Fol.* V, p. 593; *Works*, VI, p. 577); Thomas Dent, instituted, 1690. (Private communication, 1965, from the Rev. F. A. O. Sanders, Rector of Stalbridge.)

Vulcan,* with a Dessein to make[a] the rest perish by the same Fate, when they came within his Power; tho amongst them were many serious Copys, & one long one amongst the rest, against Wit Profanely or Wantonly employ'd; those two Vices being euer perfectly detested by him in others, & religiously declin'd in all his Writings.

About this Time also[b] Philaretus began to [c]–be taught–[c] some skill in the Musick both of the Voyce/song/& hand;† but the Discouragement of a Bad voyce soone[d] persuaded him to Desist.

It was now the Spring of the Yeare when[e] newes was brought to Stalbridge, of the Approach of his Sister the Lady Goring, & in her Company two of his Brothers (the Lords of[f] Kinalmeakye & of Broghill:) then newly return'd from their Three Yeares Trauells.‡ In their Company arriu'd one Mr Marcombes, a French Gentleman who had been their Gouernor;§ & behau'd himselfe so handsomly in that

---

[a] Interlined above *let* deleted. [b] Followed by *he* deleted. [c–c] Interlined above *learn to* deleted. [d] Interlined above *quickly;* no deletion. [e] Followed by *Philaretus* deleted. [f] Followed by *Broghill & the Lord of* deleted.

---

* This reference to the sacrifice of his poems on coming of age indicates that the account was written after 25 Jan 164$\frac{7}{8}$. The use of the expression 'during his Minority' in the title of the account suggests the same conclusion.

One of RB's poems that has survived, though not necessarily from this period, is entitled 'The Deist's Plea, answered by the Honourable, *Robert Boyle* Esq.' and was published shortly after his death in *Miscellany Poems upon Several Occasions.* Sm 8vo. London, 1692, edited by Charles Gildon. (I. MacAlpine and R. A. Hunter. 'Robert Boyle—Poet'. *J. History of Medicine and Allied Sciences,* (1957) **XII,** 390).

Another trifle ascribed to RB, when a youth, is recorded in B.M., Add, MSS. 4229, fol. 39.

† One wonders if his instructor were John Birchenshaw (?–1681). In early life he was in the family of the Earl of Kildare. He left Ireland at the rebellion, and settled in London, where he was a teacher for the voice and viol. We find him mentioned in connection with experiments on sound by members of the Royal Society. (Birch, *History*). Of particular interest is the fact that Birchenshaw wrote a treatise on music 'for the use of the Honourable Robert Boyle Esq' (B.P. **XLI**). Unfortunately it is undated.

‡Lewis and Roger Boyle arrived back in London from their foreign travels, 4 March 163$\frac{8}{9}$, and arrived at Stalbridge ten days later. (*I. Lismore.* **V**, pp. 78, 80).

§ To the character sketch given by RB, the following information may be added. Isaac Marcombes originated from Marsenac in Auvergne, and had been recommended to the Earl of Cork, as a suitable governor to Lewis and Roger Boyle, by Sir Henry Wotton, Philip Burlamacchi and William Perkins, all of whom gave favourable reports about him. Shortly before May 1637, he married Madeleine Burlamacchi, widow of Antoine Drelincourt, and so became a nephew by marriage of the famous pastor Jean Diodati, thereby acquiring connection with important protestant families with interests in England and the continent. Marcombes established his home in Geneva, of which town he and two of his sons were admitted Bourgeois de Genève in June 1647. He had a family of at least seven children. I have not found the date of his death. His wife died in 1665, aged 56 years, but he had died before 23 August 1654. (R. E. W. Maddison. 'Studies in the Life of Robert Boyle, F.R.S. Part VII. The Grand Tour.' *N. & R.* (1965) **20,** 51. Further details and references are given in this article.)

Relation; that the Old Earle remou'd[a] Philaretus (his brother lying sick at a Doctor's house) to his owne House againe, &[b] entrusted his whole[c] Education with this Gentleman. He was a man whose Garbe, his Mine [*sic*] & outside, had very much of his Nation: hauing been diuers Yeares[d] a Traueller & a souldier; he was well fashion'd, & very well knew,[e] what belong'd to a Gentleman. His Naturall were much better then his acquired Parts; tho diuers of the Latter he possess't, tho not in an Eminent, yet in a very competent Degree: Schollership he wanted not, hauing in his[f] greener Yeares been a profess'd Student in Diuinity; but he was much lesse Read in Bookes then Men. And[g] hated Pedantry as much as any of the seauen Deadly sins. Thrifty he was extreamely, & very skilfull in the Slights of Thrift; but[h] Lesse out of Auarice then a iust Ambition, & not so much out of Loue to Mony, as a Desire to liue Handsomely at last. His practicall Sentiments /Op[inion]s/ in Diuinity were most of them very sound; & if he were giu'n to any Vice himselfe, he was carefull by sharply[i] condemning it, to render it vninfectious; being industrious, whatsoeuer he were himselfe, to make his Charges, Vertuous. Before Company he was always[j] very[k] Ciuill to his Pupills; apt to Eclipse their Failings, & sett off their Good qualitys to the best aduantage: but in his Priuate Conversation he was cynically[l] Dispos'd; & a very[m] nice Critick both of Words & Men; which Humor he vs'd to exercise so freely with Filaretus; that at last he forc'd him to a very cautious & considerate way of expressing himselfe; which after turn'd to his no small aduantage. The Worst Quality he had, was his Choller, to[n] Excesses of which he was excessiuely[o] prone: & that bei[n]g the onely Passion to which Philaretus was (much) obseru'd to be inclin'd: his Desire to shunne clashing with his Gouernor; & his accustomednesse to beare the sudden sallys of his impetuous humor, taught our Youth, so to subdue that Passion[p] in himselfe, that[q] he [r-]was soone able[-r] to gouerne[s] it, habitually & with ease. The Continuance[t] of which Conquest he much acknowledg'd to that Passage of St. James; *For the Wrath of Man*[u] *worketh not the righteousnesse of God* [*In marg.* Jam. 1. 20.]. And he was euer a strict obseruer of that Precept of the Apostles, *Let not the Sun go downe vpon Your wrath* [*In marg.* Eph. 4. 26.]: for Continu'd Anger turnes easily to Malice; which made him,

[a] Followed by *his* deleted. [b] Followed by *assig* deleted. [c] Interlined. [d] Followed by *eit* deleted. [e] Altered from *knowing*. [f] Followed by *Younger* deleted. [g] Followed by *professt to* deleted. [h] Followed by *not* deleted. [i] Interlined. [j] Interlined. [k] Followed by *respect* deleted. [l] Interlined above *stoically;* no deletion. [m] Followed by *strict* deleted. [n] Followed by *an* deleted. [o] Followed by *pro enclin'd* deleted. [p] *Anger* interlined above; no deletion. [q] Followed by *at last* deleted. [r-r] Interlined above *came* deleted. [s] *bridle* interlined above; no deletion. [t] *Enioyment* interlined above; no deletion. [u] Followed by *God* deleted.

vpon occasion of this sentence of St Paull; to say that[a] Anger was like the Jewish Manna, which might be wholesome for a Day or so; but[b] if 'twere kept [c]–at all–[c], 'twud presently[d] breed Wormes, & corrupt.*

With this new Gouernor our Philaretus spent the Greatest Part of the Summer, partly in reading[e] & interpreting the vniuersall history written in Latin† & partly in a familiar Kind of [f]–Promiscuous Discourse–[f] in French, which Philaretus found equally Diuerting & Instructiue, & which was as well consonant to the humor of his Tutor, as his owne. About this time his eldest Brother the Lord of[g] Dungaruan, hauing at[h] his owne Charges rais'd a Gallant Troope of Horse for the King's seruice in the Scotch Expedition; His Father sent[i] two other of his sons, (Kinalmeaky[j] & Broghill) to accompany him in that seruice,‡ & design'd Philaretus for the same Employment, but[k] the[l] Sicknesse of his Elder Brother, (Mr F.) whom he was to go along with in that Voyage, Defeated all our Young man's greedy hopes. During his stay at Stalbridge, all that Summer, His Father, to oblige him to [be] Temperate, by freely giuing him the Opportunity to be otherwise; trusted[m] him with the Key of all[n] his[o] Garden & Orchards. And indeed Philaretus; was little giuen to greedinesse either in Fruite or sweetmeats; in the latter he [p]–had no fancy to at all–[p]; & in the former he was very Moderate: so valuing such

---

[a] Followed by *Mann* deleted. [b] Preceded by *if* deleted. [c–c] MS has *all it.* [d] Followed by *corrupt &* deleted. [e] Followed by *the* deleted. [f–f] Interlined above *Conversation;* no deletion. [g] Interlined. [h] Interlined above *vpon* deleted. [i] Followed by *the* deleted. [j] Preceded by *the Lords of* deleted. [k] Followed by *his* deleted. [l] Followed by *Desperate lon* deleted. [m] Followed by *w* deleted. [n] Interlined. [o] Followed by *Orchar* deleted. [p–p] *was almost abstemious* interlined above; no deletion.

---

* Exodus xvi. 20.

† Perhaps part of Giovanni Botero, *Relazioni universali* (various Italian editions), which had been translated into Latin. (Bompiani, *Opere*, **VI**, p. 175).

‡ The Earl of Cork's diary contains the following entries:

4 *May* 1639. "This day my son Dongarvans waggon and carriadges began their Jorney from Stalbridge to goe to his Ma[ts] Camp at Newcastle, or yorck."

"Dick power, Dongarvans Cornett of his horse troop, arrived at Mynhead owt of Ireland yesterday, and cam this night late to Stalbridge, leaving the horsemen and horses to come hether after him."

7 *May* 1639. "Cornett Richard power, with some horses and horsemen, cam to Stalbridge owt of Ireland."

9 *May* 1639. "This daie . . . my three sons, the Lord Dongarvan, Kynalmeakie and Broghill, with the troop of one hundredth horse that I supplied my son dongarvan with moneis to raise, buy, and arme, to serve his Ma[ty] withall against his rebillows Scottish Subiects, departed from me at Stalbridge."

24 *June* 1639. "My son Broghill came poste from the campe, & attayned to Stalbridge on Midsomer daie at two of the clock in the morning, and brought hether the firste happie newes that his Ma[ty] had concluded an honorable peace with the scotts, and dissolved his army, before thearle of Barrymore landed owt of Ireland, with his yrish Regiment of 1000 foot."

(*I. Lismore*. **V**, pp. 87–89, 96).

Nicetys & Daintys, that tho he enjoy'd them with Delight;/Ple[asure]:/ he could want them without the least Disquiet/Regret/. During this[a] Pleasing season, when the intermission of his Studys, allow'd Philaretus Leasure for Recreations; he would very often steale away from all Company, & spend 4 or 5 howres alone in the fields, to [walk] about,[b] & thinke at Random; making[c] his delighted Imagination[d] the busy[e] Scene, where some Romance or other was dayly acted: which tho imputed[f] to his Melancholy, was in effect[g] but an vsuall Excursion[h] of his [i]-yet vntam'd-[i] Habitude of Rauing; a Custome (as his[j] owne Experience often & sadly taught him) much more easily contracted, then Depos'd.

Towards the End of this summer, [k]-the Kingdome-[k] hauing now attain'd a seeming settleme[n]t by the King's Pacification with the Scotts, there arriu'd at Stalbridge Sr Thomas stafford, Gentleman-vsher to the Queen, with his Lady[l] to visit their old Friend the Earle of Corke, with whom, ere they departed, they concluded a Match betwixt his Fourth sonne, Mr F. B: & Mrs E. K.[m] Daughter to[n] my L.S. by Sr K: & then a Maid of Honor[o] both Young & Handsome.* To make his Addresses to this Lady, Mr F. was sent (& Ph. in his Company) before vp to London:† whither within few weekes they were

[a] Followed by *time what* deleted. [b] Followed by *& Raue* deleted. [c] Preceded by *acti* deleted. [d] *Fancy* interlined above; no deletion. [e] Interlined. [f] Interlined above *ascrib'd* deleted. [g] Interlined above *deed* deleted. [h] Interlined above *Effect* deleted. [i-i] Interlined above *ramage;* no deletion. All interlined above *Ill centred* deleted. [j] Followed by *so Ex* deleted. [k-k] Interlined above *the Kingdom affaires* deleted. [l] Followed by ([Blank] *Earle of Corke*, [Blank]) deleted. [m] Followed by *th* deleted. [n] Followed by *th* deleted. [o] Followed by *of* not deleted and by *about much cry'd vp for her* deleted.

* Under date 6 August 1639 the Earl of Cork wrote in his diary:

> "My Ladie stafford and I conferred privately between ourselves towching our children & concluded." (*I. Lismore*. V, p. 101).

Elizabeth Killigrew, the wife chosen for Francis Boyle, was the daughter of Sir Robert Killigrew, (1579–1633), Vice-Chamberlain to Queen Henrietta Maria, by Mary, daughter of Sir Henry Wodehouse. Lady Killigrew married as her second husband Sir Thomas Stafford.

Because of Francis's youth and his impending departure to travel abroad, the Earl of Cork wished for no more than a marriage contract at this time. He was out-manoeuvred by Lady Stafford, who, through her intercession with the King and Queen, secured a speedy marriage of her daughter.

(*I. Lismore*, V, pp. 106, 111: *II. Lismore*. IV, p. 90. Letter of Charles I to the Earl of Cork dated 4 September [1639]: B.M., Add. MSS. 19832, fol. 48v: *II. Lismore*. IV, p. 91. Letter of Lady Stafford to the Earl of Cork dated 4 September 1639: *ibid*. p. 92. Letter of Earl of Cork to Lady Stafford dated 19 September 1639.)

† Francis and Robert Boyle were preparing for their foreign travels in the latter part of August 1639. (*II. Lismore*. IV, p. 86). The Earl of Cork's diary subsequently records:

> "This Thursday being the xix of 7ber, 1639, my two yongest sons, Frances & Robert, with their governor and shannowey [Chanouis], their French servant, departed from Stalbridge towards London. And I gaue Mr Marcombes xli and delivered him other xxli to bear their chardges and to give me an accompt of at London . . . And I gave to franck & Robin 40s a peec." (*I. Lismore*. V, pp. 107–108).

follow'd by the[a] Earle & his Family; of which a greate part liued at (the Lady Staffords house), the[b] Sauoy;* the rest (for his Family was much encreased by the Accession of his Daughters the Countesse of Barrymore & the Lady Ranalagh, with their Lords & Children;) were lodg'd in the Adiacent[c] Houses; but [d]–tooke their Meales–[d] in the Sauoy; where the Old Earle kept so plentifull a house, that in [Blank] months his accompts for bare house-keeping exceeded [Blank] pound.

Not long after his arriuall, Philaretus his Brother hauing been succesfull /hauing prosper'd/ in his Addresses to his Mistris, at last in the Presence of the King & Queen, publickely marry'd her at Court;† with all that Solemnity that vsually attends Matches with Maids of Honor. But to render this Joy as short as it was Greate,[e] P. & his Brother, were within 4 dayes after commanded away for France,‡

[a] Altered from *their Fa.* [b] Preceded by *in* deleted. [c] Followed by *Lod* deleted. [d–d] Interlined above *liued* deleted. [e] [*In marg.*] He was so suddenly depriued of his Joy; that he had scarcely [1]–found that it was reall–[1], before he was forc't from it.

[1–1] *leasure to consider it* interlined above; no deletion.

* The Earl of Cork with his retinue left Stalbridge, 1 October 1639, to journey towards London, where he arrived safely three days later, and

"where Sir Thomas stafford had prepared his howse of the Savoye bravely furnished withall things, except Lynnen and plate, which I brought with me, and in a moste noble and frendly manner lent it me freely to wynter in." (*I. Lismore.* V, p. 111).

† The Earl of Cork records the event thus:

24 October 1639. "This day my Fourth son, Frances Boyle, was married in the Kings chapple at Whitehall to M[rs] Elizabeth Killoghgreev. . .The Kinge with his own Royall hand gaue my son his wife in marriadge, and made a great Feast in court for them, wherat the Kinge and queen were both presente, and I, with 3 of my daughters satt at the King and queens table, amongst all the great Lords and Ladies. The King took the bryde owt to dawnce; and after the dawncing was ended, the King led the Bryde by the hand to the bedchamber, where the queen herself, with her own hands, did help to-vndress her. And his Ma[ty] and the queen both staied in the bedchamber till they saw my son & his wife in bed together; and they bothe kissed the bride and blessed them, as I did: and I beseech god to bless them." (*I. Lismore.* V, p. 112).

‡ The travellers left London 28 October 1639; and the Earl of Cork put in his diary

"My sons Frances and Robert Boyle (the god of heaven ever bless, guide, and protect them), with M[r] Marcombes, shannoway, and the French boy, being 5 in company, M[r] Marcombes having, when they departed stalbridge, receaved 30[li] of me, and more at their departure from London, other 250[li] for their half yeares maintenance before hand at geneva, departed London this day, having his Ma[ts] licence vnder his hand and privy signett for to continew abrode 3 yeares: god guide them abrod and safe back." (*I. Lismore.* V, p. 112).

The King's licence was in the following terms:

"A Licence to Frances Boyle and Robert Boyle Esq[rs] two of the Sonns to the Earle of Corke, to travaile beyond Seas, w[th] six Servants, 50[li] in money & theire necessarie utensills, there to remaine, 3. yeares frō the tyme of theire departure. His Ma[ts] pleasure signified by M[r] Secretarie Coke and by him pr[oc]ured Dat. 23° Octob. 1639." (Public Record Office. Signet Office Docquets. Index 6811).

& after hauing Kiss't their Maiesties hands; they tooke a differing[a] farewell of all their Friends; the Bridegroome extreamely afflicted to[b] be so soone depriued of a Joy; which he had tasted but iust enuf of, to encrease his Regrets, by the Knowledge of what he was forc't from; but Philaretus as much satisfy'd; to see[c] himselfe in a Condition to[d] content a Curiosity to which his Inclinations did passionately addict him. With these differing resentments of their Father's Com[m]ands; accompany'd by their Gouernor, two French Seruants, & a Lacquay of the same Cuntry; vpon the End of October 1638 [*sic*],* they tooke Post for Rye in Sussex; where the next day hiring a Ship, tho the Sea were not very smooth, a Prosperous [e–]Puffe of Wind[–e] did safely[f] by the next morning, blow them into France.†

The Third Section

After a short refreshment at Deepe, our Philaretus trauell'd through[g] Normandy to the cheefe City of it, Rouën;[h] but by the way[i] receiu'd

[a] Interlined above *sad* deleted. [b] Followed by *ha* deleted. [c] Interlined above *see* altered to *send* deleted. [d] Followed by *see & learne new things* deleted. [e–e] Interlined above *Gale,* deleted. [f] Followed by *blow them* deleted. [g] Interlined. [h] Preceded by *Roan* deleted. [i] Followed by *u* deleted.

* A slip of the pen by RB. It was 1639; see note ‡ on p. 25.

† An account of RB's travels on the continent has been given by the present writer elsewhere; see reference in note § on p. 21.

The travellers arrived at Rye 29 October 1639, and as the wind was favourable, Marcombes hired a barque in which the company put to sea the same evening. However, the wind changed, and they were obliged to return to Rye, where they spent the night. (*Lismore MSS.* **XX,** No. 117, 118. Letters of Marcombes to Earl of Cork, 29 and 30 October 1631, the latter wrongly dated 'y[e] Last of 8[ber] 1639'.)

> They sailed a second time the next morning (30 October), "and had uery fauorable winde all y[e] day long and untill twelue of y[e] clocke of y[e] night following, about which Time we came in sight of Diepe, but because then y[e] tide was quit spent we were forced to cast an ancher and to be att Sea all night where we were tossed in a most strange manner, and ye next morning about nine of y[e] clocke we had much adoe to gett Diepe for y[e] sea was uery rough and y[e] winde mighty uiolent, and y[e] seamen of Diepe worse then they both, for they came to us perforce and tooke our portemantels and our personns in spit of our teeth and made us pay what they would, but they told us that they doe not use others better, for my part I was neuer used worse of y[t] kinde of people then att Rye and att Diepe, y[e] Donkercks after gaue us a little aprehention, but thankes be to God we are arriued into this towne in perfect health onely M[r] francis and M[r] Robert were a little wery, and because of that I did desire them to supp sooner than ye rest of ye companie, and soe they are gone to bed and sleepe all ready soundly, Tomorrow God willing because we are ten or twelue in companie, and y[t] ye wayes are some thing dangerous, we intend to Goe to Rohan [Rouën] where we shall rest ourselues one day or two and see ye rareties of y[t] cittie. . .I tooke a barke a purpose att Rye because there was no passengers but onely two or three other gentlemen y[t] would not be att soe great charges as I, yet they payd something for theyr share." (*Lismore MSS.* **XX,** No. 119. Letter of Marcombes to Earl of Cork, 13 October 1639).

aduertisement of a Robbery freshly committed in a Wood, he must[a] trauerse by Night; but iudging the feare of being apprehended would deterre the Robbers from a suddaine return to the same Place after so recent a Crime, the Company quietly continu'd on their Journy to Rouën, & arriu'd safely at it: where amongst other singularitys, Philaretus tooke much Notice of a Greate Floating Bridge,* which rising & falling as the Tyde-water[b] dos, he vs'd to resemble to the vaine Amorists of[c] outward Greatnesse, whose spirits[d] resent all the floods & Ebbes[e] of that Fortune it is built on. From Roane† they pass't to Paris; & hauing spent some time in visiting that vast Chaos of a Citty;‡ they shap't their Course for Lyons;§ where, after 9 dayes vnitermitted trauell they arriu'd: hauing by the way (besides diuers considerable Places) passed by the Towne of Moulins; (here fam'd

[a] Interlined above *did;* no deletion; preceded by *mus* deleted. [b] Followed by *did* deleted. [c] Followed by *wo* deleted. [d] Followed by *do* deleted. [e] Followed by *the fickle Fortune* deleted.

* This *Pont de Batteaux* attracted the attention of most travellers. Cf. Evelyn (de Beer) **II**, p. 122 and n.5.

At Rouen the party caught up with the sons of Sir William Steward, who wished to continue the journey with Marcombes, but he had no inclination to wait for them, as the Stewards had not kept their promise to travel with him from Rye. "We had very good companie of french gentlemen which we would not Loose, because there was great danger of being robbed between Rohan and Paris." (*II. Lismore.* **IV**, p. 96. Letter of Marcombes, to the Earl of Cork dated 10 November 1639). Presumably 'the Stewards' were sons of Sir William Stewart (?-1647). (G.E.C. *Baronetage.* **I**, p. 251: W. Playfair. *British Family Antiquity*, Vol. **IX**, London, 1811. Appendix, p. clxxix.)

They probably left Rouen, 3 November 1639, and would have passed through Fleury, Magny (where doubtless they spent the night), Pontoise, on to Paris, where they arrived, 4 November 1639. (*II. Lismore.* **IV**, p. 96.)

† A slip of the pen for *Rouën.*

‡ They spent about 10 days at Paris. RB gives no account of what he saw there, though Marcombes says they saw 'the most part of ye varieties of this famous Cittie' (*ibid:* cf. Evelyn (de Beer) **II**, pp. 91-135). Amongst the English gentlemen they met at Paris was Francis Boyle's brother-in-law, Thomas Killigrew (1612-1683), who later caused trouble by his false reports, which implied that Marcombes was neglecting his own obligations. (*Lismore MSS.* **XXII**, No. 39. Letter of Francis Boyle to the Earl of Cork dated 20 July 1641: *II. Lismore.* **IV**, p. 231. Letter of Marcombes to the Earl of Cork dated 20 December 1641).

Marcombes tells us that

> "M[r] Francis att his departure from London was so much troobled because of your Lordships anger against him that he could neuer tell us where he put his sword and y[e] kaise of pistoles that your Lordship gaue them, so that I haue been forced to buy them here a kaise of pistolles a peece, and because it is so much y[e] mode now for euery gentleman of fashion to ride with a kaise of Pistoles, that they Laugh att those that haue them not. I bought also a Sword for m[r] francis and when M[r] Robert saw it he did so earnestly desire me to buy him one, because his was out of fashion, that I could not refuse him that small request." (*II. Lismore.* **IV**, p. 96).

In the afternoon, 10 November 1639, they paid the customary visit to the English ambassador extraordinary, Robert Sidney, 2nd Earl of Leicester (1595–1677).

§ The route from Paris to Lyons would have taken them through Montargis, Nevers, Moulins, and Roanne.

enuf for the fine Tweezes it supplys vs with;* a part of the [a]–French Arcadia, the–[a] pleasant Pays de Forest; where the Marquis d'Vrfé† was pleas'd to lay the Scene of the Aduentures & Amours of that Astrea, with whom so many Gallants are still in loue so long after both his & her Decease: being also by the way[b] vsefully diuerted by the Company of two Polonian Princes, who had[c] as well a right vnto that Title, by their Vertu & their Education, as their Birth.

After some stay at Lyons,‡ (a towne of great Resort & no less Trading, but fitter for the Residence of Marchands then of Gentlemen,) they trauers't those lofty mountaines that formerly[d] belonging to the Duke of Sauoy, were now by stipulation (in exchange of the Marquisat of Saluzzo (deuolu'd to the French King;§ & hauing by the Way beheld that famous Place, where the Swiftest & one of the Noblest Riuers of Europe, the Rhosne, is so streightned betwixt to neighbour Rockes, that 'tis[e] no such[f] large stride to stand on both his Bankes;‖ the third Day after their Departure from Lyons, they arriu'd safely at Geneua;¶ a little Common-wealth which their early embracing & constant Profession of the Reformed Religion, together[g] with that peculiar[h] Care of Prouidence in so long & so vnlikely a Preseruation,

---

[a]–[a] Interlined. [b] Followed by *hi* deleted. [c] Followed by *a* not deleted and by *better* deleted. [d] MS. has *fortmerly*. [e] Interlined. [f] Interlined. [g] Preceded by *has* deleted. [h] Followed by *w* deleted.

---

* Moulins was famed for its cutlery. Cf. Evelyn (de Beer), **II**, p. 155.

† Honoré d'Urfé (1567–1625), Comte de Chateauneuf, Marquis de Valromery. Author of *L'Astrea*, a pastoral romance in five parts published between 1607 and 1627, which long enjoyed popularity on the continent during the seventeenth century. Bompiani, *Opere*, **I**, p. 294: Bompiani, *Autori*, **III**, p. 743.

‡ Cf. Evelyn (de Beer), **II**, p. 156–158: J. Lough. *Locke's Travels in France 1675–1679*. Cambridge, 1953, pp. 4–8.

§ By the Treaty of Lyons, 1601, Henry IV of France ceded Saluzzo to Charles Emmanuel I, Duke of Savoy, in exchange for Bresse and other places in Savoy.

‖ *La Perte du Rhône* is described by H. B. de Saussure in *Voyages dans les Alpes*, Neuchatel, 1803–1796, Vol. **II**, p. 61.

RB in his writings in later years not infrequently referred to things he had seen during his travels on the continent. In the *New Experiments Physico-Mechanicall*, published in 1660, he says:

> " I thought it might be said, that motion alone, if vehement enough, may, without sensible heat, suffice to break water into very minute parts, and make them ascend upwards, if they can no where else more easily continue their agitation."

After recalling his visit to *La Perte du Rhône*, he adds in support of this opinion:

> " that rapid stream dashing with great impetuosity against its rocky boundaries, doth break part of its water into such minute corpuscles, and put them into such a motion, that passengers observe at a good distance off, as it were a mist arising from that place, and ascending a good way up into the air." (*Works Fol.* **I**, p. 35. *Works*. **I**, p. 53).

¶ The travellers must have left Lyons, 25 November 1639, in order to arrive at Geneva three days later. (*I. Lismore*. **V**, p. 117). Marcombes' first letter from Geneva to the Earl of Cork in which he gave an account " of all ye passages and circomstances of our peinefull but (thankes to God) very happy journey from London to Geneua " has not been found. (*II. Lismore*. **IV**, p. 98). The receipts and disbursements in the course

has rendred very much the Theame [a]not only[a] of Discourse, but some Degree of Wonder.

Philaretus his Instructions[b] commandi[n]g him a long stay in this Place, his Gouernor, (hauing both[c] Wife & Children in the Towne) prouided him Lodgings & Entertainment in his owne house;* & at

---

[a–a] Interlined. [b] Followed by *tying* deleted. [c] *a* interlined above *both;* no deletion.

---

of this journey were as follows:

	£ s d
"Mr Marcombes disbursments for Mr Francis and Mr Robert Boyle in their trauells from London to Geneua as by his acct appears is	106 - 9 - 0
More his reconing to paie his debts	020 - 0 - 0
More that his wife was damnefied in wine & Lodging	013 - 15 - 0
Totall	140 - 4 - 0
Delivered Mr Marcomul by Wm Chettle the 19° September 1639	30 - 0 - 0
More he received for Mr Francis and Mr Robert Boyle there halfe years allowance the 26° October 1639"	250 - 0 - 0

(*Lismore MSS.* **XX**, No. 140). The Earl of Cork protested that this was an 'extravagant accompt'. (Earl of Cork's Letter Book, 24 April 1634–19 July 1641). In the same letter, dated 18 January $16\frac{39}{40}$, the Earl of Cork exhorted Marcombes to "provyde all things necessary, & nobly for my Children, to be mayntayned in such Fashion, & degree, as becomes my Sons, acording the great trust J repose in you."

* The two boys, as stated here, lodged in the house of Marcombes, and not, as has sometimes been said, at the Villa Diodati outside Geneva. (Flora Masson. *Robert Boyle*. London, 1914, p. 82). The Villa Diodati at Cologny did not exist before 1710. (W. S. Clark. 'Milton and the *Villa Diodati*.' *The Review of English Studies* (1935), **11,** 51). Their two brothers, Lewis and Roger, had indeed stayed with Jean Diodati. It is safe to say that Francis, and Robert were received into his house, particularly in view of the family connection with Marcombes, and that they attended the church of which Diodati was pastor. Diodati may well have been a formative influence in the development of RB's deeply religious outlook.

The Earl of Cork allowed £500 a year for the maintenance of the two boys, and Marcombes was always anxious to assure him that they were being well cared for in his home. Fewer informative letters are available concerning the stay of Francis and Robert on the continent than in the case of their elder brothers Lewis and Roger. We do, however, obtain some glimpse of their domestic affairs. In his letters to the Earl of Cork, Marcombes says:

> "I haue made a complet blacke satin sute to Mr Robert; ye cloake Lined with plush, and I allow them euery moneth a peece ye value of neare two pounds sterlings for their passe time." (*II. Lismore*. **IV**, p. 99. Letter of 7 January $16\frac{39}{40}$).

> "This Leter end of ye winter is mighty cold and a great quantity of snow is fallen upon ye ground, but that brings them to such a stomacke that your Lordship should take a great pleasure to see them feed. I doe not giue them daintys, but I assure your Lordship that they haue allwayes good bred and Good wine, good beef and mouton, thrice a weeke good capons and good fish, constantly two dishes of frut and a Good peece of cheese; all kind of cleane linnen twice and thrice a weeke and a Constant fire in their chamber wherein they haue a good bed for them, and another for their men." (*II. Lismore*. **IV**, p. 100. Letter of 12 February $16\frac{39}{40}$).

> "I haue fournished Mr francis and Mr Robert with three sutes of Clothes a peece and with all kind of good Linnen; I haue made also my prouision of Corne, wine, wood and Candles for all this winter; but I protest unto your Lordship

set howres taught him[a] both Rhethoricke & Logicke, whose Elements (not the Expositions) Philaretus wrote out with his owne hand; tho afterwards he esteem'd both those[b] Arts (as they are vulgarly handled[c]) not only vnseasonably taught, but[d] obnoxious to those (other) Inconveniences & guilty of those Defects, he dos fully particularize in his Essays. After these slighter studys he fell to learne the Mathematickes, & in a few months grew [e-]very well acquainted with[-e] the most Practicke[f] Part of Arithmetick, Geometry (with it's subordinades [*sic*]) the[g] Doctrine of the Sphere, [h-]that of the Globe[-h], & Fortification, in all which being instructed by a Person that[i] had [j-]a gre[a]ter[-j] [regard] of his Scholler's Proficiency, then the Gaines he might deriue from the common tedious & dilatory way of teaching, he quickly grew so enamor'd of those delightfull studys, that they were often prou'd both his Bisness & his Diuersion in his Trauels; & he afterwards improu'd his Opportunitys to the attainment of a more then ordinary skill in diuers of them.* He also frequently converst

---

[a] Followed by *the* deleted. [b] Followed by *sciences* deleted [c] Interlined above *taught* deleted. [d] Followed by *guilty* deleted. [e–e] *attain'd a competent skill in* interlined above, and deleted. [f] Interlined above *vsefull;* no deletion. [g] Followed by *Sph* deleted. [h–h] Interlined above *Geography* deleted. [i] Altered from *the.* [j–j] Interlined above *more;* no deletion.

---

that all things here are deare aboue all measure; and of that your Lordship shall iudge by this, I pay for a paire of great bootes for Mr Robert the ualue of fiue and twenty shelings and for a paire of shoes for him the ualue of fiue shelings, for a quarteron of wine which is two quarts of London, we pay ye ualue of a sheling and as much for a pound of Candels. The wood also is dearer here then in any other place of ye wordle that I know, to the great greef and Misery of ye poore people; and the reason thereof is because they transport all kinds of Commodities into Germany or into the frenche Conte of Burgondy." (*II. Lismore.* **IV**, p. 162. Letter of 16 November 1640).

* RB's account of the instruction he received in Geneva can be amplified from the letters of Marcombes to the Earl of Cork.

"And as for their Learning . . . we speake all and allwayes french, wherein Mr Robert is perfect allready and Mr Francis able to expresse him selfe in all companies; besides euery morning I teach them ye Rhetorike in Latin, and I expound unto them Justin from Latin into french, and presently after dinner I doe reade unto them two chapters of ye old Testament with a brief exposition of those points that I think that they doe not understand; and before supper I teach them ye history of ye Romans in french out of florus and of Titus Livius, and two sections of ye Cateshisme of Caluin with ye most orthodox exposition of the points that they doe not understand; and after supper I doe reade unto them two chapters of ye new Testament, and both morning and euening we say our prayers together, and twice a weeke we goe to church." (*II. Lismore.* **IV**, p. 100. Letter of 12 February $16\frac{39}{40}$).

Instruction in mathematics and fortification continued throughout the year. RB's letters during this period are so badly mutilated as to be impossible to reconstruct, but they report his progress in these studies. In later life RB referred to various aspects of the tuition he received with Marcombes. Apart from his comments, here, on rhetoric and logic,

with a Voluminous, but excellent French Booke, that call'd, *Le Monde*,* which so iudiciously informes its Readers, both of the Past & present Condition of all those States that now possesse[a] our Globe & by a delectable & instructiue variety not only [b]–satisfys men's Curiosity–[b], but so copiously supplys them with matter, both of Curious & serious Discourse, that he vs'd to say of[c] that Booke, that it was worth it's Title which meanes, *The World*.

But to employ his Body, as well as his Mind, because P. his Age was [d]–yet vnripe–[d] for so rude & violent an Exercise as the Greate horse, he spent some Months in Fencing, & 10 or 12 in learning to Dance;† the former of which Exercises he euer as much affected as he contemn'd the latter.

His Recreations[e] during his Stay at Geneua were sometimes Maill, Tennis (a Sport he euer passionately lou'd;) & aboue all the Reading of Romances; whose Perusall did not only extreamely diuert him; but (assisted by a Totall Discontinuance of the English tongue[f]) in a short time taught him a Skill in French[g] (somewhat) vnusuall to Strangers. In effect, before he quitted France, he attain'd a readinesse in the language of that Country, which enabled him, (when he made Concealement his Dessein) to passe for a Natiue of it, both amongst them that were so; & amongst Forreiners also: & in all his writings whilst he was abroad; he still[h] made vse of the French Tongue, not out of any intention[i] to improue his Knowledge in it; but because it was that he could expresse himselfe best in.

---

[a] Followed by *th* deleted. [b–b] Interlined above *diuerts them;* no deletion. [c] Followed by *it* deleted. [d–d] *not yet* interlined; no deletion. [e] Interlined above *Diuersions* deleted. [f] Followed by */lang/ taught* deleted. [g] Followed by *no* deleted. [h] Interlined. [i] Followed by *of* not deleted and by *im* deleted.

---

in his essay 'Of the Usefulness of Mathematicks to Natural Philosophie', (1671), he says of these studies:

> "I have often wished, that I had employed about the speculative part of geometry, and the cultivating of the specious algebra I had been taught very young, a good part of that time and industry, that I spent about surveying and fortification (of which I remember I once wrote an entire treatise) and other practick parts of mathematicks." (*Works Fol.* **III**, p. 156; *Works*. **III**, p. 426).

* This work is undoubtedly that by Pierre Davity. *Le monde ou la Description Générale de ses Quatre Parties avec tous ses Empires, Royaumes . . .* 7 vols. Folio, Paris, 1637. There were several subsequent editions.

† On these exercises Marcombes says:

> "Mr Robert Loues his booke with all his heart; ye onely thing yt vexes him now is, that ye fencing Maister of this cittie is yet soe sicke yt he can not giue him Lessons till ye Latter end of this moneth; and that exercise he is soe desirous to Learne yt I am almost afraid yt he should haue left a quarell unperfect in England: but for his danse, his Maister assures me yt he shall doe admirable well." (*II. Lismore*. **IV**, p. 98. Letter dated 7 January 16$\frac{39}{40}$).

But during Ph. his Residence at Geneua there hapned to him an Accident which he euer[a] vs'd to mention as the Considerablest of his whole life. To frame a right apprehension of this, yow must vnderstand; that tho Philaretus his Inclinations were euer Vertuous, & his Life [b]-free from-[b] Scandall & Inoffensive; yet had the Piety he was master of allready, so diuerted[c] him from aspiring vnto more; that Christ who long had layne asleepe in his Conscience (as he once did in the Ship vpon [Blank],) must now (as then) be wak't by[d] a Storme. For on a night which (being the[e] very heart of summer) promis'd nothing lesse; about that time[f] of[g] Night that adds most terror to such accidents; P. was suddenly waked in a Fright with such loud Claps of Thunder (which are oftentimes very terrible in those hot Climes & Seasons;) that he thought the Earth would owe an ague to the Aire: & euery clap was both[h] preceded & attended with Flashes of lightning so numerous /fr[equent]/ & so dazling, that P. began to imagine them the Sallyes of that Fire that must consume the World. The long Continuance of that /a/ dismall Tempest where the Winds were so loud as almost drown'd the noise of the very thunder; & the showres so hideous, as almost quench't the Lightning ere it could reach his Eyes; confirm'd P. in his Apprehensions, of the Day of Judgment's being at hand /come/: Wherevpon the Consideration of his Vnpreparedness to welcome it; & the hideousnesse of being surprized by it in an vnfitt Condition, make him Resolue & Vow, that if his Feares were that night disappointed all further additions to his life shud be more Religiously & carefully[i] employ'd. The Morning come, & a[j] serener cloudlesse sky return'd he[k] ratify'd his Determination so solemnly that from that Day he dated his Conversion; renewing now he was past Danger, the vow he had made whilst he fancy'd[l] himselfe to be in it: that tho his Feare was (& he blush't it was so) the Rise/occasion/ of his Resolution of Amendment; yet at least he might not owe his more deliberate consecration/ded[ication]/ of himselfe to Piety, to any lesse noble Motiue then that of it's owne Excellence. Thus[m] had this happy storme an operation vpon P. resembling that it had vpon the Ground: for the Thunder did but terrify/fright/ & Blasted not: but with it fell such kind & geniall showres, as water'd his parch't & almost wither'd Graces/vertues/& reuiuing their Greenenesse, soone render'd them both Flourishing &

[a] Interlined above *always;* no deletion. [b-b] Interlined above *voyd of* deleted. [c] Altered from *diuerting*. [d] Interlined above *vpon;* no deletion. [e] Followed by *mi* deleted. [f] *dead* interlined above; no deletion. [g] Followed by *eue* deleted. [h] Interlined. [i] Interlined above *watchfully;* no deletion. [j] Interlined above *the* deleted. [k] Followed by *so* deleted. [l] Interlined above *beleeu'd;* no deletion. [m] Followed by *was Christ borne in* deleted.

Fruitfull.* And tho his boyling[a] Youth did[b] often very earnestly sollicit to [be] employ'd in those culpable Delights that[c] are vsuall in & seeme so proper for that season, & haue repentance adiourn'd till/to/ Gray haires/old Age/ yet did it's importunitys meete euer with Denyalls/repulses/P. euer esteeming that Piety was to be embrac't not [d]–so much–[d] to Gaine heau'n as[e] to serue God with. And I remember that being once in company with a Crew of mad Young Fellowes; when one of them was saying to him, What a fine thing it were if men could sinne securely all their Life-time, by being sure of leasure to Repent vpon their Death-beds; P. presently reply'd, that truly for his part he shud not like/accept/ of sinning, (tho) on those Termes; & wud not [f]–all that while–[f] depriue himselfe of the satisfaction of [g]–liuing vertuously–[g], to enioy so many Yeares Fruition of the World.† In effect, 'tis strange that men shud [h]–make it–[h] for an Inducement to an Action, that they are confident/certin/ that they shall/must/ repent it. But P. himselfe[i] hauing sufficiently discours't[j] that Point of Early Godlinesse/Piety/ in the sixth Treatise of his *Christian Gentleman*,‡ I shall, [k]–at present–[k] onely adde to the [l]–Arguments/reasons/–[l] yow may find recorded[m] There (alleadg'd); that he vs'd to say, that 'twas a kind of Meannesse in Deuotion,[n] to consider the very Joyes of t'other Life, more as a Condition then a Recompense; & that true Gallant Christians (in their Dutys to their Maker) looke vpon Heau'n it selfe as Gallant Louers do on the Portion in their seruices/add/ to their Mistresses; for they consider it as[o] a /property/ Consequent of their Fruition, not[p] the[q] Motiue of their Loue.

But (as when [r]–in sommer–[r] we take vp our[s] Grasse-horses into the Stable & giue them store of oates; 'tis a signe that we meane to trauell

[a] Followed by /[Blank]/. [b] Followed by *ea* deleted. [c] Followed by *see* deleted. [d–d] Interlined above *onely;* no deletion. [e] Interlined above *but;* no deletion. [f–f] Interlined. [g–g] Interlined above *seruing God;* no deletion. [h–h] Interlined above *take it;* no deletion. [i] Altered from *himselves.* [j] *-ours't* interlined above *-uss't;* no deletion. [k–k] Interlined above *now;* no deletion. [l–l] Interlined above *Motiues* deleted. [m] Interlined. [n] *paying Dutys to our Maker* interlined above; no deletion. [o] Followed by *the* deleted. [p] Preceded by *no* deleted and by /[Blank]/. [q] Followed by *sole* deleted. [r–r] Interlined. [s] Followed by *som* deleted.

* RB's conversion presumably occurred before his first communion the last Sunday of the year 1640. (*Lismore MSS.* **XXI**, No. 78. Letter of Marcombes to the Earl of Cork, 23 December 1640.) RB had scrupled to take communion at Easter of that year. (*II. Lismore.* **IV**, p. 114).

† Royal Society MS. 196 contains three essays, the first of which, entitled *Of Sin*, expresses the same sentiment in section 12.

‡ Perhaps the three essays, *Of Sin*, *Of Piety*, *Of Valour*, referred to in the previous note are part of the treatise called *Christian Gentleman*. There is no surviving treatise with this title. The subject of 'death-bed repentance' is touched upon also in the *Christian Virtuoso*, Second Part, Proposition VI, which is a collection of fragments, and admittedly an early work by RB.

them) our P. soone[a] after he had receiued this new strength, found a new weight to support.* For spending some of the Spring in a Visit to Chambery, (the Cheefe Towne of Sauoy) Aix (fam'd for its Bathes) Grenoble the head Towne of Dauphiné & Residence of a Parliament,† his Curiosity at last lead him to those Wild Mountaines where the First & Cheefest[b] of the Carthusian Abbeys dos[c] stand seated; where the Deuil taking aduantage of that deepe, rauing Melancholy, so sad a Place; his humor, & the st[r]ange storys & Pictures he found[d] there of Bruno the Father/Patriark/ of that order,‡ suggested such strange &

[a] Altered from *no sooner*. [b] Followed by *Abbeys* deleted. [c] Interlined. [d] Followed by /[Blank]/.

* RB does not mention that about the middle of June 1640 "we went into Sauoye for to see a little Contrey with two other Gentlemen: y[e] one English and y[e] other french. Wee were two days abroad and were neuer so merry in our Liues." (*II. Lismore.* **IV**, p. 116. Letter of Marcombes to the Earl of Cork dated 23 June 1640). The names of the 'two other Gentlemen' remain unknown. We know very little about the persons whom RB must have met during his initial stay in Geneva, which lasted till the autumn of 1641. Francis and Robert signed their names in the armorial of the Académie de Genève, as their brothers, Lewis and Roger, had done before them. (A. Chopard 'Genève et les Anglais.' *Bull. Soc. d'Histoire et d'Archéologie de Genève.* (1940) **7**, 238). There is no proof that they studied at the Academy for other students and visitors of note were allowed to add their autographs. Thomas Killigrew, whose sister was married to Francis Boyle, was in Geneva from the latter part of March 1640 until about the middle of April following. It was he, who on returning to England, spread reports that Francis and Robert were not being properly maintained, and in consequence Marcombes was put to some trouble to exonerate himself with the help of supporting statements from the two boys. (*Lismore MSS.* **XXII**, No. 39. Letter of Francis Boyle to the Earl of Cork dated 20 July 1641: *II. Lismore.* **IV**, p. 202. Letter of Marcombes to the Earl of Cork dated 14 July 1641: *ibid.* p. 235. Letter of the same to the same dated 20 December 1641). In November 1640 a certain Mr Baker, Francis Lysson (both unidentified), and Robert Coventry, younger son of the former Lord-Keeper Thomas, Baron Coventry (1578–1640), were in Geneva. (*II. Lismore.* **IV**, p. 161. Letter of Marcombes to the Earl of Cork dated 16 November 1640). In May 1641 Bernard Stuart (killed at the battle of Rowton Heath, 26 September 1645) and John Stuart (killed at the battle of Alresford, 29 March 1644), the younger brothers of the 4th Duke of Lennox (1612–1655), together with their tutor Mr Buchanan were in Geneva on their return from Italy. (*Lismore MSS.* **XXII**, No. 25. Letter of Marcombes to the Earl of Cork dated 30 May 1641). In July 1641 RB in a letter to his father said, "among other good company we haue here my Lord of Hertford's two sonnes." (*II. Lismore.* **IV**, p. 193. Letter dated 6 July 1641). They were William Seymour (1620?–1642), styled Lord Beauchamp, and Robert Seymour (1624?–1646). A little later, at the end of August 1641, they met Mr Nightingall and Mr Northon (both unidentified) on their return from Milan. (*Lismore MSS.* **XXII**, No. 56. Letter of Marcombes to the Earl of Cork dated 1 September 1641).

† This excursion was made in April 1641. The route taken was probably from Geneva through Cruseilles, Annecy, Rumilly, Aix-les-Bains, Chambéry, along the Isère to Grenoble, Vienne, Lyons, and thence back to Geneva. Francis Boyle tells us 'we haue seene very good company together with many rare things.' (*Lismore MSS.* **XXII**, No. 33. Letter of Francis Boyle to the Earl of Cork dated 6 July 1641). RB relates, as the only significant event during this excursion, his visit to La Grande Chartreuse near Grenoble.

‡ St Bruno (?–1101), founder of the Carthusian monastic order.

hideous thoughts, & such distracting Doubts of some of the Fundamentals of Christianity/Religion;/ that tho his lookes did little betray his Thoughts, nothing but the Forbiddenesse[a] of Selfe-dispatch, hindred his acting it. But after[b] a tedious languishment of many months in this tedious perplexity; at last it pleas'd God one Day he had receiu'd the Sacrament, to[c] restore vnto him the withdrawne[d] sense of his Fauor. But tho since then Philaretus euer look't vpon those impious suggestions, rather as Temptations to be suppress't /reiected/ then Doubts to be satisfy'd[e]; yet neuer after did these fleeting Clouds, cease now & then to darken/obscure/ the clearest serenity of his quiet: which[f] made[g] him often [h]–say that–[h] Iniections of this Nature were[i] such a Disease to his Faith as the Tooth-ach is to the Body; for tho it be not mortall, 'tis very troublesome. Howeuer,[j] (as all things worke together to them that loue God,) P deriu'd[k] from this Anxiety the Aduantage of Groundednesse in his Religion: for the Perplexity his doubts created oblig'd him [l]–(to remoue them)–[l] to be seriously inquisitiue of the Truth of the very fundamentals of[m] Christianyty: & to peruse[n] what both Turkes, & Jewes, &[o] the cheefe Sects of Christians cud alledge for their seuerall opinions: that so tho he beleeu'd more then he could comprehend, he might not beleeue more then he cud proue; & not owe the stedfastnesse[p] of his Fayth to so poore a Cause as the Ignorance of what might be obiected against it. He say'd (speaking of those /Persons/ that want not[q] Meanes [r]–to enquire &–[r] /Conv[enience]/ &[s] Abilitys to iudge) that 'twas not a greater happyness to inherit a good Religion; then 'twas a Fault[t] to haue it only by inheritance; & thinke it the best because 'tis generally embrac't; rather[u] then embrace it because we know it to be the best. That tho we cannot[v] often give a Reason for/of/ What we beleeue; we shud be euer able to giue a Reason Why we beleeue it. That it is the greatest of Follys to neglect any diligence that may preuent the being[w] mistaken, where it is the greatest of Miserys to be deceiu'd. That how deare soeuer things taken vp on the score are sold; there is nothing worse taken

---

[a] Followed by /[Blank]/. [b] Followed by *many* deleted. [c] Followed by *shine on* deleted. [d] Followed by /[Blank]/. [e] Interlined above *resolu'd;* no deletion. [f] Preceded by *Am* deleted. [g] Interlined above *oblig'd* deleted. [h–h] Interlined above *to compare* deleted. [i] Preceded by *to the* deleted. [j] Interlined above *And;* no deletion. [k] Followed by *this Aduanta* deleted. [l–l] Interlined. [m] Altered from *in.* [n] Interlined above *consider* interlined above *heare;* no deletion. [o] Followed by *m* deleted. [p] Followed by /[Blank]/. [q] Interlined above *neither* deleted. [r–r] Interlined. [s] Interlined above *nor* deleted. [t] Followed by /[Blank]/. [u] Preceded by *not* deleted. [v] Followed by *euer* deleted. [w] Followed by *decei* deleted.

vp vpon Trust then Religion, in which he deserues not to meet with the True one[a] that cares not to examine whither or no it be soe.

And now P. hauing spent one & 20 months in Geneua, about the Middle of September (164) departed towards Italy;* & hauing trauers't Swizzerland, & by the Way seene Lausanne (an Academy seated vpon the Greate Geneuan lake) Zurich & Saleurre, (the heads of Cantons weari[n]g the same Name; & wander'd seuen or 8 dayes amongst those hideous Mountaines wh[e]re the Rhosne takes it's source & he saw the Rhine but a Brooke;† he at length arriu'd at the

---

[a] Interlined.

---

* Marcombes had started making plans for the journey into Italy toward the end of 1640. There was much correspondence between him and the Earl of Cork respecting the route to be taken, money payments, passports, companions, and the dangers of travelling in Roman Catholic territories. Marcombes had intended to start for Italy early in March 1641, but the Earl's consent was so long delayed that he postponed his departure until September of that year. (*Lismore MSS.* **XXI,** Nos. 78, 87, 82, 89. *I. Lismore.* **IV,** p. 170. Letters of Marcombes to the Earl of Cork dated respectively 23 December 1640, 24 January 1641, 2 January [error for February] 1641, 9 February 1641, 20 January 1641: *ibid.* p. 165. Letter of Earl of Cork to Marcombes dated 13 January 164$^{0}_{1}$). This delay in going to Italy was a great disappointment to Francis and Robert; so by way of compensation and as a change from their studies they made the excursion into France, which has already been mentioned (see note †, page 34. *Lismore MSS.* **XXII,** Nos. 2, 14. Letters of Marcombes to the Earl of Cork dated respectively 28 March 1641, 5 May 1641).

The Earl of Cork's reluctance to give permission for the journey into Italy was mainly the result of fear of the danger of persecution, when travelling in Roman Catholic territory, as a possible reprisal for the severe anti-catholic measures then being enforced in England. However, he finally allowed Marcombes to use his own discretion in the matter. (*II. Lismore.* **IV,** p. 205. Letter of the Earl of Cork to Marcombes undated, but about August 1641). Marcombes requested the Earl of Cork to obtain 'letters in Latine from y[e] King to all Kings, princes, Republickes &c in y[e] behalfe of your sons Trauells. . .' 'and that you would be pleased also to get us from his Majesty a speciall Licence to trauell into Rome, Least your Lordship or your sonnes should be questioned hereafter.' (*Lismore MSS.* **XXI,** No. 78: *II. Lismore.* **IV,** p. 173. Letters of Marcombes to the Earl of Cork dated respectively 23 December 1640, 20 January 164$^{0}_{1}$). The Earl attended to this request, for Giovanni Giustinian, the Venetian ambassador in England, informed Francesco Erizzo, the Doge of Venice, that the king had informed him of the intended visit of the Earl of Cork's sons to Venice and had requested letters of introduction. The ambassador supplied such letters on behalf of Francis, and Robert, Boyle to the Rectors of Brescia, as well as to certain members of important Venetian families. (*Calendar of State Papers: Venice, 1640–1642.* London, 1924. p. 204. Despatch dated 28 August 1641).

As a result of the long delay in making their departure, we do not know if Marcombes travelled with extra company; it was his original intention to travel with Messrs Coventry, Baker, Lyssons, and some Dutch gentlemen. (*Lismore MSS.* **XXI,** No. 78. Letter of Marcombes to the Earl of Cork dated 23 December 1640).

† The route from Geneva to Lausanne was certainly by way of Nyon, Rolle and Morges; thence through Moudon, Payerne, Murten, Aarberg to Solothurn; then through Olten, Aarau, and Baden to Zürich; from there the recognized route to Chur was along the northern shores of the Zürchersee and Walensee through Männedorf, Schmeriken, Weesen, Walenstadt and Maienfeld. (Verdier. *Le Voyage de France.* Paris,

Valtollina, a spacious[a] Valley wall'd round with the Steepe Alpes;* but so delicious, & (especially in[b] that season) so crown'd with all that Ceres & Bacchus are able to present; that it deseru'd to be the Motiue, but not the Stage, of those late Warres it has occasion'd betwixt the Riuall Crownes of France & Spaine. There Ph: had the Curiosity to visit the Place on which stood Piur, a Pleasant little Towne; once esteem'd for it's Deliciousnesse; but now much more & more meritedly famous for it's Ruine, which happn'd[c] some 2 dozen of yeares since; by the sudden & vnexpected fall of a neighbouring Hill vpon it, which strucke the whole Towne so deepe into the Ground; that no after-search by digging has euer preuail'd to reach it.†. Hauing visited the singularitys of this Earthly Paradice; P. & his company began to climbe that Mountaine of the Alpes; which[d] denominated by[e] a Towne seated[f] vpon it's foote[g], is vsually call'd La Montagna di Morbegno. The Hill was 8 miles in ascent & double that Number downewards. It was then free from snow; but all the Neighbouring hils (where store of Christall's digg'd) like perpetuall Penitents, do all the Yeare weare white. Vpon the top of this Hill (which is[h] entirely vninhabited) P. had the Pleasure to see the Clouds (which they[i] pass'd

---

[a] Interlined. [b] Interlined above *at;* no deletion. [c] *fortun'd* interlined above; no deletion. [d] Followed by *to* deleted. [e] Interlined above *from;* no deletion. [f] Interlined above *that sits;* no deletion. [g] *—ee—* interlined above *—oo—*; no deletion. [h] Followed by *u* deleted. [i] Followed by *trau* deleted.

1673, pp. 394–396). From Chur there are three possible routes into Italy, (i) via the Splügen Pass, (ii) via the Septimer Pass, (iii) via the Berniner Pass. We are not explicitly told which of these was taken, but RB's reference to 'those hideous mountains, where the Rhosne takes its source, and he saw the Rhine but a brook', leads to the assumption that the Splügen Pass was selected as being the shortest, even though the most arduous.

In one of his *Physiological Essays*, published in 1661, RB, in discussing the effect of different soils upon plants, recalled

> "that travelling divers years since from Geneva towards Italy, I was in my passage through Switzerland by a gentleman of those parts (whose brother had been formerly my domestick) invited to his castle, and entertained among other things with a sort of wine, which was very heady, but otherwise seemed to be sack; and having never met with any such liquor during my long stay in those parts, I was inquisitive to know whence it was brought: and being answered that it grew amongst those mountains, I could not believe it, till they assured me, that growing on a little spot of ground, whose entrails abounded with sulpher, it had from the soil acquired its inebriating property, and those other qualities, which made it so differing from the wine of the rest of the vineyards of that country"

(*Works Fol.* I, p. 210; *Works.* I, p. 327).

* From Chur the route would be Thusis, the Splügen Pass, Chiavenna.

† The catastrophe whereby the little town of Piuro had been completely buried by the fall of a nearby hill occurred in September 1618. There were only four survivors out of an estimated population of 3,600. (*Newes from Italie: A prodigious, and most lamentable Accident, latelie befallen: Concerning the swallowing up of the whol: Citie of Pleurs:* . . . Edinburgh, 1619).

thorough in their desce[n]t) [a]-beneath them-[a] darkning the Midde of the Mountaine; which on the Top they had/enioy'd/ a cleare serentiy.* But notwithstanding the fairenesse of the Day they spent it all in Trauersing this Hill, at[b] the height[c] of which [d]-they left-[d] the Grisons Territory, & at the Bottome enterr'd a Village belonging to (that of the) Venetians; but hauing pass'd ouer[e] such a Purgatory as the Alpes to their Italian Paradice/to that Paradice call'd Italy/, I cannot but suppose them somwhat weary: & so [f]-my Pen-[f] oblig'd to[g] let them (& it selfe) take some short rest.

The [*Fourth Section*]

Philaretus being thus entred into the vast & delicious Plaines of Lombardy, trauers't the greatest part of that rich Prouince, & hauing stay'd the Stomacke of his Curiosity with the Obseruables of Bergamo, Brescia, Verona, Vincenza & Padoa (a famous Vniversity, but more peculiarly deuoted to Esculapius then Minerua's skills[h])† he gaue it a full meale at Venice; where the greate Concourse of forreine Nations numerously resorting thither [i]-for Trade or Nobler Bisnesse,-[i] presents the senses with a no lesse Pleasing then constant Variety. From Venice, returning through Padoa, & passing[j] through Bologna & Ferrara (Townes, whose Names, allowe me to spare their Characters) he at last arriu'd at Florence‡ with a[k] Determination (hauing dispos'd of

---

[a-a] Interlined; *beneath* altered from *below*. [b] Preceded by *of whi* deleted. [c] Interlined above *Bottome* deleted. [d-d] Repeated in MS. [e] Interlined above *thorough;* no deletion. [f-f] Interlined. [g] Followed by *g* deleted. [h] Interlined above *Arts;* no deletion. [i-i] Interlined. [j] *passion* in MS. [k] Followed by *Resolu* deleted.

---

* This experience on crossing the hill at Morbegno is referred to again by RB in his *General History of the Air*, published in 1692. (*Works Fol.* **V**, p. 177; *Works.* **V**, p. 710).

† It is regrettable that RB has not enlarged upon the 'observables' in the course of his journey to Padua, and given us his experiences or impressions. It can only be assumed that the letters of introduction to the Rectors of Brescia were used to advantage. (See note * on p. 36). His remark on the medical renown of Padua makes one wonder if he was present at a dissection there. (E. A. Underwood. 'The Early Teaching of Anatomy at Padua. . .' *Annals of Science*. (1963, published 1964) **19**, p. 1). His reference in the *Usefulness of Experimental Philosophy*, published in 1663, to 'the veins, arteries, and nerves of a human body, laid out in their natural situation upon three boards, by the pains and skill of an accurate anatomist of Padua' (*Works Fol.* **I**, p. 469; *Works.* **II**, p. 74), may relate to recollections of what he himself saw at Padua, or to the set of tables bought by Evelyn when he was in Padua in 1646, and which he brought to England. (Evelyn (de Beer) **II**, p. 475 and note 5).

‡ They probably arrived at Florence about the end of October 1641. The letters written by Marcombes, Francis and Robert from Florence to the Earl of Cork on the 2nd and 20th November 1641 are missing. They were acknowledged by the Earl, who said, 'those two pacquets are all that I haue receaued from you since you departed Geneva'. (*II. Lismore.* **V**, p. 19. Letter of the Earl of Cork to Marcombes dated 9 March 164$\frac{1}{2}$).

the Horses he rid on from Geneua thither,)* to passe the Winter there. Florence is a Citty to which Nature has not grudg'd a Pleasant[a] situation, & in which Architecture has been no niggard either of Cost or[b] Skill: but[c] has so industriously & sumptuously improu'd the Aduantages liberally conferr'd by Nature, that both[d] the[e] seate & Fabrickes/buildi[n]gs/ of the Towne, (abundantly) iustify the Title the Italians[f] haue giuen it of Faire. Here P. spent much of this Time in learni[n]g of his Gouernor (who spake it perfectly[g]) the Italian Tong, in which he quickly attain'd a natiue accent, & knowledge enuf[h] to vnderstand both Bookes & Men; but to speake & expresse himselfe readily, in that language, was a skill he euer too little aspir'd to /car'd for/ acquire. The[i] rest of[j] his spare howres he spent in reading the Moderne history in Italian,† & the New Paradoxes of the greate Star-gazer Galileo;‡ whose ingenious Bookes,[k] perhaps[l] because they could not be so otherwise were[m] /confuted/ by a Decree from Rome; [n]–his H.–[n] the Pope, it seems presuming [o]–(& that iustly)–[o] that the Infallibillity of his Chaire extended [p]–to determine (as vnerringly) points in Philosophy as in Religion;–[p] & loath to haue the stability of that Earth question'd, in which he had established his Kingdome; /Greatnesse/. Whilst[q] Philaretus liu'd at Florence, this famous Galileo dy'd within a league of it; his memory being honor'd with a celebrati[n]g Epitaph, & a[r] faire Tomb erected for [him] at the Publicke Charges/Cost/.§ But before his Death being long growne Blind; to certaine Friers (a Tribe whom for their Vices & Impostures

---

[a] *-ant* interlined above *-ing* of *Pleasing;* no deletion. [b] *of* in MS. [c] Preceded by *Here Phi* deleted. [d] Interlined. [e] Followed by *Towne* deleted. [f] Followed by *ascri* deleted. [g] Interlined above *like a Natiue* deleted. [h] Followed by *readily* deleted. [i] Preceded by *Some Rabb* deleted. [j] Followed by *the* deleted. [k] Interlined above *Opinions;* no deletion. [l] Interlined above *perhaps for* deleted. [m] Interlined above *be ruin'sd* deleted. [n–n] Interlined. [o–o] Interlined, [p–p] Conjectured. MS is confused; it has *as* interlined above *vnerringly* deleted. and *as well* deleted following *Philosophy*. [q] Preceded by *This Famous Galileo* deleted. [r] Followed by *rich* deleted.

---

* Marcombes had told the Earl of Cork that 'there they must keepe theire Coache and Linen Galantly for it is a Court where English Gentlemen are much Considered and observed.' (*Lismore MSS.* **XXI**, No. 78. Letter dated 23 December 1640).

† Compare note † on p. 23.

‡ Galileo Galilei. *Dialogo sopra i due massimi sistemi del mondo.* Florence, 1632. In *Seraphick Love*, published in 1659, RB refers to 'Galileo's Optick Glasses, (with one of which Telescopioes that I remember I saw at Florence, he merrily boasted that he had, *Trovato la Corte a Giove*).' (*Works Fol.* **I**, p. 168; *Works.* **I**, p. 264).

§ Galileo died, 8 January 1642 N.S., at Arcetri in the vicinity of Florence. The proposals to honour Galileo by a public monument were delayed by the great opposition of the Church of Rome, and were not realised for nearly a century. (J. J. Fahie. *Memorials of Galileo Galilei.* Leamington and London, 1929, p. 142).

he long[a] had hated) that reproacht him with his Blindnesse as a Just Punishment of heau'n, incens'd for being so narrowly pry'd into by him, he answer'd, that he had the satisfaction of not being blind till he had seene in heau'n what neuer mortal Eyes beheld before.* But to returne to P. the Company of certaine Jewish Rabbins lodg'd vnder the same roofe with him, gaue him the opportunity of acquainting himselfe with diuers of their Arguments & Tenents, & a Rise of further Disquisitions in that Point. When Carneuall was come (the season when Madnes is so generall in Italy that Lunacy dos for that time loose it's Name) he had the Pleasure to see the Tilts maintain'd by the greate Dukes Brothers;† & to be present at the Great men's Balls. Nor did he sometimes scruple, in his Gouernor's Company, to visit the famousest Bordellos; whither resorting out of bare Curiosity,[b] he retain'd[c] there an vnblemish't Chastity[d], & still[e] return'd thence as honest as he went thither. Professing that he neuer found any such sermons against them, as they were against themselues. The[f] Impudent Nakednesse of vice, clothing it with a Difformity, Description cannot reach, & the worst of Epithetes cannot but flatter. But tho P. were noe Fewell for forbidden Flames, he prou'd the Obiect of vnnaturall ones. For being at that Time [g-]not above 15[-g], & the Cares of the World hauing not yet faded[h] a Complexion naturally fresh enuf; as he was once vnaccompany'd diuerting himselfe abroad, he was somewhat rudely presst[i] by the Preposterous Courtship of 2 [j-]of those[-j] Fryers, whose Lust makes no distinction of Sexes; but that which it's Preference of their owne creates; & not

---

[a] Interlined above *euer* deleted. [b] Followed by *he still brou* deleted. [c] Altered from *return'd;* followed by *thence* deleted. [d] Interlined above *honesty* deleted. [e] Interlined. [f] Followed by *Impudence of* deleted. [g-g] Interlined above *in the Flower of Youth;* no deletion. [h] Interlined above *stain'd;* no deletion. [i] Interlined above *storm'd;* no deletion. [j-j] Interlined.

---

* It is a coincidence that John Milton, who visited Galileo in his old age, was similarly taunted by Salmasius and others. In his reply, Milton denied that loss of sight could be considered as a judicial punishment for his sins. The fallacy that 'all calamitous Providences are to be considered as Judgments' is exposed in the amusing story to which Milton's misfortune has given rise. ([P. du Moulin.] *Regii Sanguinis Clamor ad Coelum.* Hagae-comitum, 1652. Dedication purporting to be by A. Vlac, but by C. Salmasius. D. Masson. *Life of John Milton.* 7 vols. London, 1859–1894. Vol. **IV**, p. 467: J. Milton. *Angli Pro Populo Anglicano Defensio Secunda.* London, 1654, p. 41: *The Prose Works of John Milton.* Bohn Edition. 5 vols. London, [n.d.] Vol. **I**, pp. 222, 235, 237: N. L. Riffe. 'Milton and Charles II.' *N. & Q.* (1964), **209**, 93.

† Prince Leopold de' Medici (1617–1675, afterwards cardinal.

Giovanni Carlo de' Medici (1611–1663), afterwards cardinal.

Mattias de' Medici (1613–1667).

The eldest brother, Grand Duke of Tuscany, was Ferdinand II (1610–1670). (G. F. Young. *The Medici.* 2 vols. London, 1911).

without[a] Difficulty, & Danger, forc't a scape from these gown'd Sodomites. Whose Goatish Heates, seru'd not a little to[b] arme Filaretus against such Peoples specious Hypocrisy; & heightn'd & fortify'd in him an Auersenesse for Opinions, which now the Religieux discredit as[c] well as the Religion. [*In marg.* Place here the story of the Duell with Coates of Maile.]

P. hauing thus spent the Winter in Florence,* towards the End of March began his Journey to Rome; & hauing[d] past thorough & seen the singularitys of Siena, Montefiascone & some other[e] remarkable Places in his Passage,† at the End of 5 dayes he safely arriu'd at that imperious[f] Theame of Fame, which destinated to some kind or other

---

[a] Followed by *Danger* deleted. [b] Followed by *impar* [?] *heihten that Auersenesse* deleted. [c] Followed by *mu* deleted. [d] Followed by *by the Way* interlined and deleted. [e] Followed by *obs* deleted. [f] Interlined above *vast* deleted.

---

* RB gives no account here of the 'rareties' in Florence. From statements in his subsequent published works it is certain that he visited the Palazzo Vecchio and the Palazzo Pitti, for in *Certain Physiological Essays*, published 1661, he refers to the loadstones of prodigious size that he had seen in Tuscany (*Works Fol.* **I**, p. 222; *Works.* **I**, p. 346); and one of the *Occasional Meditations*, published 1665, is entitled 'Upon the sight of a branch of coral among a great prince's collection of curiosities' (*Works Fol.* **II**, p. 223; *Works.* **II**, p. 455). Both these rareties were remarked upon by other travellers of the period (Evelyn (de Beer) **II**, pp. 185–200, 411–419 for a description of Florence). Probably it was here that he saw a large piece of crystal which was transparent and green like an emerald at the vertex, and shaded to colourless crystal at the base. This item is mentioned in his *Essay about the Origin and Virtues of Gems*, published 1672 (*Works Fol.* **III**, p. 222; *Works.* **III**, p. 525). At the Palazzo Vecchio he would undoubtedly have seen the large collection of portraits of famous persons (now known as the Gioviana Collection), little thinking that at a later date his own likeness would be added thereto (R. E. W. Maddison. 'The Portraiture of the Honourable Robert Boyle, F.R.S.' *Annals of Science.* (1959. Published 1962) **15**, 157). In one of his Moral Essays, written about 1647–1648, which is directed against cruelty to animals, RB writes, 'I remember at Florence in the Great Duke's excellent Stables, I saw a horse who for some eminent piece of service he had done this Prince, was kept there with a liberall assignment of Provender, & priviledg'd with an exemption from further service for his whole life. This I mention to Yow onely to insinuate that Greate & wise men have esteem'd horses Objects of thankfulnesse, as well as Compassion.' (B.P. **XXXVII**). In another of his Moral Essays, written about the same time, entitled 'Against Confidence' [=presumption, impudence], he writes, 'whilst I liv'd in Italy I had frequently occasion (tho with as much scorne as Wonder) to observe the Italians to be so much of that Roman's Opinion[1], that oftentimes, to secure their Women from the suspition of Dishonesty, they enslave them to an impossibility of being dishonest. Particularly I remember that in Florence I spent divers Months without ever being permitted to see the Mistris of the house above twice, or speake with hir alone once; which (restraynt) I must Confesse was the onely way in making me take notice of that kind of custome to make me forgive it.' (*ibid*).

† For the singularities of Siena, Montefiascone and other places on to the way to Rome see Evelyn (de Beer) **II**, pp. 201–212.

---

[1] The reference is to Caius Julius Caesar, who gave as his reason for divorcing his wife Pompeia: *meos tam suspicione quam crimine judico carere oportere*. (Suetonius. *Caesar*. 6; 74.)

of vniversall Monarchy, is now no lesse considerable by it's present superstition then formerly by its Victorious Armes; the Moderne[a] Popes bringing it as high a Veneration as the Antient Cesars; & [b]–the Barberine Bees flying as farre–[b] as did the Roman Eagles.* The more conveniently to see the numerous raritys of this Vniversall Citty, P. to decline the distracting[c] Intrusions & importunitys of English Jesuits, past for A frenchman, which neither[d] his Habit nor[e] Language much contradicted. Vnder this Notion he restlesly[f] & delightfully pay'd his Visits, to what in Rome [g]–& the Adiacent Villas–[g] most deseru'd;† & amongst other Curiosityes & Antiquitys, had the Fortune[h] to see the Pope:[i]‡ at Chapell with the Cardinalls, who seuerally appearing[j] mighty Princes, in that Assembly,[k] look't like a Company of Common Fryers. Here P. cud not chuse but smile, to see a Young Churchman after the seruice ended, vpon his knees carefully with his feet sweepe into his handkercheefe the Dust his holynesse's (Gowty) feet, had [l]–by treading on it–[l] consecrated, as if it had been some Miraculous Relique. Nor was P. negligent to procure the Latin & Tuscan Poëms of this Pope;§ whose Name Vrbanus, his Actions did not belye; he hauing more[m] of the Gentleman in him then his Pontificall Habit, would /might/ seeme to let him weare. A Poet he was & a Mat. . .[n] & tho neither of them in perfect. . .[n] much, that he delighted. . .[n] [P. said] that he neuer[o] found the Pope lesse valu'd then in Rome, nor his Religion fiercelyer disputed against then in Italy; &[p] sometimes added, that he ceas'd to wonder that the Pope shud forbid the sight of Rome to[q] Protestants, since nothing could more confirme them in their Religion.

---

[a] Interlined above *Miter being* deleted. [b–b] Preceding page in MS has *the* [Followed by *Ro* deleted] *Barberens* [Altered from *Barberinian*] *Bees flying as far;* not deleted. [c] Followed by *&* interlined and deleted. [d] Interlined above *both* deleted. [e] Interlined above *&* deleted. [f] Interlined above *visited busily* deleted. [g–g] Interlined. [h] Followed by *at the Chappell amongst all the Cardinalls* deleted. [i] Followed by *one that* [?] deleted. [j] Interlined above *seeming* deleted. [k] Followed by *appea* deleted. [l–l] Interlined. [m] Interlined above *much* deleted. [n] Paper torn away affecting about one-third of each line. [o] Followed by *heard* deleted. [p] Followed by *wo* deleted. [q] Followed by *the* deleted.

---

* A reference to the coat of arms (*Azure, three bees or, two and one*) of the powerful Barberini family; and to the Roman standard of an eagle borne upon a spear.

† For the sights of Rome at this period consult Evelyn (de Beer) **II**, pp. 212–317, 355–405.

‡ Maffeo Barberini (1567–1644), cardinal, became Pope Urban VIII in 1623.

§ *Poesie Toscane*. Rome, 1637. *Poemata*. Paris, 1642; Oxford, 1726. In his *Considerations touching the Style of the Holy Scriptures*, published in 1661, RB tells us that he was delighted to read some of Urban's reflections upon the scripture 'in the handsome paraphrases of his pious muse'. (*Works Fol.* **II**, p. 133; *Works*. **II**, p. 315).

P. hauing in a short Time suruey'd the Principal Raritys,* of this proud Mistris of the World, was vnwillingly driven thence by his Brothers Disability to support th'encreasing Heates; which there[a] proue often insupportable to strangers, (the neighbouring Cuntry being very scorch't[b] barren & vninhabited.) Wherefore he tooke his way backe towards Florence, by that delicious Valley that ennobles Perugia, & passing by Pistoya,† came to Florence; where after a short repose, they descended the riuer Arno vnto Pisa,‡ [*In marg*. Place here the Accident about his Teeth.§] & from thence to Liuorno, || where in a

---

[a] Followed by *seeme* interlined above *doe* [?] *often;* all deleted. [b] Followed by & deleted.

---

* John Aubrey records that he had 'oftentimes heard him [RB] say that after he had seen the antiquities and architecture of Rome, he esteemed none anywhere else.' (J. Aubrey. *Brief Lives*. Edited by A. Clark. 2 vols. Oxford, 1898. Vol. **I**, p. 120).

There is no record of any visit being paid by the travellers to the Venerable English College, for their names do not appear in the Pilgrim Book.

One thing RB failed to see in Rome was the extraordinary hydraulic organ of Athanasius Kircher (1601?–1680). In *New Experiments Physico-Mechanicall* RB quotes Kircher's description of this 'musical engine', and wished,

> " I had had the good fortune when I was at Rome, to take notice of these organs; or that I had now the opportunity of examining of such an experiment. For, if upon a strict enquiry I should find that the breath that blows the organs doth not really upon the ceasing of its unusual agitation by little and little relapse into water, I should strongly suspect that it is possible for water to be easily turned into air."

This was mentioned in connection with a discussion of the properties of water vapour, for RB had 'found by experience that a vapid air, or water rarefied into vapour, may at least for a while emulate the elastical power of that which is generally acknowledged to be true air.' (*Works Fol.* **I**, p. 34; *Works*. **I**, p. 52: Athanasius Kircher. *Musurgia universalis sive Ars Magna Consoni et Dissoni in X libros digestos*. 2 vols. Rome, 1650. p. 309 and plate XVIII).

One wonders if, whilst in Rome, a visit was paid to Cassiano dal Pozzo (?–1657), antiquarian and naturalist, who had been a member of the Accademia dei Lincei, which existed from 1603–1630. Dal Pozzo had taken into his own residence the musuem and other collections of the academy. (Michele Maylender. *Storia delle Accademie d'Italia*. 5 vols. Bologna, 1926–1930. Vol. **III**, p. 476).

A surviving catalogue of the museum illustrates the Cartesian diver (C. N. Bromehead. 'Bottle Imps'. *Antiquity*. (1947), **21**, 105), which Boyle described in 1673 in connexion with experiments on the influence of atmospheric pressure on bodies in water. Boyle's account gives no indication that he had previously seen these objects, so he may have invented them independently. (*Works Fol.* **III**, p. 307; *Works*. **III**, p. 656). The travellers probably left Rome about the end of April 1642.

† For Pistoia see Evelyn (de Beer) **II**, p. 410.

‡ For Pisa see *ibid*. **II**, p. 179.

§ RB had read the *Lives of the Philosophers* by Diogenes Laertius (*Works Fol.* **I**, p. 228: *Works*. **I**, p. 355; B.M., Add. MSS. 4229 fol. 60: Bompiani. *Opere*. **VII**, p. 846). "The sect, which then struck him most, was that of the Stoics; and he tried his proficiency in their philosophy by enduring a long fit of the tooth-ache with great unconcernedness." (*Works Fol.* **I**, *Life*. p. 15; *Works*. **I**, p. xxvi). Perhaps the occasion of this fortitude is what is referred to here.

|| For Livorno=Leghorn see Evelyn (de Beer) **II**, pp. 183–184.

Felouca* with a good Winde, they ventur'd, for expedition sake,[a] some 15 or 16 miles into the sea; & coasting along the country, still neere the shore for feare of sudden stormes, they each night lay in some Towne, drawing their Boat ashore (which was not vneasy in regard of the Inconsiderable Tides of the Medditeranean there) & soone (tho not without Danger) reach't proud Genoa†. . .[b] As if she meant to giue Italy a Good Deboire;‡ & indeed[c] the Cuntry could no way haue left P. a better Relish & Idea of hir, then by[d] making the last Place he stop't[e] at in hir, one of the Pleasantest that could be seene. The next day Filaretus prosecuted his Journey, & passing by Monaco (a very strong Place, then newly betray'd by the Prince of it to the French, §) & by Mentone, a little Principality belonging to the same Prince, & [f]⁻stopping a While⁻[f] at Nizza, a Place extreamely & meritoriously famous for that strength, which Nature & Art, haue emulously giuen it, by night they[g] landed at Antibe, one of the Townes of France,[h] that most approaches Italy. ||

The Fifth Section.

The Morning that succeeded P. his Arriuall at Antibe, ¶ he left it to crosse the Cuntry to Marseilles.** But he was welcom'd into France by an Accident which was very hazardous & might haue prou'd

[a] Followed by *to* deleted. [b] Bottom of sheet missing. [c] Followed by *P.* deleted. [d] Followed by *letting* deleted. [e] *saw* interlined above; no deletion. [f–f] Interlined above *calling in* deleted. [g] Followed by *arriu'd* deleted. [h] Final *s* deleted.

* Felouca = felucca. A coasting vessel formerly much used in the Mediterranean. It is propelled by oars and a lateen sail, and is capable of great speed. The ship is not decked.

† On nearing Genoa Marcombes must have had sad recollections of his former visit to that town, when in charge of Lewis and Roger Boyle, and Boyle Smyth. All three fell ill of the smallpox, and Boyle Smyth died 30 December 1637 N.S. Marcombes encountered insuperable difficulties over burial of the body; the inquisition would have sent it to be buried with the horses and dogs because Smyth died without confession; the cost of embalming was prohibitive; mariners would not take the corpse. Finally, he was obliged to put the body in a trunk, hire a boat, and cast the body into the sea 10–12 miles from land. (B.M., Add. MSS. 19832 fol. 41v. Letter of Marcombes to the Earl of Cork dated 16 January 1638 N.S.)

In the *History of Firmness*, when discussing the properties of mortar, RB recalls having seen the breakwaters at Genoa and elsewhere (*Works Fol.* **I**, p. 267; *Works.* **I**, p. 421).

‡ RB has clearly used this word to mean *good impression*, which is the opposite of its usual figurative sense.

§ The Prince of Monaco at that time was Honoré II, who had placed himself under the protection of Louis XIII by a treaty of 1641. (G. Saige. *Documents historiques relatifs à . . . Monaco.* 3 vols. Monaco, 1888–1891. Vol. **III**, p. cc.)

|| For the places mentioned here on the journey from Genoa to Antibes see Evelyn (de Beer) **II**, pp. 168–171.

¶ The last city on French territory; a garrison town called the key of France.

Probably the route taken was through Cannes, Brignoles, past Rouziers, St Baume, Marseilles. (*Francis Mortoft: His Book.* Edited by M. Letts. Hakluyt Society, 2nd Series, No. LVII. 1925. pp. 34–35). For Marseilles see Evelyn (de Beer) **II, p. 164.

Tragicall.[a] During the whole time of his being a Traueller or Resident in Italy he had religiously adher'd to his. . .[b] Englishmen that [c-]. . .before Philar.[-c] thinking as much better as safer to take of their hats then to venture their heads, complimented with the Crucifixe: but P. without the least Act of superstition, tho not with ill Words, & worse Menaces, [d-]ventur'd & past[-d] boldly thorough them all: as euer[e] resoluing that the Soule shud not more transcend the Body in it's owne Value, then in his Esteeme. Thus Danger thus happily scap't, P. continues his Way to Marseilles, where the third Day he arriu'd, with intent there to expect [bills] of Exchenge promis'd to be then sent thither, to enable him[f] to prosecute his future Trauells. His Dentention here was shortned by his Visits of so excellent a Harbor for Gallyes & small vessels; which as 'tis compriz'd in the Towne, is euery Night, like it, assur'd[g] with Locke & Key. Here P. had the pleasure to see the French King's Fleet of Galleys put to Sea, & about 2000 poore[h] Slaues tugge at the Oare to row them.*†

[End of *An account of Philaretus* written by Robert Boyle.]

It was at Marseilles that the travellers first became aware of the disastrous consequences to the Earl of Cork of the outbreak of the Great Irish Rebellion. The bills of exchange that would have supplied their next quarter's allowance of £250 had not arrived; in fact, Mr Perkins had not sent them. When making arrangements for the transmission of this sum, the Earl of Cork had said,

> " good M[r] Perkins let me entreate you to vse your vthermost care & endeauors, that M[r] Michael Castell‡ may not be disappointed, and thereby my creditt endangered; & my sonns with their

---

[a] Followed by *Vpon his Determ* deleted. [b] Bottom of sheet missing. [c-c] Interlined above . . . *the Way* deleted. [d-d] Interlined. [e] Followed by *bele* deleted. [f] Interlined. [g] Interlined above *shut vp* deleted. [h] Interlined.

---

* For an account of the galleys see Evelyn (de Beer) **II**, p. 165. In the *History of Firmness* Robert regrets that he had no opportunity to visit the coral fisheries in the neighbourhood of Marseilles in order to have 'an ocular satisfaction', whether or no the tradition is strictly true 'that coral grows soft at the bottom of the sea, but when it is brought up into the open air, though it retains its bulk and figure, it hardens into a stony concretion'. (*Works Fol.* **I**, p. 276; *Works.* **I**, p. 435).

† Attention is directed to the bibliographical notes at the end of volume **II** of Evelyn (de Beer) relating to contemporary guide books, travel diaries, and maps covering the places visited by Evelyn and RB during their continental travels. Although both were abroad at the same time, they did not meet.

‡ Marcombes had drawn bills on Castell.

Tutores necessities vnsupplyed; for I put my whole trust heerein into your hands. And god knowes with what difficulty I haue gott those moneyes together to make good my reputacon, & supply my Childrens occasions. And therefore as you loue me, be soe carefull to doe all things so seasonably and conveniently, as neither they nor I suffer thereby, for yf it concerned my lyfe I would entrust itt in your hands. And it much perplexeth me, that I can neither heare of Mr Marcombes, nor my two sonnes, nor where they are, nor how they doe; and Therefore yf you haue any intelligence from them, send it me to youghall."*

Instead of carrying out the Earl's instructions, Mr Perkins himself being in dire want of money used the £250 to repay himself for expenditure on the parliament robes of the Earl's sons and Lord Kinalmeakys' debts. He does not seem to have realised the seriousness of the situation and the great inconvenience caused by his action.† The Earl was greatly annoyed, and instructed a third person to write as follows:

" Mr Perkins, I am sorrie you should forgett yrselfe so much as to deceive the trust my Lord imposed vpon you to make ouer by bill of exchange the some of 250li to his Lordship's two sonnes, Mr Frauncis and Mr Robert Boyle (they being strangers in a forrin cuntrie): that you should keepe the said 250li and disapoint them of the allowance my Lord was pleased to imploie you for the makeing it ouer to them, by which neglect and deceiuing the trust imposed vpon you, you put two gentlemen to a greate streight, and hath wrought such an impression in my Lord of your ill carriage therein, that his Lordship cannot with patience write vnto you."‡

After long delay the Earl of Cork received letters from his two sons and Marcombes written whilst they were in Florence. In his letter of acknowledgment the Earl described the sad state of Ireland as a result of the rebellion, and said,

" Necessitie compells me to make you and them [his sons] know, the dangerous and poore estate wherevnto by gods prouidence I am at this instant reduced. . .I am deeply indebted, and haue neither money, revenue, nor stock left me, nor can longer subsist for want of means. . .And the 25th of January last, I scraped

---

* *II. Lismore.* IV, p. 257. Letter of Earl of Cork to Mr. Perkins dated 27 January 164½.

† *Lismore MSS.* XXIII, No. 13. Letter of W. Perkins to the Earl of Cork dated 14 April 1642.

‡ *II. Lismore.* V, p. 117. Letter from a third person on behalf of the Earl of Cork to W. Perkins endorsed 5 November 1642.

together, and with much difficulty gott together two hundred and fifty pounds by selling of plate (wherein I lost the fashion), and haue made it ouer to be paid by Mr Perkins in London to Mr Castell. And I am confident it was either paid the 1st of March, or before the 10th. . .hitherto I haue punctually paid your Bills; which longer to doe I am noe waies able, neither must this 250^{li} now paid, be by my sonnes and you expended to maintaine you till the first of June next, but it must be employed either to bring my two sonnes into Ireland, out off some meet port in France to land either at dublin, Corke, or Youghall, (for all other Cities and Sea Townes are possessed by the enemy) or else my two sonnes, till this generall rebellion and waste contynues, must of necessitie vpon receipt of these my lettres, presently begin their iourney from the place where these lettres shalbe deliuered you, and travaile into Holland, and putt themselues into entertaynement vnder the service and conduct of the Prince of Orange; for they must henceforward maintayne themselues by such entertaynements as they gett in the warres; for with inward greefe of soule I write this truth vnto you, that I am no longer able to supply them with any more moneyes for their maintenance. . .I am no longer able to giue them exhibition. . .It wilbe a worke worthy your Consideracōn, how you gouerne my two sonnes, and how you with their owne consents will dispose of them, either for Ireland or Holland. I know you are too generous to leaue them, till you see them shipped for Ireland, or well entred in the warres of Holland, as their owne desires, with your good advice, shall lead them; for into England I will not consent they shall yet come, till Ireland be recouered, for I haue neither present money nor meanes to defray their expences there: And for them that haue been soe well maintayned to appeare there with out money, would deiect their spirits, and grieue and disgrace me, and draw contempt vpon vs all. And therefore I can onely pray, that God, in his divine providence will guide them and you well, and my prayers and best blessings shall goe along with you; for as now you shall behaue and carry your selfe towards them at the parting blow, soe shall you oblige me and them euer to value and esteeme you. But in any case, I pray be very circumspect how you spend this last 250^{li} now made ouer vnto you; but grow less and decline in your vsuall expences, and putt all vnnecessary servants and dependants from you, and so husband this little remaine of money, as at your departure they may haue some competency either to bring them out of France into Ireland, or to setle them in the warres in the low

Countries; for I know you are too noble to lett them departe from you with empty purses; for I am confident God will so dispose of me and them, as wee shall haue meanes and opportunity to testify our thankfulness, and reward your faithfulnes towards vs. Yf not God will reward you, and the world will speak honorably of you. I must now forbidd you to giue forth any more Bills of Exchange, or to charge me with the payment of any more moneys, for I am no longer able to satisfy them, and to protest them, wilbe my disgrace and yours. And I am well assured this Caution will prevent both. For conclusion I pray you beleeue this great truth from me, that I putt a very high value and esteeme vpon you, your honesty, and integrity, that I affect you, and the many good parts that are in you cordially, otherwise I would not haue trusted you with my two yong sonnes, that are so deare vnto me, nor with the disburseing of a thousand pounds a yeare for their maintenance and expence, without accompt, as hitherto I haue done. And therefore I coniure you by all the protestac̄ons of loue and fidelity, that you haue made to me and my Children, that when you shall leaeu them, you will leaue them in such a manner, as may oblige them vnto you in all hearty thankfulnes and me to speake your goodnes in all places, and vpon all occasions, for your good and advancement. And with this Coniurac̄on I conclude my selfe to be your faithfull & assured frend, R. Corke."*

The news must have greatly perturbed the travellers, quite apart from the failure of Mr Perkins to transmit funds—a failure of which the Earl was quite unaware when he wrote the above letter. The result of their earnest deliberations is contained in a letter written by Robert to his father from Lyons, whither they had gone after leaving Marseilles, probably by way of Aix and Avignon. The projected tour of France before returning home had perforce to be abandoned.

"Hauing according to your Lordship's order and directions, seriously pondered and considered the present estate of our affaires, we haue not thought [it] expedient for divers reasons that my Brother will tell your Lordship by word of mouth, that I should goe into Holland; for besides that I am already weary and broken with a long Journey of above eight hundred miles, I am as yet too weake to vndertake so long a voyage in a strange country, where when I should arrive I know no body and have little hope by reason of my youth to be receaued among the troopes, and

* *II. Lismore.* **V**, pp. 19–24. Letter of Earl of Cork to Marcombes dated 9 March $164\frac{1}{2}$.

withall Mr Perkins hauing not sent vs the 250 pound starling that your Lordship had ordained for our present Journey, I could not part from hence in any good equipage: wherefore Mr Marcombes having offered me to bring me to Geneva, where I shall want no necessary thing and stay there untill it please your Lordship to give me some more particular order and directions, or till it please God to change the face of the affaires; I haue accepted his offer, hoping that one day God will give me the meanes not onely to render him what he shall haue layed out for me, but also to requite him for his greate friendship and courtesy. I. . .then to Geneva (with God's blessing) to repose my selfe after so long a. . .to gather some force and vigour to serue and defend my Relig. . .my King and my Country according to my little power. . .employ my time so well, that I shall render my selfe. . .ploy whereunto your Lordship may put me: I beseech. . . betweene this and the end of August to thinke vpon. . .I may honorably gaine my liuing, assuring you that. . .bour and good endeauour to render my selfe capable thereof. . .if your Lordship hath need of me in Ireland I beseech your Lordship to acquaint me therewith and to beleeve that I have never beene taught to abandon my parents in adversity, but that there and in all other places I will alwayes striue to show my selfe an obedient Sonne: as for my Brother Francis, . .he is now ready to take horse to goe towards Ireland to secoure your Lordship according to his power; wherefore being confident that he will informe you more largely of all our affaires, I most humbly take my leave. . . "*

Francis Boyle then parted from his brother to return home. Whether he did so alone, or in company with other travellers going from Lyons, we do not know. We may be certain that Robert Boyle and Marcombes took the direct route from Lyons through Amberieu and Nantua to Geneva, where they were settled before the end of July 1642. In the meantime Francis had arrived safely in Ireland, and written to Robert telling him of the exploits of their brother Lewis, Lord Kinalmeaky, against the Irish rebels, whereupon Robert wrote to Lewis:

. . ." I beleeve that by this time yow have seene my Brother Francis, who hath (as I esteeme) acquainted yow with the reasons that caused our separation: wherefore I'le content myselfe to assure yow now, that I am at Geneva in very good health at Mr Marcombe's lodging, Where I want nothing but some confortable

* *II. Lismore.* V, pp. 71–72. Letter of RB to the Earl of Cork dated 25 May 1642 (N.S.). The letter is mutilated.

letters out of England or Ireland, and where I dayly expect fresh orders and money from my father. I lately receaued with a greate deale of contentment and pleasure the glad newes of your generous and fortunate Combats against the Irish Rebels. . ."*

Lewis never received this letter, for he was killed fighting the rebels at the Battle of Liscarrol, 3 September 1642. The letter was received by the Earl of Cork, 13 October following.

After a 'long sylence' Robert heard from his father.

> " . . .I receaved therein a Commandement from your Lo: to stay heere with Mr Marcombes till the Spring time, which order . . .I am resolued to obey, and to employ my selfe so diligently to my Studyes and Exercises, that at my comming home I may render my selfe capable of some employment in the affaires of Ireland. I am no lesse astonished than your Lordship at the baseness and infidelity of Perkins, nor no lesse sensible to the obligations of Mr Marcombes, whom I should altogether despaire ever to requite if your Lo: in your Letter had not promised to take your selfe that matter in hand: he alwayes continues his care of me, and lets me want nothing that is necessary either for the body or for the mind; so that I am now (God be thanked) in as perfect health as ever I haue beene and am ready (with his helpe) to spend my time as well as euer I did in my life. . ."†

No further letters between Robert Boyle and his relations during his second residence at Geneva are known. The Earl, solicitous as ever for the welfare of his youngest son, wrote as follows to his son Richard, Lord Dungarvon:

> " . . .and I pray you wryte unto yr brother Robin & deliver your lrs to Mr Burlemacke [Philip Burlamacchi] & know of him whether he hath sent Robin that money I assigned him, wch yf hee hath not already done, I pray you presse him to doe it speedily."‡

Probably the last gift that the Earl made to Robert is recorded in his diary under date 5 February $164\frac{2}{3}$: " . . .my choice dvn mare I sente to Lismore to be kept and drest carefully for my son Robert when god shall send him home from his Forreign travailes; and appointed Mr pomfrett to oversee and to ryde her gently."§ Father

---

* *II. Lismore.* V, p. 97. Letter of RB to Lord Kinalmeaky dated 1 August 1642 (probably O.S.). The letter is mutilated.

† *II. Lismore.* V, p. 114. Letter of RB to the Earl of Cork dated 30 September 1642 (undoubtedly O.S.).

‡ B.M. Egerton MSS. 80. fol. 14. Letter of Earl of Cork to Lord Dungarvon dated 21 November 1642.

§ *I. Lismore.* V, p. 221.

and son did not meet again, for the Earl died, 15 September 1643, at Youghal.

Occasional references to observations made by Robert at some time or other during his travels on the continent are scattered through his subsequent published works. It was in Italy that he saw conservatories or pits wherein snow and ice were stored throughout the summer. When he wanted to give a detailed description of them in his *Experimental History of Cold* (1665), he found it necessary to consult John Evelyn, whose account he printed.* In the *Usefulness of Experimental Natural Philosophy* he discusses the effects of thunder and earthquakes on fermented liquors, and says: "I remember, that when I returned out of Italy through Geneva, there happened in that place an earthquake; upon which, the citizens complained, that much of their wine was soured, though I, that lodged in the highest part of the town, saw nothing to make me believe, that the bare succussion of the earth was capable to produce so great and sudden an alteration in the wine".† He makes passing reference to the same earthquake in his communication to the Royal Society dealing with an earthquake near Oxford.‡ *In the General History of Air* he says: "I remember, that being one night at a town built almost upon the great lake of Geneva, anciently called *Lacus Lemanus*, the thunder was so violent, as much frighted the inhabitants; though (by reason of the neighbourhood of the high mountains of Savoy and Switzerland) thunders be very frequent there: and the next day I had a great complaint made me, of the strong stink of sulphur, produced by the thunder, that fell hard by into the lake, and was ready to overcome, by its smell, even a soldier, that stood centinel near it".§ In the *Experimental History of Colours* (1663) when describing the changes that occur in vegetable juices, he says: "I could add many other instances, of what I formerly noted touching the emergency of redness upon the digestion of many bodies, insomuch that I have often seen upon the borders of France (and probably we may have the like in England) a sort of pears, which digested for some time with a little wine, in a vessel exactly closed, will in not many hours appear throughout of a deep red colour (as also that of the juice, wherein they are stewed becomes)".‖ Again, in the *Experimental History of Cold*, with reference to the duration of snow, he says: "I shall take notice, that in

* *Works Fol.* **II**, p. 307; *Works.* **II**, p. 584.

† *Works Fol.* **I**, p. 534; *Works,* **II**, p. 171.

‡ *Philosophical Transactions* (1665–1666). **I.** No. **XI, p. 179.** *Works Fol.* **II.** p. 448; *Works.* **II**, p. 797.

§ *Works Fol.* **V**, p. 123; *Works.* **V**, p. 636.

‖ *Works Fol.* **II**, p. 61; *Works.* **I**, p. 756.

the Alps, and other high mountains, even of warmer climates, though the snow doth partly melt towards the end of summer; yet in some places, where the reflexion of the sun-beams is less considerable, the tops will even then remain covered with snow, as we among many others have in those countries observed ".* In the same work we read: " When I was in Savoy, and the neighbouring countries, which have mountains almost perpetually capped with snow, I heard them often talk of a certain white kind of pheasants to be met with in the upper parts of the mountains, which for the excellency of their taste were accounted very great delicacies."† When writing " About the surfaces of contiguous Fluids ", Boyle refers to " . . .what I haue seen in Savoy near Geneva, where the Rhoane and the Arue run together a pretty way w[th]out manifestly mingling their waters; whose distinction, by reason of the great clearnes of the former, and the muddines of the later, is easily discerned."‡

Whilst in Geneva, most likely during his second stay in that town, Robert met François Perreaud (1572–1657), a protestant divine, who wrote a book entitled *Démonographie, ou traité des démons* (Geneva, 1653). Part of the book was translated by Peter du Moulin (the younger, 1601–1684) into English at the request of Robert Boyle, who in the letter prefixed to the English version§ relates that " the conversation I had with that pious author during my stay at Geneva, and the present he was pleased to make me of this treatise before it was printed, in a place where I had opportunities to enquire both after the writer, and some passages of the book, did at length overcome in me (as to this narrative) all my settled indisposedness to believe strange things ".||

Robert Boyle's home-coming can be told in his own words.

> " At length, though I could not receive any supply at all from my friends, yet being unable to subsist any longer, I was forced to remove thence; and having upon his [Marcombes's] credit made shift to take up some slight jewels at a reasonable rate, we made sale of them from place to place, and by their help, at last,

---

* *Works Fol.* **II**, p. 307; *Works.* **II**, p. 584.

† *Works Fol.* **II**, p. 330; *Works.* **II**, p. 621.

‡ B.P. **XXVII.**

§ F. Perreaud. *The Devill of Mascon*, Oxford, 1658. See Madan. **III**, No. 2407. p. 87.

In the ' Autobiographie de Pierre du Moulin ', (the elder, 1568–1658), we read: " Passants par Mascon, nous visitasmes Monsieur Perreaux, ministre, dont la maison avoit esté travaillée par l'espace de six semaines par un esprit malin. Dieu enfin le délivra de cette affliction." *Bulletin de la Société de l'Histoire du Protestantisme Français.* (1888) **7**, 470.

|| *Works Fol.* **I**, *Life.* p. 141; *Works.* **I**, p. xccxi.

by God's assistance, we got safe into England towards the middle of the year 1644".* The use of 'I' and 'we' in this letter suggests that Marcombes accompanied Robert from Geneva in order to ensure his pupil's safe arrival back in England, thereby properly discharging his obligations as guardian. After Robert was returned into England, "he posted away from Dover to London, where he found by Inquiry, for he had no full address to her, where his sister the Lady Catherine Jones who was fled thither out of Ireland, was lodg'd. And coming into the house in his french Dresse, he pass'd by several of his younger Relatives, who gaz'd at him wthout knowing him, as he on ye other side knew them but by guess, but advancing to the innermost part of the House, and meeting no servants of whom to make enquiry, he spoke to a Person, that was going down a pair of stairs, your Back turn'd towards him, and enquir'd for such a Lady. This question made her come up to see who ask'd it, and at first she look'd upon him wth surprize, because she knew not at all that he was in England, whence he had been absent so many years, but afterwards having upon an attentive view, discern'd who he was, she cry'd out, Oh! 'tis my Brother, when wth the joy & tendernes of a most affectionate sister, she saluted & imbrac'd him when many questions & answers were interchang'd between them, she took care privately to haue an apartment provided for him in her House, & would by no means suffer, he should lodge any where else whilst he staid in London, wch was four months & a half.

"This abode wth so excellent a Person in a faire large House, where there liv'd also a sister in Law of hers, marry'd to one of the chief members of the then House of Commons, who were both of them eminently Religious.† [Robert] many times wth thankfulness to God, acknowledg'd as a seasonable Providence to him for more than one Reason, since first in the heat of his youth, it kept him constantly in a Religious family, where he heard many pious discourses, & saw great store pious examples, whereas if a gracious Providence had not detain'd him here, he had lost the benefit of all these, & instead of them being expos'd to ye manifold & great temptations of a Court & an Army, where tho' there were besides the excellent King himself, and diverse

---

* *Works Fol.* I, *Life*. p. 15; *Works*. I, p xxvii. Letter of RB to the steward of his manor at Stalbridge, undated, but last quarter 1644.

† I suggest that Lady Ranelagh was living in the house of Sir John Clotworthy in Holborn. She was sister-in-law to Sir John and his wife Margaret. Sir John was one of the leading members of the Presbyterian party in the House of Commons.

eminent divines, many worthy Persons of several ranks, yet the generality of those he would have been oblig'd to converse wth were very debaucht & apt, as well as inclinable to make others so. Besides by this means [Robert] grew acquainted wth several Persons of power & interest in ye Parliam^{t} and their party, wch being then very great, & afterwards the prevailing one, prov'd of good use, & advantage to him, in reference to his estate and concerns both in England and Ireland '.*

After an absence of almost five years Robert Boyle had returned safely into England to be reunited with his family. One period of his life had closed; yet it was a period, we may be sure, in which he profited by every occasion during the Grand Tour " to enrich the mind with knowledge, to rectify the judgment, to compose the outward manners, and to form the complete gentleman ".†

The Great Earl of Cork, who died 15 September 1643, had throughout his years of prosperity made generous gifts of money and property to his children. The Earl's diary records the numerous acquisitions of property he made for the benefit of his son Robert. Under date, 1 March 163$\frac{6}{7}$, we read:

" This daie I did examyn and caste vp that faier & Lardge Revenew. . .which I haue in, and by my half yeares Rentall book, For our Ladie day, 1637, apporc̄oned, and distributed amongste those my Five sons, which god in his great mercie hath also blest me withall: . . .And my yongest son, Robert Boyle, his half yeares rent, then also to be due, out of the lands and estate by me assigned vnto him, will amount vnto 1079^{li} 9^{s} 3^{d} ster. . ."‡

By the Septpartite Indenture of 1636 the Earl vested all his estates on behalf of his sons in four trustees.§ Robert's intended residence was to be at Mallow, co. Cork, Ireland, and his lands in Tipperary and Kildare. The unfortunate death of Lewis Boyle, Viscount Kinalmeaky, necessitated new arrangements, which were embodied in a fresh will signed by the Earl, 4 November 1642.||

Apart from the disposal of lands, etc., there are two matters in the will that are of particular interest in regard to Robert. The Earl's sons and daughters had all, with one exception, been married to important members of the nobility. The one exception was Robert;

---

* B.M. Slo. MSS. 4229. fol. 68.

† Thomas Nugent. *The Grand Tour*. 4 vols. London, 1749. Vol. **I**, p. viii.

‡ *I. Lismore*. **IV**, p. 228.

§ Townshend. Appendix II.

|| Townshend. Appendix III, which prints the will *in extenso*.

and for him, the Earl had designed him to marry Anne Howard, daughter of Edward, Lord Howard of Escrick. The Earl's diary records for 11 September 1641:

> "I and dongarvan rod to Hatfeild to take leave of thearle of salisburie; I gaue the servants v^{li}, and to my Robins yonge M^{rs}, my Lo. Edward howards daughter, a smale golde Ring with a dyamond;"*

and for 16 November 1642:

> ". . .I delivered to my daughter Broghill in gold, v^{li}, to bwy a Ringe besett rownd with diamonds, to be by her presented from me to M^{rs} An Howard, daughter to the Lorde Edw: Howarde."†

His will contains the following conditional bequest:

> "I do give grant legate and bequeath to Mrs Ann Howard Daughter of ye Lord Edward Howard my Silver Cistern weighing 680 ounces, my Silver Kettle or Pot weighing 162 ounces, my Silver Ladle weighing 27 ounces, whereon my own Coat of Arms is engraven, which three pieces of Plate I bought of Sir Thomas Jermin the Younger Knight, and paid him for them £274 18s. 6d. besides what I paid for Engraving my Arms, if the said Mrs Ann Howard shall be marryed to my said Son Robert, But in case she shall not be marryed unto and become ye wife of my said Son, Then I give the said three pieces of Plate to my said Son Robert if he live and attain unto ye age of twenty one years, Otherwise I give and bequeath them to my son Dungarvan's eldest Son Charles Boyle."‡

The Earl's hopes were not fulfilled: Anne Howard married her cousin Charles Howard, who became the 1st Earl of Carlisle.§ It is generally considered that the following remarks in a letter written by Katherine, Viscountess Ranelagh, to her brother Robert refer to this marriage.

> "You are now very near the hour, wherein your mistress is, by giving herself to another, to set you at liberty from all appearances you have put on of being a lover; which though they cost you some pains and use of art, were easier, because they were but

---

* *I. Lismore*. V, p. 190.

† *ibid*. p. 216.

‡ Townshend. p. 485.

§ G.E.C. He died 24 February $168\frac{4}{5}$, aged 56 years.

Anne Howard was probably born, 1628, or perhaps later. Her marriage was undoubtedly towards the end of 1645, for her son was about three years old in September 1649. R. E. Chester Waters. *Genealogical Memoirs of the Extinct Family of Chester of Chicheley*. 2 vols. London, 1878, Vol. **I**, pp. 149–151: *Autobiography of Anne, Lady Halkett*. *Camden Society*. (1875), N.S. **13**.

appearances. It is well, if she put not herself by that act of bounty into more slavery than she gives you liberty; but now she must perfect making the venture she has so far proceeded in."*

No doubt numerous attempts were made in later years to provide a wife for so elegible a bachelor. One such attempt was made by John Wallis in 1669 when he proposed as 'an excellent wife for esquire Boyle' the Lady Mary Hastings, daughter of Ferdinando, 6th Earl of Huntingdon.† However, Robert never married.

The second matter in the Earl's will in regard to Robert Boyle reads:

> "And I do desire my Son Robert that in regard of the Fidelity and Trust that I have found in the said William Chettle (who waited upon me in my Chamber and Carried my Purse for above 26 years) that he will entertain him into his Service in the same Place and Condition wherein he hath so long served me, whom I hereby seriously commend to my said son for a faithfull true and honest Servant that will Carry his Purse and keep his Accounts faithfully and Justly as he hath done mine."‡

The statement that Robert Boyle studied at Leyden is still occasionally encountered, even though its inaccuracy was pointed out long ago.§ It originally appeared in Anthony Wood's *Fasti* (1691–1692),|| and has been repeated in numerous works since that time. Wood most probably took it from *The Minutes of Lives* by John Aubrey, the manuscript of which he had borrowed.¶

---

* *Works Fol.* V, p. 565; *Works.* **VI**, 534.

† *Works Fol.* V, p. 515; *Works.* **VI**, p. 459. Letter from J. Wallis to RB dated 17 July 1669.

‡ Townshend, p. 501.

§ *Biogr. Brit.* Vol. **II**, (1748), p. 913, note A.

|| Vol. **II**, col. 838. Bliss edition, Vol. **II**, col. 286.

¶ *Aubrey's Brief Lives*, edited by O. L. Dick. London, 1949. pp. ci, 36.

II

The Stalbridge Period

(1645–1655)

The manor, advowson and lands of Stalbridge, co. Dorset, were purchased by the Great Earl of Cork from James Touchet, 3rd Earl of Castlehaven,* for £5000 in 1636.† Other properties around Stalbridge were also acquired subsequently at a further expenditure of £2085.‡ It was the first estate to be purchased by the Earl in England, and he saw it for the first time, 6 August 1638.§ Alterations and improvements to the property had already been started, for it was in great decay.||

A part payment of £20 was made to "the plomer for carrying the water in Leden pipes to Severall vnder Roomes in my howse of Stalbridg".¶ The Earl of Bristol's free mason, Gregory Brimsmead, was engaged

> "to pave the way from my owt-moste gate to my hall door at Stalbridg, and the Tarras before my howse, with Rangerpavies of Free stone, owt of the same vayne of hard Free stoan [as] that [of] thearles Court going into his howse at Sherborn, and in the same manner that the same court is paved. Hee is to dig, and lay those stones even, neatly even and worckmanlyke in all respects, saving the carriadg of the stones from the quarry, which I am to do; for which worck I am to give him three pence halfe penney the foot . . ."**

---

* (1617?–1684) (*D.N.B:* G.E.C.)

† *I. Lismore.* **IV**, pp. 192, 199, 202, 203, 206; *idem.* V, pp. 11, 12, 21, 67, 68, 126: John Hutchins. *The History and Antiquities of the County of Dorset.* 4 vols. Folio, London, 1861–1873. Vol. **III**, p. 671.

‡ *I. Lismore.* V, p. 107.

§ *ibid.* p. 57.

|| *The Earl of Cork's Letter Book* 24 *April* 1634—19 *July* 1641. pp. 144, 168, 170, 182. Letters of Thomas Crosse, Steward of Stalbridge, to Earl of Cork dated 17 December 1636, 24 December 1636, 24 January 163$^{6}_{7}$, 28 March 1637 respectively.

¶ *I. Lismore.* V, p. 60. 9 October 1638.

** *ibid.* p. 67. 18 December 1638.

Forty shillings were paid

> " for a beam of 24 foot long to lay in the Flowr vnder the great chamber at Stalbridge."*

An agreement was made with another free mason, Walter Hyde, of Sherborn,

> " to make and sett vp for me in my new intended bwylding over the great sellar in Stalbridge 4 wrought chymneis, with Figures answerable or better then the chymney in my own bed-chamber, to be made of the stoan of Marnell quarrey; which [he] is to fynishe in all respects neatly and worckmanlyke, with halfe pares of hewed stone at his own chardges in all respects, except carriadge, which I am to provide, and to pay him for the Fowre, Forty marks. . . . He is to doe no other worck till this be don. . . "†

To the same mason

> " I paid 29 december laste. . .5^{li} ster: in part payment for eight Rownd Pillars with capitalls and Bases of hambdon stone a foot long, answerable to those in my stayrcase at Stalbridge, which he is to make, bring home, and sett vp in my new cellar there, to lay beames of my firste Flowers vppon, which 5^{li} I am to make vp 7^{li} for that peec of worck, when it is fynished, without any other trowble or chardg to me, then the payment of the said vij^{li} in all. Alsoe he is to digg owt of Hambden quarry soe much good stone, as will make sufficient Rayles, bases, Balasters and peddistalls, answerable in every respect to those of my stairecase at stalbridge, as will serve between my owtwarde gate and my hall door, and the Tarras before my hall door, from wall to wall, being in all that is soe to be rayled in about 300 foot: but be it more, or less, I am to pay him 4^{s} a foot runing measure, he to be at all chardges for making, carriadg, setting vp, with the cramps of iron and lead to fasten them together. The balasters are to be 3 inches higher then those of my staircase and to be sett soe neer and close together, that a dog cannot creep between. He is to keep at the least six masons still at worck, and to ply them still to this worck, and not to take any other worck in hand till myne ar fynished, and to be paid his money as the worck goe forward. more paid him 13 Feb^{r}, 1638, 12^{li}, making in all 22^{li} ".‡
>
> " I paid Hynde [or Hyde], my Free mason, xx^{s}, and makes vp vj^{li} for the six Round pillers of Hambdon stone that he made, brought home, and sett vp at his own chardges in my sellar at Stalbridge.

---

* *ibid.* p. 67. 19 December 1638.
† *ibid.* p. 68. 21 December 1638.
‡ *ibid.* p. 70. 2 January 163^{8}_{9}.

And for the 4 tonnels and tips with balles of Hambledon stone to be sett vp in my new bwilding at Stalbridge, I am too paye 24^{s} a peec in the quarry, (besides carriadg and setting vp, which I am to vndergoe); and haue this day paid him one halfe thereof, being 48^{s}."*

Gregory Brimsmead, the other mason, was

"paid vj^{li}, in part of xxxj^{li}, for making the staire case owt of my little Court in Stalbridge."†

and John Dean, the plumber of Sherborn, received £40

"in full satisfac̄on for all the Lead, sawder, and worckmanship of my plump cestern, Boylers, gwtters, pipes, daie labors and all other demaunds for the plomer worcks he hathe hethertoo don at my howse of Stalbridge."‡

"I haue agreed with Xtofer watts, free mason & carver, who dwels in horse street, in Bris[tol], to make me 2 such chymnes as are in my bedchamber at Stalbridge, or better; which he is to sett vp & make half pares and fynishe at his chardges, (all saving carriadge which I am to paie for); for which I am to paie him 4^{li} a chymney, and he is to make me a very fair chymney also for for my parler, which is to reach vp close to the seeling, with my Coat of Armes compleate, with crest, Helmett, Coronett, Supporters and mantling, and foot pace, which he is to sett vp and fynesh with a foot pace, all at his own chardges, fair and graceful in all respects; and for that chymney I am to pay him x^{li}, and I am to fynde carriadg also. He is also to make me 12 Fygures, each 3 foot highe, to sett vpon my staircase, for which he demaunds 20^{s} a peece, & I offer him 13^{s}. 4^{d}, and he is presently to cutt one of them with the figure of pallace with a sheeld; one with my coat with a corronet is to be cutt for a tryall."§

"I purchased of Sara Eyres, of Stalbridge, widdo, in W^{m} chettles name, to my vse, for v^{li}, her estate for lyfe [in] the Cottage & garden wherein the sope boyler dwels, adioyning to my orchard wall, to lengthen the walke, or carreer by me newly made and sett with yonge elmes, called Wm Sidnams walck."||

The Earl's expenditure in respect of building up to 14 September 1639 amounted to about 1000 marks.¶

---

* *ibid.* p. 79. 13 March 163$\frac{8}{9}$.
† *ibid.* p. 81. 16 March 163$\frac{8}{9}$.
‡ *ibid.* p. 82. 27 March 1639.
§ *ibid.* p. 84. 14 April 1639.
|| *ibid.* p. 101. 13 August 1639.
¶ *ibid.* p. 107. 1 mark = 13*s.* 4*d.*

Later in the year,

"I gaue 40li in part of 50li to haue a bowling green made me at Stalbridge, by a noseles man, named Thomas ford, and I am to provide all materials, and carriadge."

The total cost came to more than £100.*

John Hopkins, the Earl's mason of Stalbridge, was paid

"xxs for adding one Round Staier more at the lower part of my stone staires ascending to my owtward gate there; as also other vli in earnest, and in part payment, for making me a dry stone wall of 2 foot thick and 9 foot highe, from my brew howse to the gate of the Lane going to sherborn; for which I am to make it vp 9d the perch, he digging & quarreing the stones, and I drawing them from the quarry to the worck, and he making the stone-wall neatly, strong, & worckmanleyke."†

Stalbridge House was assessed at 30 hearths in 1662–1664, so it ranked as the fifth largest manor house in Dorsetshire.‡

This was the stately home that the Great Earl by indenture dated 30 November 1640 conveyed to

"Edward Lord Howard of Escricke, and Robert Lord Digby of Geshill in Ireland to the use of Robert Boyle Esqr his youngest sonne dureing his naturall life the remaynder to his first sonne & the heyres males of his body. . .The remaynder to Francis Boyle another sonne of the sd Earle of Corke for Terme of his life, wth the like remaynders to his sonnes. . ."§

The Earl tells us

"that I haue putt into my Iron cheste which I haue lefte in my howse at Stalbridge the ancient pattents, deeds, wrytings, evidences, and papers which I haue, that any waie Conce[rn] that Mannor, and my purchace thereof, exceptinge on deed which I made thereof to the Lo. Edward Howarde, & the Lo. Robert digbie of Geshill; which orrigenall deed of myne being faste sealed vp with my own seale, I delivered in a black Boxe to Sir Wm Howard, to be by him given vp to his brother the Lo. Edward Howard to be Safely kept; and the cownterpart thereof I haue delivered to the Safe Keeping of my cozen Mr Peeter Nayler, an Attorney in the King's bench, whose chamber is in Newe In,

---

* *ibid.* pp. 114, 115, 137. 2 November, 15 November 1639; 13 May 1640.

† *ibid.* p. 163. 24 October 1640.

‡ C. A. F. Meekings. *Dorset Hearth Tax Assessments 1662–1664.* Dorchester, (For the Dorset Natural History & Archaeological Society), 1951. pp. xxviii, 57.

§ B. L. IV, 85. Letter of John Nicholls to RB dated 12 September 1657.

London, and that is also sealed vp in another boxe with my own seale."*

For the subsequent history of the entailed estate see note *, p. 259.

After his return to England about the middle of 1644, Robert Boyle would certainly have needed some time in which to become accustomed to the contemporary conditions. He had left England as a boy of eleven years of age. During his residence abroad his father had died; the Great Irish Rebellion had broken out bringing disastrous consequences to the fortunes of various members of his family; and the Civil War was in progress. On account of the disturbed state of the country it was nearly four months before he could get to Stalbridge,† so it is most unlikely that he was there on 8 October 1644, when King Charles I dined and stayed the night at Stalbridge.‡

As in the case of other families some of the Boyle family supported the King, whilst others supported the Parliament. It was by good fortune, as he himself recognized, that Robert found his sister Katherine, when he arrived in London, for she had valuable and influential relationships with the parliamentarian party, whereby he secured protection for his English and Irish estates.§ Other members of the family had not been so lucky; the estate of his eldest brother had been sequestered.||

From the point of view of their undoubted influence on Robert Boyle, it is of considerable interest to consider the friends, acquaintances, and certain family relationships of his sister Katherine, Lady Ranelagh, at this time. She was the niece of Mrs Dorothy Moore, who married John Dury¶ in 1645; she was also the sister-in-law of Sir John Clotworthy (afterwards viscount) and his wife Margaret. These four individuals had been known to Samuel Hartlib** for some years, and so Robert Boyle's acquaintanceships became extended into Hartlib's circle. Benjamin Worsley†† is another who was known to Boyle at an early date. How this contact between the two came about

---

* *I. Lismore.* V, p. 196. 28 September 1641.

† *Works Fol.* I, *Life.* p. 15: *Works.* I, p. xxvii. Letter of RB to the steward of his manor at Stalbridge, undated.

‡ Thomas Manley. 'Iter Carolinum' in John Gutch, *Collectanea Curiosa.* 2 vols. Oxford, 1781. Vol. II, p. 425: Richard Symonds. *Diary of the Marches of the Royal Army during the Great Civil War.* Camden Society. 1st Series. (1859), 74.

§ *Works Fol.* I, *Life.* p. 15; *Works.* I, p. xxvii.

|| *C.J.* III, p. 307. 10 November 1643.

¶ Footnote is on page 62.

** Footnote is on page 62.

†† Footnote is on page 63.

is not clear, because Worsley's movements from 1642 to 1646 are not

---

¶ Further ramifications of the Boyle family are shown by the following tables, which are not complete.

Sir John KING—
- —Dorothy KING = (1) Arthur MOORE
  = (2) John DURY
- —Sir Robert KING
- —Mary KING = William CAULFIELD, 2nd Baron Caulfield.

Garret or Gerald MOORE, 1st Viscount Moore of Drogheda ———
- —Sir Thomas MOORE = (1) Sarah BOYLE = (2) Robert DIGBY, 1st Baron Digby
- —Arthur MOORE = Dorothy KING
- —Frances MOORE = Roger JONES, 1st Viscount Ranelagh
  - —Arthur JONES 2nd Visc. Ranelagh = Katherine BOYLE
  - —Margaret JONES = John CLOTWORTHY, 1st Visc. Massereene
  - —Mary JONES = John CHICHESTER

(Based on G.E.C.: Lodge, Vol. **IV**, p. 302: Turnbull, *H. D. & C.* p. 247.)

** (?–1662) came from Poland. Described by John Beale as " the zealous solicitor of Christian peace amongst all nations, the constant friend of distressed strangers, the true hearted lover of our native country, the sedulous advancer of ingenious arts and profitable sciences." (J. Beale. *Herefordshire Orchards a Pattern for all England.* London, 1657. In Dedication to Hartlib.)

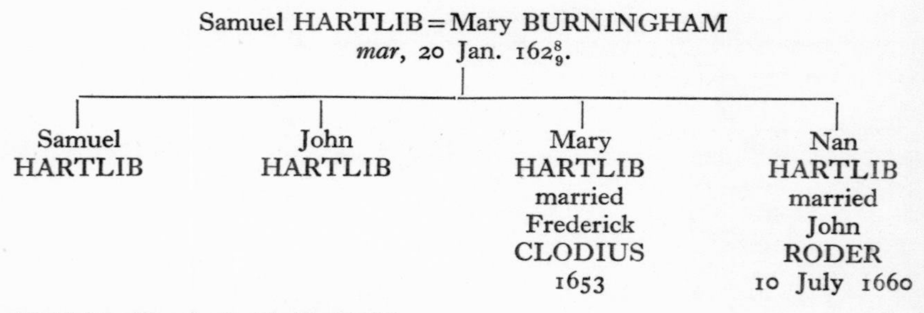

(*D.N.B.:* Turnbull, *H. D. & C.*)

known to me. He had been surgeon-general to the army in Ireland in 1641–1642, so, in consequence of this position, he may have become known to some member of Lady Ranelagh's group. The present writer has, elsewhere,* given the *latest* dates by which certain individuals became known to, or started to correspond with, Hartlib and Boyle. In every case, with the one possible exception of Worsley, the date is later in respect of Boyle. It is reasonable to conclude that Robert Boyle's circle of acquaintances on his return from abroad stemmed from his sister Katherine, and was extended through Hartlib and his acquaintances. Some of them had a strong religious outlook, which undoubtedly helped to intensify Robert's own natural disposition, and which found expression in his early writings. I believe, but have no proof, that Benjamin Worsley was responsible for stimulating Robert Boyle's interest in medical and chemical pursuits. It is perhaps significant that the earliest available sheet of philosophical notes (1650) made by Boyle mentions only Mr Worsley and Dr Boate (Arnold or Gerard), and that the former supplied two-thirds of the items.†

Any notion that Boyle went into seclusion by going to Stalbridge is unjustified, for during this period he frequently visited friends and relations, besides making visits to Ireland in connection with his estates there. Apart from attending to the requirements of his property at Stalbridge, he spent some of his time during 1645 and 1646 in writing the *Ethics*, two versions of which are still extant.‡

---

†† Benjamin Worsley (1618?–1677) is not mentioned in any standard work of reference. Variously described as Doctor of Physic, merchant. Surgeon-general of king's army in Ireland, 1641. Petitioned for a patent for saltpetre. 1646. In Amsterdam, 1649. Connected with Council of Trade, 1650, Secretary to Commissioners in Ireland, 1651. Surveyor General of Forfeited Estates in Ireland, 1654. Married Lucy Cary, 1656. Took farm of the Post Office, 1659. Nominated Commissary General of Musters, 1659. In government service, 1663. Petitioned for a patent relating to production of senna, 1666. Member of Council of Trade, 1668. Secretary and treasurer of combined Council for Trade and Plantations, 1672. His library sold, 1678. (*N. & Q.* (1943) **185**, 123: R. P. Bieber, 'The Plantation Councils of 1670–4.' *Engl. Hist. Rev.* (1925) **40**, 100: *L. J.* **7**, 401, 424, **8**, 574: *C. S. P. Domestic: C. S. P. Ireland: C. S. P. Colonial (America):* Turnbull, *H. D. & C.:* Hooke, *Diary:* T. Larkom. *History of the Down Survey*, London, 1851: *Sale Catalogue of Library of Benjamin Worsley*, London, 1678.)

* *N. & R.* (1963), **18**, 104.

† B.P. **XXVIII.**

‡ Royal Society. MS. 192. *The Ethical Elements;* and MS 195. *The Aretology or Ethicall Elements of Robert Boyle. Begun At Stalbridge, The of*

Further evidence of Boyle's predilection in these early years for writings of moralistic or theological tenor is provided by a list of titles drawn up in his own hand.

" Materialls & Addenda
Design'd towards the Structure
& compleating of Treatises
already begun or Written.
January the 25th 16^{49}_{50}. . . .

> Theodora: Dayly Reflection: Publicke-Spiritidnesse: The Antagonist of Romances: My Opionion of Romanecs: the Swearer Silenc't: The Appendix to it: Seraphicke Love: Consolation to T's Mother: Occasionall Reflections: Essays. (1) Of Time & Idlenesse (2) Of Essays (3) Of Naturall Philosophy & Filosofers (4) Of Chymistry & Chymists (5) Of Universality of Opinions & of Paradoxes (6) Of Authority in Opinions (7) Of Antiquity & New-light (8) Of Cold (9) Of Atoms (10) Of Dungs (11) Of Good Language (12) Of the Holy Scriptures (13) Of the Mechanicks (14) Of Inkes & Writing (15) A Censure of some Preachers & Hearers of Sermons (16) Of the Weapon-salve & occult Qualitys (17) Of Heresies: & how Little Knowledge is necessary in all Persons to Salvation (18) Of Spotts (19) Of true Christianity (20) Of Reasoning & Discourse (21) Of the Ethickes (22) An Inquisition concerning Hapynesse (23) Of the Interpretation of Scripture (24) Of Complaisance (25) Of Censuring & Censurists (26) Of the Vanity & Partiality of Fame."*

It will be noted that less one third of these items has any scientific content. Theodora was not published till 1687, when it appeared anonymously as *The Martyrdom of Theodora and of Didymus;* Publicke-Spiritidnesse† appeared in a modified form in 1655 as *An Invitation to a free and generous Communication of Secrets and Receits in Physick;* The Swearer Silenc't, with the appendix, appeared posthumously in 1695 as *A Free Discourse against Customary Swearing;* Seraphick Love was used as the running title to *Some Motives and*

[*Blank*] 1645. *That's the tru Good that makes the owner so.*

Works Fol. **I**, *Life*. pp. 17, 20; *Works*. **I**, pp. xxx, xxxiv. Letters of RB to Lady Ranelagh dated 30 March 1646, and to Marcombes dated 22 October 1649 respectively.

* B.P. **XXXVI.**

† Mentioned in RB's letter to Lady Ranelagh dated 2 August 1649. *Works Fol.* **V**, p. 237; *Works*. **VI**, p. 48.

Incentives to the Love of God published in 1659; *Occasional Reflections* appeared in 1665. Essays 1 and 21 alone have survived.*

Francis Boyle published during his brother's lifetime a volume of *Moral Essays and Discourses* (1690) in which he has made very free use of Robert's *The Swearer Silenc't*. It is highly probable that Francis used other youthful writings of his brother in the same work.†

The number of letters surviving from Robert Boyle's correspondence in these early years is less than when Birch wrote his *Life*. It is possible to trace some of his movements, as well as to throw light on his occupations and interests. At the beginning of May 1645 he was at Bristol;‡ for what purpose we do not know. In August of the same year he probably went to France again for a short while, and the assumption is that he did so in order to settle his indebtedness with Marcombes.§ According to Birch, Boyle was at Cambridge in December following,|| and this was perhaps the occasion that enabled Boyle to make his friendship with Francis Tallents,¶ Fellow of Magdalene College, who had tutored certain of the sons of the Earl of Suffolk, whose daughter Margaret was married to Lord Broghill. Boyle was in London during February 1646, when he wrote to Arthur Annesley,** giving him authority to handle the affairs of his estates in Connaught and elsewhere.††

The hazards of travelling at this time under the disturbed conditions prevailing in the country as a result of the continued conflict between Royalist and Parliamentarian forces are described in a light-hearted manner in the letter Boyle wrote to his sister, Katherine, at the end of March 1646. He left London in company with his brother Roger on the 26th of that month, dined at Egham, and stayed the night at

---

* Essay 1 is in B.P. **XIV**, fol. 15. It is mentioned in *Occasional Reflections. Works Fol.* **II**, p. 147; *Works*. **II**, p. 336.

The appendix to the *Sceptical Chymist* (1680) has mention in the author's preface of 'an essay that I wrote many years since, *Of the Usefulness of Chymistry to the Empire of Man*,' which may have some relation to Essay 4. *Works Fol.* **I**, p. 374; *Works*. **I**, p. 591.

† R. E. W. Maddison. 'The Plagiary of Francis Boyle.' *Annals of Science*. (1961, published 1963), **17**, 111.

‡ B. P. **XXXVII**. A literary effusion 'To my Mistress', written "but to please a Person of my owne sexe that desir'd it."

§ *Works Fol.* **I**, *Life*. p. 15; *Works*. **I**, p. xxvii. Letter of RB to his brother, Lord Broghill, dated 25 August 1645.

|| *ibid.*

¶ (1619–1708) Divine; author of various religious works.

** (1614–1686), afterwards Viscount Valentia, and Earl of Anglesey. He was sent to Ulster by authority of parliament in 1645 as a commissioner to defeat Ormond's negotiations with the Scots. (*Biogr. Brit: D.N.B.:* G.E.C.)

†† Bodleian MSS. Clarendon 27, fol. 67. Catalogue No. 2132. Letter dated 28 February $164\frac{5}{6}$.

Farnham. The next day they dined at Winchester, and stayed the night at Salisbury. Boyle arrived at Stalbridge on Saturday night, 28 March 1646, intending to go to Bristol when the weather permitted. " My stay here, God willing, shall not be long, this country being generally infected with three epidemical diseases (besides that old leiger* sickness, the troop-flux) namely the plague, which now begins to revive again at Bristol and Yoevil six miles off, fits of the committee, and consumption of the purse. . . "†

It is most noticeable that Boyle throughout his life avoided partisanship in political affairs, and it is not often that he writes a news-letter. Publication of the *Articles of Surrender* following the capitulation of the Royalists at Oxford in June 1646 gave him cause to write from London to inform his brother Richard, Earl of Cork, about compounding for his estate.‡ One of Boyle's rare news-letters describing contemporary events is that dated 22 October 1646 written from London to Marcombes, who was hoping for further employment.§ He says:

> " An employment fit for you we cannot yet procure, because all our nobility stands at a gaze, to see whether the issue of the treaties now in debate will be either peace or war; in either of which cases it is probable, that a good many of them will make visits to foreign climates."
>
> " The sadness of your condition I very much resent, and would offer you my assistance to sweeten it, if I did not think the proffer superfluous; but truly I believe it would less afflict you, if you were a spectator of our miseries here, where every day presents us with much more unusual dispensations of providence, where I myself have been fain to borrow money of servants, to lend it to men of above £10,000 a year."

This same letter reveals the difficulty of Robert Boyle's position in view of the fact that he had relatives supporting one or other of the political factions.

> " I was once a prisoner here upon some groundless suspicions, but quickly got off with advantage. . . . I have been forced to observe

---

* =fixed, stationary.

† *Works Fol.* I, *Life*. p. 17; *Works*. I, p. xxx.

‡ Oxford capitulated 20 June 1646. The *Articles of Surrender* are printed at large in Joshua Sprigg, *Anglia Rediviva*, Oxford, 1854 (Original edition London, 1647). Holders of estates were given six months in which to compound for them. The Earl of Cork's fine was fixed in 1646 at one-tenth =£1,631.

Works Fol. V, 235; *Works*. VI, p. 44. Letter of RB to Richard Boyle dated 14 July 1646.

§ *Works Fol.* I, *Life*. p. 17; *Works*. I, p. xxx.

a very great caution, and exact evenness in my carriage, since I saw you last, it being absolutely necessary for the preservation of a person, whom the unfortunate situation of his fortune made obnoxious to the injuries of both parties, and the protection of neither. Besides I have been forced to live at a very high rate, (considering the inconsiderableness of my income) and, to furnish out these expences, part with a good share of my land, partly to live here like a gentleman, and partly to perform all that I thought expedient in order to my Irish estate, out of which I never yet received the worth of a farthing."*

This particular letter is of importance because it mentions for the first time the vexed question of 'The Invisible College', concerning which so much has been written.† Boyle tells Marcombes about his literary efforts, and discloses his study of "natural philosophy, the mechanics, and husbandry, according to the principles of our new philosophical college, that values no knowledge, but as it hath a tendency to use;" and he requests Marcombes "to enquire a little more thoroughly into the ways of husbandry, &c., practised in your parts; and when you intend for England, to bring along with you what good receipts or choice books of any of these subjects you can procure; which will make you extremely welcome to our *Invisible College*, which I had now designed to give you a description of, but a gentleman, whom I have been forced to keep talk with all the while I was writing this, together with the fear of having too much already trespassed upon your patience, call upon me to end your trouble and this letter together." This unknown gentleman little knew what discussions he would provoke amongst historians of science as a result of his interruption, whereby we have been deprived of a first-hand account of the Invisible College.

---

* The straitened circumstances of his sister, Katherine, about this time are shown by the fact that Parliament ordered the sum of six pounds per week to be paid to her "for the present Support and Relief of the extreme Necessities of herself and her Children, out of the Treasury for Sequestrations at Guildhall, London." (*L.J.* 8, 710. under date 6 February 164$\frac{6}{7}$.) The matter of her pension, and payment of arrears was raised at a later date. (B.M., Add. MSS. 5501 fol. 46. Memorandum dated 25 April 1650.)

† The most important and valuable articles dealing with this problem are the following:

R. H. Syfret. 'The Origins of the Royal Society.' *N. & R.* (1948), **5**, 75.

G. H. Turnbull. 'Samuel Hartlib's Influence on the Early History of the Royal Society.' *ibid.* (1953), **10**, 101.

D. McKie. 'The Origins and Foundation of the Royal Society of London.' *ibid.* (1960), **15**, 1.

In connection with the recent work of M. Purver, *The Royal Society: Concept and Creation.* London, 1967, see the essay reveiw by C. Webster in *History of Science.* (1967), **6**, 107–128.

The next mention is in Boyle's letter written from London, 20 February 164^{6}_{7}, to Francis Tallents at Cambridge. He says: "The best on't is, that the corner-stones of the Invisible, or (as they term themselves) the Philosophical College, do now and then honour me with their company." He extols their knowledge, broad outlook, and universal good-will; and regrets "there is not enough of them."*

The third and last mention occurs in Boyle's letter, dated 8 May 1647, written to Samuel Hartlib. "You interest yourself so much in the Invisible College, and that whole society is so highly concerned in all the accidents of your life, that you can send me no intelligence of your own affairs, that does not (at least relationally) assume the nature of Utopian."†

These three are the only occasions on which we find any mention of the Invisible College, and they are restricted to the period 22 October 1646 to 8 May 1647. No other contemporary mention of the Invisible College is known.

Thomas Birch's association of this Invisible College with the groups of virtuosi who were meeting in London or Oxford from 1645 onwards, and who were later incorporated as the Royal Society, was qualified by the words 'probably' (1744) and 'seems to be' (1756);‡ but with the passage of time the supposition has become positive identification. On various grounds, it is improbable that Boyle, a young man of 20 years, was then on such close terms with the 'corner-stones' at Gresham College.§ In fact, it seems rather unlikely that the individuals in question were known to Boyle at this early date. The present writer has, elsewhere,‖ given dates at which they became known to Hartlib and Boyle. Although the true dates are unquestionably earlier than those given, yet the order in which the individuals present themselves is probably near the truth as well as being more significant. The emerging pattern shows that they became known to Boyle at a later date than to Hartlib. Indeed, William Petty, who was introduced to Boyle by Hartlib in 1647, is the only one of them concerning whom there is any surviving record, as yet known, of having been in contact with Boyle before his (Boyle's) departure for Ireland in 1652.

The present writer is not alone in the opinion that the Invisible College or the Philosophical College, mentioned only three times by

* *Works Fol.* **I**, *Life*. p. 20; *Works*. **I**, p. xxxiv.

† *Works Fol.* **I**, *Life*. p. 24; *Works*. **I**, p. xl.

‡ *Works Fol.* **I**, *Life*. p. 25; *Works*. **I**, p. xlii: Birch *History*. **I**, p. 2.

§ D. McKie. 'The Origins and Foundation of the Royal Society of London.' *N. & R.* (1960), **15**, 22.

‖ R. E. W. Maddison. 'The Stalbridge Period, 1645–1655, and the Invisible College.' *N. & R.* (1963), **18**, 109 *et seq.*

Boyle during 1646–47, has no connection with those virtuosi who are mentioned by Wallis and Sprat in their accounts of the origin of the Royal Society relative to the years 1645 or 1648–49.* Pertinent remarks on the absence of Hartlib's name in these accounts are made by Dircks.† The name Invisible College is perhaps a *jeu d'esprit* on the part of Boyle to describe a small group of persons interested in the utilitarian aspects of natural philosophy, and calling themselves the Philosophical College. Many proposals for academies or colleges for the purpose of acquiring or dessiminating knowledge were put forward in this period; and it is sufficient to name Bacon, Hartlib, Comenius, Dury, Hall, Petty, Cowley, Evelyn in this connection. It has been suggested, (but not convincingly), that the name Invisible College was a conceit inspired by Boyle's stay in Italy.‡

There is no documentary evidence as to the membership of the Invisible College, which most likely had no formal constitution at all. The present writer suggests that besides Robert Boyle, Benjamin Worsley, whose interests and activities were both utopian and utilitarian, and who most probably in these early years played a significant part in the development of the former's scientific interests, was another. John Dury may have been in the group, for he had an interest in some mining projects, and corresponded with Worsley in 1648–49 with a view to co-operation.§ The conjunction of Boyle, Worsley and Dury in the activities of George Stirke (*vide infra*) is perhaps not without significance in this connection.|| Could it be that this Philosophical College of 1646–47 comprised Boyle, Worsley, Dury and some others who were interested in utilitarian projects because of their potentially profitable nature; and was it Invisible because such activities were not compatible with their other publicly better known utopian schemes, even if the profit from those activities were to be used in furtherance of the utopian schemes? Such duality of outlook is not impossible, for the rumour at a later date that Mrs Dury contemplated engaging in retail trade, when the Durys were in rather straitened circumstances, provoked surprise if not disapproval that she could stoop so low.¶

---

* John Wallis. *A Defence of the Royal Society* . . . London, 1678.
Thomas Sprat. *The History of the Royal Society of London*. . . London, 1667.

† Henry Dircks. *A Biographical Memoir of Samuel Hartlib*. London, [1865], pp. 20–22.

‡ R. F. Young. 'The Invisible College.' *The Times Literary Supplement*. 12 December 1936. No. 1819, p. 1035.

§ Turnbull. *H. D. & C.* p. 262: B. L. VII. Copy of a letter from J. Dury to B. Worsley dealing with a furnace and an antimony mine.

|| See p. 77 also.

¶ Turnbull. *H. D. & C.* p. 261.

Evidence for a well-established degree of acquaintanceship between Boyle and Worsley at this time is provided by a letter* written by the former. Although undated, it can be confidently assigned to a date shortly after 21 November 1646, because of the reference to Worsley's saltpetre patent.† The opening salutation is the intimate one of 'Dear Mr. Worseley'. Boyle says: "My grand employment, in my spare hours, is to catechise my gardener and our ploughmen concerning the fundamentals of their profession:" and that he has not forgotten "his promise to transmit to you any thoughts or experiments of mine, that I shall judge conducible to the furtherance of your great design, and the enabling you to do for the great world, what the chairman of the physicians has done for the little, publish a discourse *de usu partium*."‡ So, Boyle was already interested in experiments of some kind or other—probably those having a bearing on medicinal curative value. His growing interest in such activity is shown by the equipping of a laboratory. "... at present the unluckly and unexpected lingring of the waggon, which is to bring me all my Vulcanian implements & my necessitated idlenesse till they come; makes me wish it drawne by Pegasuses & thinke it drawne by snayles; & makes me sadly walke up & downe in my laboratory.... This unseasonable disaster ... [enforces] ... an unlooked for backwardness in my preparations ..."§ Shortly after, he bemoans to his sister, Katherine,

> "... That great earthen furnace, whose conveying hither has taken up so much of my care, and concerning which I made bold very lately to trouble you, since I last did so, has been brought to my hands crumbled into as many pieces, as we into sects; and all the fine experiments, and castles in the air, I had built upon its safe arrival, have felt the fate of their foundation. Well, I see I am not designed to the finding out the philosophers stone, I have been so unlucky in my first attempts in chemistry. My limbecks, recipients, and other glasses have escaped indeed the misfortune of their incendiary, but are not, through the miscarriage of that grand implement of Vulcan, as useless to me, as good parts to salvation without the fire of zeal. Seriously, madam, after all the pains I have taken, and the precautions I have used, to prevent this furnace the disaster of its predecessors, to have it transported a thousand miles by land, that I may after all this receive it

* *Works Fol.* V, p. 232; *Works.* VI, p. 40.

† *L. J.* VIII, p. 574.

‡ Galen. *De usu partium corporis humani.*

§ *Works Fol.* V, p. 231; *Works.* VI, p. 39. This undated draft without superscription I suggest was written for Worsley, about the end of February 1647. RB was in London on 20 February and at Stalbridge on 6 March, by which date some of his apparatus had arrived.

broken, is a defeat, that nothing could recompense but that rare lesson it teaches me, how brittle that happiness is, that we build upon earth . . ."*

A little later he tells us, "I often divert myself at leisure moments in trying such experiments, as the unfurnishedness of the place, and the present distractedness of my mind, will permit me."†

Samuel Hartlib was already known to Boyle; a correspondence had been started between them and was to continue for some years. The matters referred to in their letters during 1647 deal with *Real Character;*‡ the wind-gun which Boyle had read about in the works of Mersenné; § two works of J. V. Andrea, which Hartlib had sent to John Hall|| at Cambridge for translation.¶ Hall wished to dedicate his *Emblems with Elegant Figures*** to Boyle, who was cautious about giving his approval. Hartlib wrote to Boyle on 16 November 1647†† about his scheme for an office of communications, 'especially in reference to universal learning',‡‡ and at the same time introduced William Petty§§ by sending the design of the History of Trades.||||

"The author of them is one Petty, of twenty-four years of age, not

---

* *Works Fol.* **I**, *Life*. p. 21; *Works*. **I**, p. xxxvi. Letter of RB to Katherine, Lady Ranelagh, dated 6 March 164^{6}_{7}.

† *Works Fol.* **I**, *Life*. p. 25; *Works*. **I**, p. xli. Letter of RB to S. Hartlib dated 8 May 1647.

‡ Undoubtedly a reference to the work of George Dalgarno subsequently published as *Ars Signorum, vulgo character universalis et lingua philosophica*. London, 1661. RB's name is the first in the list of persons thanked by the author for their support. 'An inchoative and indigestible form' of this appeared at Oxford in 1657. (Madan. **III**, No. 2335).

§ Marin Mersenne, *Hydraulica, Pneumatica; arsque navigandi, Harmonica Theorica, Practica, et mechanica phaenomenas*. Paris, 1644. p. 149 *de sclopeto pneumatico*. On the wind-gun see J. G. Fischer, *Geschichte der Physik*, Göttingen, 1801–8. Vol. **I**, p. 456. Evelyn (de Beer) Vol. **IV**, p. 520.

|| John Hall (1627–1646), poet and pamphleteer. (*D.N.B.:* G. H. Turnbull, 'John Hall's Letters to Samuel Hartlib.' *Review of English Studies* (1953), **4**, 221.)

¶ The two works were *Christianae Societatis Imago* and *Christianae Amoris Dextera Porrecta*.

All the above matters are in RB's letters to Hartlib dated 19 March 164^{6}_{7} and 8 April 1647. (*Works Fol.* **I**, *Life*. **22**; *Works*. **I**, p. xxxviii.)

** London, [1648]. Turnbull, *loc. cit.* p. 228.

†† *Works Fol.* **V**, p. 256; *Works*. **VI**, p. 76.

‡‡ Concerning Hartlib's scheme see Turnbull, *H. D. & C.* p. 79 *et seq.*, and *passim*.

§§ William Petty (1623–1687), political economist. Professor of anatomy at Oxford, 1651; carried out the 'Down Survey' in Ireland; knighted, 1661; original F.R.S. *D.N.B.:* E. Fitzmaurice, *The Life of Sir William Petty*, London 1895: E. Strauss, *Sir William Petty*, London, 1954: I. Masson and A. J. Youngston, 'Sir William Petty, F.R.S.' *N. & R.* (1960), **15**, 79.

|||| This projected work was never written. For an outline of the contents see Marquis of Lansdowne, *The Petty Papers*, London, 1927, Vol. **I**, p. 203.

altogether a very dear Worsley, but a perfect Frenchman, and good linguist in other vulgar languages besides Latin and Greek, a most rare and exact anatomist, and excelling in all mathematical and mechanical learning, of a sweet natural disposition and moral comportment; as for solid judgment and industry, altogether masculine."

During this year (1647), whilst at Stalbridge, Boyle composed certain of the *Occasional Reflections*, some of which were subsequently published, and others of which are still in manuscript. Other much longer moral disquisitions were written, too; such as those to Mrs Dury,* his brother [? brother-in-law],† an unknown lady,‡ and Fidelia.§

From the end of February‖ to the beginning of April 1648 Boyle was in the Netherlands, whither he went not only to view the country, but also to accompany his brother Francis and sister-in-law back from The Hague.¶ His stay at The Hague provided him with subjects for some of the *Occasional Reflections*, as for example those "Upon a Courts being put into Mourning.** Hague 1648", and "Upon the Shop of an ugly Painter rarely well stor'd with Pictures, of very handsome ladies. At the Hague"; and the return journey on "Sailing betwixt Roterdam and Graves-end on Easter-day 1648", when he found "a real storm a very troublesome and uneasie thing", provided "Looking through a Perspective Glass upon a Vessel we suspected to give us Chase, and to be a Pyrat".†† At Leyden he "was

* B. L. **I**, 109. Letter of RB to Mrs. John Dury dated, Stalbridge, 15 April 1647.

† B. L. **I**, 137. Letter of RB to [?] dated Stalbridge, 18 April 1647.

‡ B. P. **XXXVII**. Letter of RB to [?] dated London, 15 August 1647. This is titled 'The Duty of Mother's being a Nurse asserted.'

§ B.P. **XXXVII**. Letter of C. K. [i.e. RB] to Fidelia dated London, 2 December 1647.

‖ *Works Fol.* **I**, *Life*. p. 27; *Works*. **I**, p. xlv. Letter of RB to Marcombes dated 22 February 164$\frac{7}{8}$ in which he mentions his intention of leaving for the Netherlands the next day.

¶ *L.J.*, **8**, 468a. Pass for Mr [Francis] Boyle, his wife and servants to go into Holland, 19 August 1646.

** It would seem that the court was still in mourning for Prince Frederik Hendrik, the stadtholder, who died 17 March 1647 N.S. The only other cause for mourning in this period would have been for Katharine Belgica, a half-sister of the former. She died 12 April 1648 N.S., but Boyle had left the country by Easter Day (12 April N.S., or 2 April O.S.). Information communicated by Dr Maria Rooseboom and supplied by the officials of the Archives of the Royal Family at The Hague.

†† *Works Fol.* **I**, p. 165; *ibid.* **II**, pp. 166, 212, 225. *Works.* **I**, p. 259; *ibid.* **II**, pp. 365, 438, 458.

brought to the top of a tower,* where in a darkened room . . . a convex glass, applied to the only hole, by which light was permitted to enter, did project upon a large white sheet of paper, held at a just distance from it, a lively representation of divers of the chief buildings in the town ".† It is most probable that Boyle visited the Anatomy School in Leyden,‡ for Hartlib in his letter of 9 May 1648 thanks Boyle for his " new discoveries in anatomy, and enquiries of other useful and ingenious knowledges,"§ which remark is more likely to refer to Boyle's report on what he had seen in Holland, than to the results of any investigations by himself. This visit to Leyden is no doubt the basis of the mis-statement that Boyle studied at the university there. At Amsterdam he visited Menasseh ben Israel,|| " the Greatest Rabbi of this Age ", whom he mentions several times in his works.¶ This was not the first time that Boyle had consorted with Jews in order to learn more about their religion: he had done the same during his visit to Florence in 1642. Among his miscellaneous observations he noted " that in Holland it is usual, as I have seen my self, to mingle sheep's dung with their cheeses, only to give them a colour and a relish."

Boyle was at Stalbridge for a while during May 1648 for he wrote to his sister Katherine from that place.** During June and July he was in London, when he told a correspondent that he had had " unexpectedly the happiness of a long conversation with the fair lady, that people are pleased to think my mistress."†† William Petty followed up his introdution to Boyle by writing him a flattering letter and dedicating to him his invention of the Double Writing Instrument.‡‡ He concluded

* Officials of the Municipal Archives at Leyden have been unable to identify this place. Information communicated by Dr. Maria Rooseboom.

† *Works Fol.* **I**, p. 194; *ibid.* **III**, p. 86. *Works.* **I**, p. 302; *ibid.* **III**, p. 312.

‡ For a short description see Evelyn (de Beer) Vol. **II**, p. 53.

§ *Works Fol.* **V**, p. 256; *Works.* **VI**, p. 77.

|| (1604–1657), Jewish theologian; petitioner to Cromwell for readmission of the Jews into England. (*D.N.B.:* C. Roth. *A Life of Menasseh ben Israel.* London, 1934.)

¶ *Works Fol.* **II**, pp. 110, 124; *ibid.* **IV**, p. 375. *Works.* **II**, pp. 280, 301; *ibid.* **V**, p. 183.

** *Works Fol.* V, p. 235; *Works*, **VI**, p. 45. Letter dated 13 May 1648.

†† *Works Fol.* V, p. 236; *Works.* VI, p. 46. Letter from RB to Lady Elizabeth Hussey dated 6 June 1648. Perhaps the ' fair lady ' was Elizabeth Carey (1631–1676) daughter of the 2nd Earl of Monmouth, for John Evelyn wrote to William Wotton saying " I have been told he courted a beautiful and ingenious daughter of Carew, Earl of Monmouth." (Evelyn (Bray) Vol. **III**, p. 349.) This episode has been romantically treated by Masson, pp. 174–179.

‡‡ J. Ward. *The Lives of the Professors of Gresham College.* London, 1740. p. 218. Petty's pamphlet *The Advice of W. P. to Mr. Samuel Hartlib. For the Advancement of some particular Parts of Learning.* London, 1648. He was granted a patent for this invention. (*L. J.* **X**, 8 March $164\frac{7}{8}$.)

his letter, saying, " For my study and ends being enquiries into nature, and useful arts, and finding how ill my abilities to make experiments answer my inclinations thereto, I knew no readier way to become fat in that kind of knowledge, than by being fed with the crumbs, that fall from your table ".* In August Boyle was at Leese Priory in Essex, the home of his sister Mary, where he completed *Some Motives and Incentives to the Love of God*, which was not published until 1659, and then dedicated to his sister Mary, Countess of Warwick. In September he was back at Stalbridge.

Visits abroad and to his relations meant that Boyle was absent from Stalbridge during much of this year. In addition, there were the disturbances resulting from the Second Civil War; so, he could not have devoted much time to any experimental work. His main interests and occupations in this year seem to have been literary. There is a noticeable absence of detailed reports or comments on current events. The Siege of Colchester receives passing mention only in an " Occasional Reflection " on the " prodigiously wet weather " that summer.† There is no record of what Boyle thought about the execution of King Charles I on 30 January 164$^{8}_{9}$.

His brother Roger, Lord Broghill, had supported the Royalist cause, and at this time was making plans to go overseas in order to join Charles II in exile. In this connection, in view of the fact that Robert had " been forced to observe a very great caution " in his conduct,‡ it is a pleasant surprise to find him writing to an unidentified lady soliciting her aid to secure from the French ambassador, Pierre de Bellièvre (1611–1683), a pass to enable his brother Roger to

> " trauell into France & To & Fro there as long as they neither Do nor Carry any thing that may forfeit the Protection they shall Live vnder . . . Yow may possibly do somebody no vnacceptable seruice, to helpe into France a Person, who elsewhere might (perchance) crosse (or at Least retard the Progresse of Desseins) & who is now dispos'd to this Retirement by Considerations that it is farre more easy for Y[our] L[adyship] to guesse then safe for me to write."§

The plan was not accomplished, for in fact Lord Broghill was later won over to the support of Cromwell.||

* *Works Fol.* **V**, p. 296; *Works.* **VI**, p. 136. Letter of W. Petty to [RB] dated 21 June 1648.

† *Works Fol.* **II**, p. 178; *Works.* **II**, p. 384.

‡ pp. 66-67.

§ Letter of RB to [?] dated 26 March 1649 and written from Marston, the seat of Lord Broghill. The draft is at The Historical Library, Yale Medical Library, New Haven, Conn., U.S.A.

|| Morrice. *Memoirs*, p. 9; *Letters*, p. 10.

Boyle's increasing interest in scientific matters becomes apparent during this year, 1649. Hartlib in one of his letters deals with the antimonial cup,* and in his *Ephemerides* mentions Boyle's use of microscopes.† In August, when he was at Stalbridge recuperating from an attack of the ague which afflicted him in July at Bath, and was " drawing a quintessence of wormwood (for my particular use) ", Boyle tells his sister, " Vulcan has so transported and bewitched me, that as the delights I taste in it, make me fancy my laboratory a kind of Elysium . . . I there forget my standish and my books ".‡ So, it is evident that by this date he had assembled some of the necessary furnishings of a laboratory, and started his experiments and investigations.

According to Birch, Boyle was in London on 15 November, and at Marston on 20 November 1649.§ He was in London during December when he wrote to his brother Roger to compliment him on his military successes in Ireland, where Lord Broghill was then serving Cromwell.|| He wrote also to his sister Alice in reference to his fourteen years' absence from his native country, and the unwelcome marriage of her son to Susan Killigrew. He counselled her to consider " making the best of those accidents that providence is pleased to dispense " . . . and to consider " whether or no, in things past remedy, it be not better, rather, by a winning gentleness to improve what you dislike to its best uses . . . than persist in testifying an implacable indignation, that cannot redress a displeasure, and may provide one."¶ Lord Barrymore's conduct as a married man called for the admonitory letter which must have been written somewhat later.**

We know very little about Boyle's activities during 1650. His list of " Materials & Addenda " (p. 64) shows that his main interest since returning to England had been for writings of a moralistic character, though subjects of a scientific nature were already engaging his attention. Another list in his hand gives further subjects of essays. Unfortunately, it is undated. Because of the predominating number

---

* *Works Fol.* **V**, p. 257; *Works.* **VI**, p. 78. Letter dated 24 July 1649.

† G. H. Turnbull. ' Samuel Hartlib's Influence on the Early History of the Royal Society.' *N. & R.* (1953), **10**, 122.

‡ *Works Fol.* **V**, p. 238; *Works.* **VI**, p. 49.

§ *Works Fol.* **I**, *Life*, p. 27; *Works.* **I**, p. xlv.

|| *Works Fol.* **V**, p. 239; *Works.* **VI**, p. 50. Letter dated 20 December 1649.

¶ *Works Fol.* **V**, p. 239; *Works.* **VI**, p. 51. Letter from RB to [Alice, Dowager Countess of Barrymore] dated 21 December 1649.

** *Works Fol.* **I**, p. 25; *Works.* **I**, p. xliii. Letter of [RB] to [Richard Barry, 2nd Earl of Barrymore]; undated, but most likely 1650. There are omissions in the printed version.

of items of scientific interest it is probably a little later than the other list dated 1650. The titles of some of these philosophical essays are worth noting as showing the direction of Boyle's thoughts at the beginning of his scientific work, and bearing on its subsequent development. We find, for example, the following:

> " Obseruations vpon some Chymicall Distinctions of salts.
> Against the Peripateticke Quaternary & the Chymicall Ternary of Elements.
> Some Assumptions about Destillation examin'd.
> Of the Efficacy of vnpromising Medecines.
> Of the Effluvia & Pores of Bodys."*

These essays, if they were indeed written, are not known. However, reference to Boyle's published works shows that these subjects were more fully developed in later years. The same paper has a list of " Experiments vnpublished " in which we note the following:

> " The Experiment of logwood & Verdigrease.
> Additions to the Experiment De Vacuo.
> The Burning of snow & Camphire.
> The Cement against Fire & Water."*

We find that Boyle is the first signatory (followed by John Sadler, John Dury, Samuel Hartlib, Benjamin Worsley, Henry Robinson) to a letter written to Cressy Dymock urging him to continue his experiments in " manure and engines of motion."† The conjunction of these names is significant in showing the composition of the group from which Boyle was deriving stimulous for his scientific enquiries. A hasty letter to Hartlib shows that Boyle was busily concerned with sorting out his private affairs,‡ in connection with which the Council of State granted him, 22 July 1650, a pass to travel with three servants into Ireland.§ His movements in the following months are uncertain.

Sometime between 29 November 1650 and 16 January 165$^{0}_{1}$ Boyle was made acquainted with George Stirke by Robert Child.|| Stirke,

---

* B.P. XXXVI.

† Turnbull. *H. D. & C.* p. 268. Concerning Dymock, see H. Dircks, *A Biographical Memoir of Samuel Hartlib.* London, [1865], pp. 91–100.

‡ *Works Fol.* I, p. 27; *Works.* I, p. xlvi. Letter of RB to S. Hartlib dated 1 May 1650.

§ *C.S.P.Domestic 1650*, p. 552.

|| S. E. Morison. *Harvard College in the Seventeenth Century.* Cambridge, Mass, 1936. Consult index under Stirke.

R. S. Wilkinson. 'George Starkey, Physician and Alchemist.' *Ambix.* (1963), **11**, pp. 121–152.

G. H. Turnbull. 'George Stirk, Philosopher by Fire.' *Publications of the Colonial Society of Massachusetts. Transactions, 1947–1951.* (1959), **38**, p. 241, note 6.

an empiric, was deeply interested in, and worked intensively on, the transmutation of metals, the preparation of the alkahest and medicines references to which occur in Hartlib's papers, and reveal Boyle's close concern with Stirke's work.*

From the early part of January 165$^{0}_{1}$ Boyle was in frequent communication with Stirke, whose activities he reported to Hartlib. Stirke prescribed for Boyle one of his (Stirke's) preparations for curing the stone in the kidneys.† Boyle reported on Stirke's preparation of the liquor alkahest saying it would take two to three months to complete.‡ Stirke was reported to be working on a new cure for consumption;§ to have a medicine for fevers;‖ and was going to refute Vaughan.¶ The remains of a letter written by Stirke later than 29 April 1651 almost certainly to Boyle describes various chemical processes, and mentions his extractions of gold and silver out of antimony and iron, which some gentlemen solicit him to continue. Worsley was one of the gentlemen who tried to persuade Stirke to turn the process to immediate profit; but Stirke was quite unwilling to collaborate with Worsley and his associates, who claimed to be able to perform similar feats.** Stirke considers that his processes are superior; that he has no need of partners, and " in a thing which I could command as a master, I would not work as an amanuensis ".†† About this time John Dury saw the extraction of antimonial gold by Stirke.** It is noteworthy in this connection that Dury, Hartlib and Clodius signed a pact—*Christianae Societatis Pactum*—on 18 August 1652 in which it is stated " that their joint efforts to promote the public good were not to injure or prejudice anyone, and that their efforts were to help Stirke and serve his honour and advantage ".‡‡ There seems to be little doubt about one of the reasons for the interest shown by Hartlib, Dury and Worsley in Stirke's processes; they hoped thereby to finance their lofty schemes for public good. A letter written in April 1652 by Dury to Hartlib confirms this; it expresses the hope that Stirke ' would set upon " the lucriferous experiments which God hath put into his hand," and so promote " his

---

* *ibid.* pp. 232, 238. G. H. Turnbull. ' Robert Child.' *ibid.* p. 21.

† G. H. Turnbull. ' George Stirk, Philosopher by Fire.' *ibid.* p. 232.

‡ *ibid.* p. 234; *Works Fol.* **I**, p. 485; *Works.* **I**, p. 97.

§ Turnbull. *loc cit.*, p. 234.

‖ *ibid.* p. 235.

¶ *ibid.* p. 238.

** Turnbull. *H. D. & C.* p. 225.

†† B.L. **VI**, 99.

‡‡ *ibid.* p. 123, section 9. G. H. Turnbull. ' George Stirk, Philosopher by Fire.' *ibid.* p. 227.

own comfort" and "the accomplishment of our joint public designs" '.*

Between the 8th and 16th January 165$^{0}_{1}$ Hartlib recorded in his *Ephemerides* Boyle's opinion on the utility of microscopes.† This year is noteworthy for the publication by Nathaniel Highmore‡ of *The History of Generation*, which is the earliest work to carry a dedication to Robert Boyle. In his dedication, which is dated 15 May 1651, Highmore addresses Boyle in these terms:

> "You have, Sir, so inricht your tender years with such choice principles of the best sorts, and even to admiration managed them to the greatest advantage; that you stand both a pattern and a wonder to our Nobility and Gentry . . . you have not thought your blood and descent debased, because married to the Arts. You stick not to trace Nature in her most intricate paths, to torture her to a confession; though with your own sweat and treasure obtained."

The fact that Highmore was a neighbour of Boyle, and practised at Sherborne, near Stalbridge, is sufficient to explain how this dedication came to be made. One feels that the dedication was inspired more by a genuine regard for Boyle than a desire for his patronage.

In November 1651 Boyle was at Twickenham, near London, evidently in connection with the marriage festivities of some young friends. His letter to John Mallet tells of "a violent & sudden fitt of Sicknesse, which unfitted me to handle a Pen till God's Blessing vpon a Strāge Remedy of a Great Chymist [? Stirke] of my Acquaintāce restor'd me to Health without the wonted Martyrdome of Physicke."§ The statements that Boyle was a resident of Twickenham, that his name occurs in the rate book and the parish register are mistaken.|| In subsequent months Boyle was moving between Stalbridge, London and Leese.

By the latter part of 1651 Stirke was working in the laboratory at St. James's Palace, London. According to the Hartlib papers he had been installed there for the purpose of distilling oils.¶ In view of the

---

* *ibid.* p. 227. Turnbull quotes here another letter dated 29 May 1652 in which Dury says much the same thing.

† G. H. Turnbull. 'Samuel Hartlib's Influence on the Early History of the Royal Society.' *N. & R.* (1953), **10**, 101.

‡ (1613–1685), M.D., Oxon; physician. (*D.N.B.*).

§ B.M., Harl. MSS. 7003, fol. 179. Letter from RB to [John Mallet].

|| D. Lysons. *The Environs of London*. London, 1792–1811. Vol. **III**, p. 578: R. S. Cobbett. *Memorials of Twickenham*. London, 1872. p. 391: E. Walford. *Greater London*. London, [n.d.], Vol. **I**, p. 93.

¶ G. H. Turnbull. 'George Stirk, Philosopher by Fire.' *Publications of the Colonial Society of Massachusetts. Transactions, 1947–1951*. (1959), **38**, 236, note 7: *George Starkey's Pill Vindicated*. . .[London, ? 1660 ?].

fact that John Dury was Library Keeper at St. James's Palace at that time,* it is reasonable to conjecture that Stirke was admitted to the laboratory facilities there through the influence of Dury. From this place Stirke wrote several letters to Boyle:† they all deal with chemical matters, and it is most regrettable that Boyle's part of the correspondence is missing. In his letter of 3 January $165\frac{1}{2}$ Stirke makes the following intriguing statement, concerning which no elucidation is forthcoming: "*Fraternitatem tuam de quâ scripsisti, non injucunde saluto, cujus fraternitatis leges si scirem, unus illorem fierem, non certerorum Fratrum, at tui gratiâ.*" His letter of 16 January $165\frac{1}{2}$ mentions a visit that day by the Earl of Newport ‡ with Lady Cork§ and Lady Ranelagh. He refers to the recent marriage of Thomas Vaughan,|| saying that he had married a clergyman's daughter of no fortune. His thanks to Boyle for " Artocreas tuum " implies that Stirke was receiving some support from him. Stirke claims that in 1652 he prepared for Boyle the *ens veneris* or *elementum ignis veneris*, which was considered to be a very valuable medicinal product.¶ In April of that year Stirke was ill, probably through working with charcoal fires and mercurial and antimonial preparations in an enclosed space. Boyle made the very sensible recommendation that Stirke remove the windows from his workroom.**

Correspondence with Hartlib and Mallet was maintained. Boyle told Hartlib that Mallet was willing to help in improving the " Legacy

* Turnbull. *H. D. & C.* p. 268.

† 3 January $165\frac{1}{2}$, B.L. V, 129; 16 January $165\frac{1}{2}$, B. L. V, 131; 26 January $165\frac{1}{2}$, B. L. V, 133; 3 February $165\frac{1}{2}$, B. L. V, 135. It is clear that other letters were written, but they are missing.

‡ Mountjoy Blount, Earl of Newport (1597?–1666). Little is known about his activities between October 1647 and December 1653. He compounded with the Commonwealth, and was probably living quietly.

§ Elizabeth (Clifford) (1613–1691). Wife of Richard, 2nd Earl of Cork, RB's eldest brother.

|| (1622–1666) alchemist and poet. His wife's christian name was Rebecca, her surname is not known. Stirke uses the phrase *Clerici eujusdam filiam* which could be taken to mean the name was Clerk. (Arthur Edward Waite. *The Works of Thomas Vaughan.* London, 1919. See the Biographical Preface.).

¶ G. H. Turnbull. ' George Stirk, Philospher by Fire.' *Publications of the Colonial Society of Massachusetts. Transactions, 1947–1951.* (1959), **38**, p. 235. ' Some Considerations Touching the Usefulness of Experimental Philsophy' *Works Fol.* **I**, pp. 510, 563; *Works,* **II**, pp. 135, 215. The *ens veneris* was prescribed and prepared by RB for Lady Conway, but it was no cure for her ailment (M. H. Nicolson, *Conway Letters*, London, 1930, pp. 113, 225, 228, 230). The substance was still in the *New Edinburgh Dispensatory* as late as 1786, (p.521).

** G. H. Turnbull. *loc. cit.*, p. 241.

of Husbandry ";* and continued his discussion with Mallet on divinity and grammars.† Birch tells us that a letter written 19 June 1652 by Boyle to his brother Roger was later modified to form the dedication to *Some Considerations Touching the Style of the Holy Scriptures*,‡ which had been composed a year or two previously at Broghill's request. §

From June 1652 to July 1654 Robert Boyle was in Ireland, || except for about three months, July to September 1653, when he was in London and Stalbridge. He arrived at Youghal from Waterford, 26 June 1652, and met his brother Richard. Both went to Castle Lyons, the seat of the Earl of Barrymore, 7 July 1652. On 15 September 1652 Richard and Robert together with their brother Roger went to Waterford " to waite vpon the Generall ¶ at whose house wee alighted, and were by him broght to our lodging at Collonell Lees house." Robert and Richard returned to Youghal, 17 September 1652. All the Boyle brothers with Lady Cork dined at Ballynetra, 11 November 1652, probably with some member of the Smyth family. Robert left Youghal to go to England, 4 June 1653. Although much of his time was taken up by visits to relatives and attention to matters affecting his estates, he certainly did not neglect his studies. On 15 April 1653 William Petty at Dublin wrote a letter to Boyle in which Boyle is severely taken to task for his continual reading, for his brothers and friends consider he does so to excess and to the prejudice of his health. Petty says it is unnecessary, as Boyle is so accomplished that he has no need for " other men's loquacity ". Boyle is reproved for " your apprehension of many diseases, and a continual fear, that you are always inclining or falling into one or the other ", this fear being in itself a disease " incident to all, that begin the study of diseases ". Boyle is indicted for " practising upon yourself with medicaments (though specificks) not sufficiently tried by those, that

---

* B.M. Add. MSS 32093 fol. 293. Letter of RB to John Mallet dated 2 March $165\frac{1}{2}$. The full title of the work in question is *Samuel Hartlib his Legacie: or An Enlargement of the Discourse of Husbandry used in Brabant and Flanders.* 1st edition, London, 1651.

† *Works Fol.* **I**, p. 30; *Works.* **I**, p. 1. Letter of RB to John Mallet dated January $165\frac{2}{3}$.

‡ Not published till 1661. This work together with *Seraphick Love* and *The High Veneration Man's Intellect owes to God* were placed on the *Index librorum prohibitorum* by decree dated 5 September 1695: they are still there.

§ *Works.* **I**, *Life.* p. 28; *Works.* **I**, p. xlvii.

|| Diary of the 2nd Earl of Cork. Lismore MSS. **XXIX.**

¶ Lieutenant-General Charles Fleetwood had landed at Waterford by 11 September 1652 to take over as Commander-in-Chief of the forces in Ireland. (*The Memoirs of Edmund Ludlow*, edited by C. H. Firth. 2 vols. Oxford, 1895. Vol. **I**, p. 330.).

administer or advise them ". Though Petty feels that none of his arguments will alter Boyle's habits, he prays that no evil consequence may come of them.*

Boyle interrupted his stay in Ireland to be in England from June to September 1653. He left Youghal for England 4 June 1653.† During this visit he met John Wilkins, then Warden of Wadham College, Oxford. It is not too much to believe that this meeting was the outcome of talks with Petty, who before going to Ireland was in 1651 anatomy professor in that university. Wilkin's letter of 6 September 1653 says: " I should exceedingly rejoice in your being stayed in England this winter, and the advantage of your conversation at Oxford, where you will be a means to quicken and direct our enquiries. And though a person so well accomplished as yourself, cannot expect to learn any thing amongst pedants, yet you will here meet with divers persons, who will truly love and honour you . . . If I knew with what art to heighten those inclinations, which you intimate of coming to Oxford, into full resolutions, I would improve my utmost skill to that purpose ".‡

On his way back to Ireland, Boyle wrote two letters from Bristol. One was to John Mallet, and dealt with the leasing of property;§ the other, in lighter vein, was undoubtedly to Frederick Clodius|| wishing

* B. M. Add. MSS. 6193 fol. 138.

† Diary of the 2nd Earl of Cork. Lismore MSS. **XXIX.**

‡ *Works Fol.* V, p. 629; *Works,* VI, p. 633.

§ *Works Fol.* **I,** *Life,* p. 32; *Works.* **I,** p. lii. Letter dated 23 September 1653.

|| Frederick Clodius; dates of birth and death unknown; evidently a native of Germany. Before coming to this country he seems to have been engaged in a buying commission for a foreign princess (*Works Fol.* **I,** p. 485; *Works.* **II,** p. 97). How he came into the Hartlib circle is not known. He was introduced by Henry More, the Platonist, to Lady Conway, for whom he was engaged to deliver a picture to her brother John Finch in Padua in 1653 (*Conway Letters,* edited by M. H. Nicolson, London, 1930, pp. 88, 90, 91). Contemporary opinions of Clodius varied greatly. He was clearly acceptable to Hartlib, for he married one of his daughters. He was on good terms with RB, as the correspondence testifies. William Brereton (later 3rd Baron Brereton) would have no one else by Clodius as his physician, though it is not known what, if any, medical qualifications he possessed (Worthington, Vol. **I,** p. 304) John Evelyn regarded Clodius with suspicion, calling him a ' professed adeptus ', who had insinuated himself into the acquaintanceship of both Hartlib and RB by the same '*methodus mendicandi*' (Evelyn (Bray) Vol. **III,** p. 391). Sir Kenelm Digby thought so highly of Clodius that he proposed to maintain him and his family for two years in addition to furnishing him with a laboratory (*Works Fol.* V, p. 265; *Works,* VI, p. 89.) Later, in 1654, Henry More regretted having introduced Clodius to Lady Conway, and his condemnation of Clodius is about as strong as it could possibly be (*Conway Letters,* pp. 94–97, 102, 104, 106). Clodius seems never to have prospered, and he probably dropped into obscurity. His ' misdemeanours ' are mentioned by Dury in 1661 (Turnbull, *H.D.C.* p. 8). He is mentioned once by Samuel Pepys.

him happiness in his marriage to Hartlib's daughter, Mary, and enclosing as a philosophical accompaniment "a way to make wine to drink the bride's health." * There is an undated letter from Clodius to Boyle which must belong to this Irish period, as it acknowledges one of Boyle's from Dublin. The letter answers several questions that had been raised concerning mercury, the liquor alkahest, and alchemical matters. Most likely it has reference to some of Stirke's activities.†

Boyle probably landed at Kinsale on 5 October 1653. Two days later, in company with his brother Francis and other members of the family, he came to Youghal from Blarney. Boyle remained in Ireland until the first week in July 1654 at least. The distressing conditions then prevailing in that country are described in one of his letters to John Mallet.‡ Correspondence, which has survived in part only, was maintained between Boyle and Hartlib, the latter of whom wrote a very informative news-letter on 28 February $165\frac{3}{4}$.§ It contains a long quotation from a letter sent to Hartlib by Worsley which discloses that there has been a dispute between Worsley and Clodius; and furthermore that Worsley has "laid all considerations in chemistry aside, as things not reaching much above common laborants, or strong-water distillers, unless we can arrive at this key, clearly and perfectly to know, how to open, ferment, putrefy, corrupt and destroy (if we please) any mineral, or metal . . . seeing a natural putrefaction and corruption is the first thing in order to be done towards the making, transmuting or multiplying of any metal. . . . In this therefore, both principally and only, I conceive learning, judgment, or wisdom to consist, either as to the knowledge of true medicine (other preparations short of this being not much to be valued) or as to the carrying on of a higher work in nature. And being led by some serious considerations to see the necessity of this (and the smallness of most chemical operations besides this) the same hand of goodness did dispose my mind to comprehend both the possibility, and somewhat of the way of it, and nature of the course concerning it . . . resolving not much to prefer a correspondency of any hands, in things chemical or medicinal, unless this, as I either may be an assistant towards it, or be assisted in it. And although I may say, I see already so much in it, as to prefer it before any other natural knowledge, or perhaps, employment; yet I can find nothing very valuable or very desirable, either in myself,

---

* *Works Fol.* V, p. 229; *Works.* VI, p. 35. Letter dated 27 September 1653.

† B.L. II, 88.

‡ B.M. Add. MSS 32093, fol. 318. Letter dated 22 January $165\frac{3}{4}$ from Youghal.

§ *Works Fol.* V, p. 257; *Works.* VI, p. 78.

or others " Worsley, " that noble and high soaring spirit ", as Hartlib calls him, leaves the chemical scene. Hartlib's letter reveals that Boyle was engaged on " a short essay concerning chemistry, by way of a *judicium de chemia & chemicis* ".* Stirke is " altogether degenerated"; for the second time has been imprisoned for debt, and is now living obscurely. " He hath always concealed his rotten condition from us; nor hath there been any communication between him and my son, as long as you have been in Ireland." Hartlib refers to the death of Arnold Boate, and expresses his concern for the completion of Gerard Boate's *Ireland's Naturall History*,† saying that, if Boyle and Child do not undertake the task, it " will never be perfected to any purpose ". It is clear that Boyle was not contented in Ireland, for Hartlib continues: " you complain of that barbarous (for the present) country, wherein you live; but if you would but make a right use of yourself, from the place where you live, towards Dr Child, Mr Worsley, Dr Petty, Major Morgan (not to mention others) they would abundantly cherish in you many philosophical thoughts, and encourage you, perhaps more vigorously that I can do at this distance and uncertainties, to venture even upon divers choice chemical experiments, for the advancement both of health and wealth." This comment points to the conclusion that Boyle did not have frequent or prolonged dealings with these individuals during his stay in Ireland. Child was not in direct content with Boyle.‡ Worsley was Surveyor-General of Lands. A survey of Ireland was undertaken after the Cromwellian conquest in order to decide on the distribution of forfeited lands in satisfaction of the claims of soldiers and adventurers. Petty was concerned in this survey, and succeeded in displacing Worsley from his position. Anthony Morgan was probably more occupied in political affairs. The letter contains much on affairs of husbandry, and closes with a reference to Clodius who is engaged on " that noble medicament of the Helmontian cinnabar ".

At the beginning of March 1654 Johann Unmussig, concerning whom almost nothing is known, but who appears to have acted in a medical advisory capacity to Lady Ranelagh's family, wrote to Boyle giving prescriptions for the stone.§ Boyle long suspected that he was

* *Cf.* p. 64, RB's " Materialls & Addenda " Essay 4.

† The first part had been published by Hartlib in 1652 with a dedication to Oliver Cromwell and Charles Fleetwood; it was reprinted in 1657 without the dedication.

‡ G. H. Turnbull. ' Robert Child '. *Publications of the Colonial Society of Massachusetts. Transactions, 1947–1951.* (1959) **38**, 44.

§ B.L. V, 153. Letter dated 1 March $165\frac{3}{4}$ from London.

afflicted with urinary calculus, but when in Ireland was assured that he did not have it.*

There is an undated letter from Boyle to Clodius which was probably written towards the end of April 1654. Boyle thanks him for sending the processes of *mercurius vitae*, and again shows his dislike of Ireland by saying: "I live here in a barbarous country, where chemical spirits are so misunderstood, and chemical instruments so unprocurable, that it is hard to have any hermetick thoughts in it, and impossible to bring them to experiment . . . For my part, that I may not live wholly useless, or altogether a stranger in the study of nature, since I want glasses and furnaces to make a chemical analysis of inanimate bodies, I am exercising myself in making anatomical dissections of living animals: wherein (being assisted by your father-in-law's friend Dr. Petty, our general's physician) I have satisfied myself of the circulation of the blood, and the (freshly discovered and hardly discoverable) *receptaculum chyli,* made by the confluence of the *venae lactae;* and have seen (especially in the dissections of fishes) more of the variety and contrivances of nature, and the majesty and wisdom of her author, than all the books I ever read in my life could give me convincing notions of." His enquiries in Ireland on behalf of Clodius concerning minerals have been "as fruitless as diligent." He sends processes of the sulphur of *stella martis*†.

During May 1654 Hartlib write at least two long news-letters to Boyle. They deal very largely with husbandry; the chemical activities of his son-in-law, Clodius, who was then in great favour with Sir Kenelm Digby; and other general news from England and the continent.‡

Boyle was certainly still in Ireland on 3 July 1654;§ he probably left that country during the same month. His movements during the following months are not known with any certainty, though we know that he was ill in London towards the end of the year.|| It appears that after his return to England he divided his time between London and Stalbridge, so his intention of going to Oxford was not immediately fulfilled. His "Philosophical Diary begun 1st January 1654/5" shows how occupied he then was with chemical matters, and that

---

* *Works Fol.* I, p. 484; *Works.* II, p. 96.

† *Works Fol.* V, p. 241; *Works,* VI, p. 54.

The discovery of the reservoir of the chyle was published by Jean Pecquet in his *Experimenta Nova Anatomica*, Paris, 1651.

‡ *Works Fol.* V, p. 261, 264; *Works.* VI, p. 83, 89. Letters dated 8 May and 15 May 1654 respectively from London.

§ Diary of the 2nd Earl of Cork. Lismore MSS., XXIX.

|| *Works Fol.* V, p. 630; *Works.* VI, p. 634. Letter from John Mallet to RB dated 27 March 1655.

much information was derived from Clodius. There is a suggestion that Stirke, too, was again in contact with Boyle*. He was at Stalbridge on 5 September 1655 when he wrote to John Mallet saying that he had suffered from distemper of the eyes, and was going to London†, where he saw Hartlib, to whom he wrote from Eton, 14 September 1655, on his way back to Stalbridge saying that he had made a visit to Oxford "on purpose to visit our ingenious friend Dr Wilkins with whom I spent a day with no small satisfaction, his entertainment did as well speak him a courtier as his discourse & the real productions of his knowledge a philosopher. But tho I were us'd with a huge deal of civility in divers colleges during my short stay at Oxford (where I continu'd not full two days) yet that which most endear'd to me my entertainment was the delight I had to find there a knot of such ingenious & free philosophers, who I can assure you do not only admit & entertain real learning but cherish & improve it, & have done & are like to do more toward the advancement of it than many of those pretenders that do more busy the press & presume to undervalue them, & during the little time I spent there I had the satisfaction to hear both the chief professors & heads of colleges maintain discourses & arguments in a way so far from servile that if all the new paradoxes have not found patrons there, 'tis not their dissent from the ancient or the vulgar opinions but some juster cause that hinders their admission, so that much of what you & I had been inform'd of concerning the servileness of & disaffection to real learning of that university may perchance have been true when the informers liv'd there but certainly now the persons are mistaken & so injur'd, they being so far from being censurable for their predecessors failings that it ought to be their praise not to have imitated them, & this I mention to you partly because you will not perhaps altogether slight the testimony of a person who has been no absolute stranger to divers foreign universities & has not yet had the ill luck to be looked upon as an enemy to useful learning & partly because what concerns natural philosophy in general must not be thought irrelative to husbandry which is but an application of it to rural subjects "‡. The tenor of this letter can be taken as supporting evidence that the London (1645) and the Oxford (1648–9) groups of virtuosi who later were original Fellows of the Royal Society did not constitute the Invisible College, and were not linked with it.

---

* B.P. VIII, 140.

† R.C.H.M. Bath, II, p. 113. Letter from RB to John Mallet dated 5 September 1655.

‡ B.P. XXXVII. Remains of a draft. Spelling has been modernized.

Boyle's "Philosophicall Diary" for January and February 1655 records the following persons from whom he received information on chemical processes or medicinal preparations:—Unmussig, Kenelm Digby, Clodius, Stirke, Smart (the operator of Dorchester House), Remé, as well as others. The various entries reveal Boyle's widening interest in these matters.* This year is of importance in that it saw the first appearance in print, albeit anonymously, of one of Boyle's works. Its title is *An Invitation to a free and generous Communication of Secrets and Receipts in Physick*, and it was one of the contributions to *Chymical, Medicinal and Chyrurgical Addresses: Made to Samuel Hartlib, Esquire*, London, 1655.†

The exact date when Boyle settled at Oxford is not known. On this point Thomas Birch has written as follows: "After his [Boyle's] return to England, which was probably, according to his intentions, in the latter end of June 1654, he went to reside at Oxford."‡ This statement, whether so intended, or not, by Birch, has been universally interpreted to mean, that Boyle settled at Oxford in the year 1654. However, it appears most likely that he was established at Oxford towards the end of 1655, or early in 1656. The tone of Boyle's letter of 14 September 1655 to Hartlib quoted above does not give the impression of being written by one who was then a resident of Oxford. Furthermore, there is the letter of Boyle's sister, Katherine, who wrote a letter dated 12 October§, in which she deals with the matter of suitable lodgings in Oxford for her brother. She describes the inconveniences of Mr Crosse's apartments, and the improvements necessary to make them comfortable. She says: "You are here much desired . . . I think you would have both more liberty and more conversation, than where you are, and both these will be necessary both to your health and your usefulness." The letter closes with her service to some members of the family and "my dear lord President and his Lady". This mention must refer to Henry Lawrence, who was given the title of "Lord President of the Council" by Cromwell

---

* B.P. **VIII.** fol. 140 *et seq.* See also RB's 'Promiscuous Observations' of September to December 1655: B.P. **XXV**.

† M. E. Rowbottom. 'The Earliest Published Writing of Robert Boyle.' *Annals of Science* (1950), **6**, 376.

R. E. W. Maddison. 'The Earliest Published Writing of Robert Boyle.' *Ibid.* (1961 published 1963), **17**, 165.

Fulton, p. 1.

‡ *Works Fol.* **I**, *Life*, p. 33; *Works*, **I**, p. liv.

§ *Works Fol.* **V**, p. 558; *Works*, **VI**, p. 523. Letter from Katherine Ranelagh to RB. No year is given in the original manuscript letter. R. T. Gunther has printed part of this letter without indicating that the original bears no year (*Early Science in Oxford*, Vol. **I**, p. 11). He has silently assigned it to 1653, presumably on the strength of Birch's statement.

16 December 1654. Chronologically, this letter dated 12 October from Lady Ranelagh fits best into the year 1655. Furthermore, the earliest surviving letter addressed to Boyle at Oxford is one from Clodius, who dates it 3 March 1655, which date is undoubtedly in accordance with Church Style, and so corresponds to the historical year 1656.

The years 1645–1655 may be called Boyle's Stalbridge Period. He did not spend them in retreat, quietly absorbed in study and experimental work. For him it was an unsettled period in unsettled times. Nevertheless, he devoted considerable time to philosophical pursuits, and was able to build up a no mean reputation. However, by 1654 the group of persons who contributed to his early development had become disrupted. Arnold Boate, Gerard Boate, Robert Child were dead; John Dury had left for the continent in 1654; George Stirke, at least, was unreliable; Benjamin Worsley was no longer interested in chemistry, both he and William Petty were now engaged in public affairs in Ireland. Only Hartlib and Clodius remained, and they seem in subsequent years gradually to have slipped from close association with Boyle, if one may judge by the surviving correspondence. Hartlib wrote regularly to Boyle up to 1659, when his letters apparently ceased. Clodius did not prosper at all, for in 1663 we find him begging for further financial assistance from Boyle, saying that thefts and misfortunes were all conspiring to his ruin.* Perhaps, finally, he was fortunate enough to find refuge with William Brereton (the 3rd baron), who had so high an opinion of him. Lord Brereton having provided himself with a convenient house at Brereton, not only declared it to be at Boyle's service, but also said, " My true Friend Cl[odius] will I hope be Resident with me."† There is an echo of him years afterwards in a letter written by Johann Christian Agricola, who had dwelt with Clodius, and in whose laboratory he had seen and waited on Boyle. Agricola, in the meantime, had become M.D., and acquired an official position in Holstein. He wrote to Boyle asking for advice in the purchase of lenses and microscopes.‡

Benjamin Worsley retained pleasant memories of his association with Robert Boyle, for years after he referred to a conjunction of friendship and intimacy during many years with him whose great virtue he emulated above most men's.§

---

* B. L. **II**, 17. Letter of F. Clodius to RB dated 12 December 1663.

† B. L. **I**, 164. Letter of W. Brereton to RB dated 9 May 1664.

‡ *Works Fol.* V, p. 640; *Works*, **VI**, p. 650. Letter dated 6 April 1668.

§ *C.S.P. Colonial.* (*1669–1674*), p. 439. Letter of B. Worsley to Sir Thomas Lynch dated 30 November 1672.

This was the end of a period.* A change was needed, and the removal to Oxford brought Robert Boyle within a sphere of brilliance eminently suited to bring him to maturity.

* RB's Stalbridge period has been treated by the present writer slightly more fully in certain aspects in *N & R.* (1963), **18**, 104. Some errors in dates on p. 116 of that account are corrected here.

III

The Oxford Period

(1656–1668)

Anthony Wood informs us that Robert Boyle, " a sojourner in the University 1659, . . lived in that house (owned then by an apothecary) next on the west side of University Coll. sometimes known by the name of Deep hall."* The apothecary was John Crosse, son of Matthew Crosse (*d.* Feb. $165\frac{5}{6}$), superior Bedell of Law;† and the house was on the south side of High Street, between University College and the ' Three Tuns ' tavern. This same house was undoubtedly formerly occupied by an ancester, Henry Crosse, Bedell of Theology, and Registrar of the University (1566–1530).‡

A view of the house, now demolished, is shown in Plate XV. A commemorative plaque, which was put in position as recently as 1965,§ marks the site. (Plate XVI). From Anthony Wood we learn that lodgings were available at this house. Lady Ranelagh stayed there

> " with design to be able to give you [i.e. Robert Boyle] from experience an account , which is the warmest room; and indeed I am satisfied with neither of them, as to that point, because the doors are placed so just by the chimnies, that if you have the benefit of the fire, you must venture having the inconvenience of the wind, which yet may be helped in either by a folding skreen, and then I think that, which looks into the garden, will be the more comfortable, though he have near hanged and intends to matt that you lay in before." ||

---

* Wood, *Life*, Vol. I, p. 290.

† *ibid.*, Vol. III, p. 460, 203.

‡ R. T. Gunther. *Early British Botanists and their Gardens. Oxford*, 1922. p. 304.

§ *Oxford Mail*, 13 May 1965.

|| *Works Fol.*, V, p. 558; *Works*, VI, p. 523. Letter of Lady Ranelagh to RB dated 12 October [1655].

The earliest positive evidence that Boyle was established at Oxford occurs in a letter from Clodius in which he acknowledges receipt of the former's letter dated 22 February from Oxford.*

Boyle's circle of acquaintances and friends was now rapidly widening. He had become acquainted with Johann Friedrich Schlezer from Hamburg, who stayed with Hartlib whilst in England during 1655 on a visit to Cromwell as the representative of the Elector of Brandenburg.† In his *Ephemerides*, early in 1656, Hartlib says that Boyle is acquainted with George Ent, the physician;‡ and soon after 20 May 1656 Hartlib was told by Boyle that Wren was likely to succeed Daniel Whistler at Gresham College.§ Boyle was now known to Ralph Bathurst of Trinity College, Oxford, as the letter written by the former proves.|| He was also known to John Evelyn with whom he dined, 12 April 1656, in company with [George] Berkeley, Dr. Jeremy Taylor, and Dr. Wilkins at Sayes Court; in the afternoon they all went to see Colonel Blount's new-invented ploughs.¶

The brief entries in Boyle's "Philosophicall Diary begun this 1st day of January 165^{5}_{6}" relate to miscellaneous chemical experiments, and mention that well-known seventeenth chemical work *Tyrocinium Chymicum* by Jean Beguin. Boyle's views on varnishes together with queries about essence of gold were communicated to Hartlib; these items find mention in Boyle's diary for the next year, the "Philosophicall Diary begun this First of Jan. 165^{6}_{7},"**

It was about this time that Boyle became acquainted with Henry Oldenburg,†† who later became the indefatigable secretary of the Royal Society. Oldenburg had been engaged by Lord and Lady

---

* RB's letter is not known to survive. The letter of F. Clodius to RB is dated 3 March 1655, which is taken to be the historical year 1656 (see comment on p. 87). B.L., **II**, 17a.

† Turnbull, *H. D. & C.*, p. 3; Worthington, **I**, p. 66. Letter of S. Hartlib to J. Worthington dated 12 December 1655. J. F. Schlezer (1609?–after 1657). M.D., Leyden. Incurred debts in England; imprisoned (November 1657).

‡ G. H. Turnbull. 'Samuel Hartlib's Influence on the Early History of the Royal Society.' *N. & R.* (1953), **10**, 108.

§ *Ibid.*, p. 112.

|| T. Warton. *The Life and Literary Remains of Ralph Bathurst, M.D.* London, 1761., p. 162. Letter of RB to R. Bathurst dated 14 April 1656 at St. James's.

¶ Evelyn (de Beer), Vol. **III**, p. 169; c.f. Letter of S. Hartlib to RB dated 25 March 1656, *Works Fol.*, V, p. 266; *Works*, **VI**, p. 91. Colonel Blount is presumably Thomas Blount; later F.R.S. *D.N.B.*

** G. H. Turnbull, *loc. cit*, p. 112: B.P. **XXV**.

†† (*ca.* 1619–1677). *D.N.B.;* A. R. Hall and M. B. Hall, 'Some hitherto unknown facts about the private career of Henry Oldenburg.' *N. & R.* (1963), **18**, 94: *idem.* 'Further Notes on Henry Oldenburg.' *Ibid.* (1968), **23**, 33.

Ranelagh in 1656 as tutor to their son Richard Jones, nephew of Robert Boyle. Oldenburg and Richard Jones were at Oxford in April 1657* awaiting instructions from Lady Ranelagh before leaving to travel on the continent. Two others of Boyle's nephews, namely, Charles and Richard Boyle, sons of the 2nd Earl of Cork, were at Oxford during this same year under the charge of Peter du Moulin.† These two nephews made their continental tour under the care of Walter Pope,‡ and it would seem that he was entrusted with this duty as a result of recommendation by Robert Boyle, for Pope wrote to the 2nd Earl of Cork saying that he was " very conscious of the trust you haue layd upon mee although I am unknowne to you; And I humbly desire your Lordship to beleeue that I shal make it my utmost endeavour, so to acquitt my selfe in this place, that I may discharge a good conscience before god, and preserue that opinion your most Learned Brother Mr. Robert Boyle is pleased to haue of me . . . "§ Pope was a member of Wadham College, Oxford, of which his half-brother, John Wilkins, was warden; so, one does not have to look far for the most likely source of Boyle's acquaintanceship with Pope.

Three books were published in 1658 with dedications to Boyle. George Starkey [Stirke], whose experiments on the alkahest, transmutation, and medicinal products had interested Boyle so much in his Stalbridge period, dedicated *Pyrotechny Asserted*‖ to him. Ralph Austen,¶ who corresponded with Hartlib during many years on the subject of gardening, dedicated *Observations upon . . . Fruit-trees, Fruits, and Flowers*** . . . to Boyle. Peter du Moulin, the younger, dedicated to him his translation of *The Devill of Mascon* by François Perreaud,†† reference to which has already made on page 52.

For the successful furtherance of his experimental work Boyle was fortunate about this time in securing the services of a remarkable

---

* *Works Fol.*, V, p. 299; *Works*, **VI**, p. 140. Letter of H. Oldenburg to RB dated 15 April 1657.

† Peter du Moulin, the younger, (1601–1684), D.D., Cantab., et Oxon; became prebendary of Canterbury. Letter of P. du Moulin to Samuel Bochart dated 15 July 1657; (Private communication from the late Dr E. Weil, 28 October 1958): B.L., **VI**, 4. Letter, partly in RB's hand, dated 3 April 1658.

‡ (?–1714) astronomer; later professor of astronomy in Gresham College: *D.N.B.*

§ Lismore MSS., **XXXI.**, No. 7. Letter dated 23 April 1659, (N.S.) from Paris.

‖ 8vo. London, 1658.

¶ ?–1676. *D.N.B.*

** Madan, No. 2374.

†† Madan, No. 2407.

inventive genius named Robert Hooke.* This young man had gone up from Westminster School to Christ Church, Oxford, in 1653. About 1655, he became known to the group of virtuosi who were holding philosophical meetings there. Concerning them Hooke says, "At these Meetings, which were about the Year 1655 (before which time I knew little of them) divers Experiments were suggested, discours'd and try'd with various successes . . ."† He became chemical assistant to Dr. Thomas Willis, and afterwards, by recommendation of the latter, he was engaged as assistant to Boyle. With the help of Hooke, Boyle was able to carry out what is probably his most important single piece of sustained, connected research work, namely his work with the air-pump. Much of Boyle's other work consists of detached experiments and observations, which have not been welded into a coherent exposition. This fact in no way derogates from Boyle's skill in making experiments, observations, or deductions therefrom.

Hooke continues his account of that time, saying: ". . . in 1658, or 9, I contriv'd and perfected the Air-pump for Mr. Boyle, having first seen a Contrivance for that purpose made for the same honourable Person by Mr. Gratorix, which was too gross to perform any great matter."‡

The Air -Pump or Machina Boyleana

Otto von Guericke§ as a result of his interest in philosophical problems carried out various experiments in pneumatics, and produced, not later than 1654, an air-pump by means of which he was able to make some very spectacular demonstrations, the most famous being that of the "Magdeburg Hemispheres." Although von Guericke's own account of his experiments was not published until 1672 in *Experimenta Nova Magdeburgica de Vacuo Spatio*, there was an earlier account by Kaspar Schott|| in 1657 in his book *Mechanica Hydraulico-Pneumatica*. Robert Boyle must have known about this account very soon, for Samuel Hartlib in a letter to him says: ' 'You speak still of the German

---

* (1635–1703.) R. Waller. 'The Life of Dr. Robert Hooke.' being the introduction to *The Posthumous Works of Robert Hooke, M.D., S.R.S.*, London, 1705: *D.N.B.:* E. N. de C. Andrade. 'Robert Hooke.' Wilkins Lecture. *Proc. Roy. Soc.* Series B., (1950), **137**, No. 887, 153: M. 'Espinasse. *Robert Hooke*. London, 1956.

† Gunther. Vol. **VI**, p. 8.

‡ Gunther., *loc. cit.*

§ (1602–1686.) Councillor to the Elector of Brandenburg, Mayor of Magdeburg.

|| (1608–1666.) Jesuit. Professor of physics and mathematics at Würzburg.

vacuum as of no ordinary beauty; but the poet says *Uritque videndo foemina*."* Boyle himself tells us that the 'public confusions' of 1659† causing him to quit "the place where my Furnaces, my Books, and my other Accomodations were, I fell afterwards upon the making of Pneumatical tryals."‡ He was stimulated to try and build, or have built, a machine similar to that constructed by von Guericke. To this end he first enlisted the help of a well-known London instrument maker, Ralph Greatorex,§ with whom Boyle had already had dealings.||

Hooke's contemptuous dismissal of Greatorex's attempt has been quoted above. The first successful air-pump as designed by Boyle and Hooke was described in the former's first published scientific work, *New Experiments Physico-Mechanicall, Touching The Spring of the Air, and its Effects*, which appeared at Oxford in 1660.¶ This book is written as a letter to Boyle's nephew, Viscount Dungarvon, son of his brother the 2nd Earl of Cork, and is dated from Beaconsfield, 20 December 1659. In the preface *To the Reader*, Boyle mentions that the distemper in his eyes prevented him not only from writing so much as one experiment himself, but even from reading over what he had dictated to others. In addition to this disability he had been afflicted with ague in April 1659.** It is clear, therefore, that Boyle must have been greatly dependent on Hooke, and perhaps others, for assistance in performing his experiments and recording them.

An account of this pneumatical engine, construction of which must have been completed early in 1659, is given in chapter VII. This engine could be used not only to rarefy air in the globe, but also to condense air into it. The main receiver, which was as large as the glassmen of the day could blow, was fashioned so that objects could be introduced into it — a facility which greatly extended the range of

---

* 7 January 165$\frac{7}{8}$. *Works Fol.*, V, p. 271; *Works*, **VI**, p. 99. The quotation is from Virgil, *Georgica*. 3. 215.

† Resulting from the death of Oliver Cromwell, and the near anarchical conditions after the overthrow of Richard Cromwell.

‡ *Works Fol.*, **I**, p. 228; *Works*, **I**, p. 356.

§ E. G. R. Taylor. *The Mathematical Practitioners of Tudor & Stuart England*. Cambridge, 1954, p. 229.

|| For example, he had communicated to RB a lute for mending cracks in iron furnaces (1655); and a way of making potashes (1656), B.P. **VIII**, fol. 143; *ibid*, **XXV**.

¶ Madan. No. 2484: Fulton, p. 9 *et seq.*: *Works Fol.*, **I**, p. 1: *Works*, **I**, p. 1. This work, together with RB's *Experiments and Considerations Touching Colours* (1664), seems to have been largely drawn upon by Samuel Johnson when compiling his *Dictionary*. (*Notes and Queries*. (1950), **195**, 517.) He it was who said of RB, 'His name is indeed reverenced, but his works are neglected.' (*The Rambler*. No. 106.)

** Worthington, **I**, p. 123. Letter of S. Hartlib to J. Worthington dated 20 April 1659.

experiments that could be performed. Nevertheless, the performance of the engine was not all that could be desired, for Boyle wrote to Hartlib, early in November, saying, "I am now prosecuting some things wth an engine I formerly writ to you of, & these things that have been already done in part are such, as would not, perhaps, be unacceptable to our new philosophers, wherever they are. But we have not yet brought our engine to perform what it should."*

Boyle's account *Touching the Spring of the Air, and its Effects* roused much interest, and was followed by many other publications either commenting on, or criticizing, his observations.

Experiment XXXV in the *Spring of the Air* mentions the capillary rise of liquids in tubes of narrow bore. An investigation of this phenomenon resulted in Robert Hooke's first published scientific work†, which he dedicated to Boyle, saying:

> "The honour you were pleased to do me, in putting me upon this enquiry, did not a little animate and encourage me to persevere in what I had begun with so happy an Omen. My good Success therefore, if any, is wholly to be ascribed to your self; as being the first Excitor and chief Abettor of it . . . I must therefore with the *Persian* offer to you, as he to the Sun, what he believes himself to have received from it. And therefore I trust my endeavouring to soar aloft with the Eagle, to enjoy the Influence of the most Glorious Light of the world, will find a Pardon, and be judged much better, than a hovering and fluttering with the silly Fly about the dim and fading Flame of a candle, that at best, will but singe those wings that raised it so high, and in stead of giving it more Vigour, will wholly disable it for the like Future Attempts . . . "

Boyle brought his air-pump to London and installed it at Chelsea‡, where Evelyn went to see experiments with the pneumatical engine.§

Boyle presented his first air-pump to the Royal Society in 1661.‖ It was one of the show pieces of the Society, experiments with it being performed for the delectation of important visitors such as royalty and foreign ambassadors. The pump was in constant use by the Society,

---

* Worthington **I**, p. 161. Letter of RB to S. Hartlib dated 3 November 1659.

† *An Attempt for the Explication of the Phaenomena, Observable in an Experiment Published by the Honourable Robert Boyle*, London, 1661. Reprinted in Gunther, **X**, (1935): Keynes, p. 6.

‡ There are frequent references to RB's staying at Chelsea. I have not identified where. It was probably a residence of one of his relatives, which he used as *pied-à-terre*.

§ Evelyn (de Beer), **III**, p. 255. 7 September 1660.

‖ Birch *History*, **I**, p. 23. The pump has not survived.

and numerous references to the experiments performed at meetings have been recorded.* They dealt with the Torricellian experiment, the effect of reduced pressure on all kinds of living creatures, Prince Rupert's drops, magnets, and so on. For all its imperfections and awkwardness in operation Boyles' Pneumatical Engine enabled many instructive experiments to be made.†

Hartlib's letters to Boyle during 1658 and 1659 throw further light on the latter's activities, besides reporting general and particular news derived from his numerous correspondents such as Worsley, Brereton, Beale, Worthington, Pell, as well as others at home and abroad. Information on all manner of topics, such as husbandry, wine making, perpetual motion, telescopes, barnacles, glass stoppers, Kuffler's inventions, the hatching of chickens, medicinal preparations, was passed on to Boyle by Hartlib, who also acted as intermediary for correspondence between Boyle and his nephew, Richard Jones, then travelling on the continent with Oldenburg. Boyle's continued interest in the Worsley-Hartlib scheme for using the revenue from forfeited estates in Ireland for the Advancement of Universal Learning:‡ is shown by his active support thereof in influential quarters of the government. Hartlib's hopes of personal benefit from the scheme were frustrated, so he believed, by the machinations of Petty against Worsley. Various other proposals for the establishment of academies of learning were made, for example by Bacon, Cowley, and others. Evelyn had one, too, which he elaborated in a long letter to Boyle, "who is alone a society of all that were desirable to a consummate felicity."§ It is obvious, that Evelyn was hoping that Boyle would provide substantial financial support to establish the philosophical college he described. Boyle also interested himself in the petition for assistance to exiled Protestant families from Bohemia.|| Boyle's life-long interest and liberality in promoting the Christian faith are exemplified at this time by his allowing £50 *per annum* to Robert Sanderson¶ to write on

---

* *Ibid., passim,* particularly 1661–1665.

† J. B. Conant. 'Robert Boyle's Experiments in Pneumatics.' *Harvard Case Histories in Experimental Science.* Case 1, Cambridge, Mass., 1950.

‡ Turnbull. *H. D. & C.*, p. 55.

§ *Works Fol.*. **V**, p. 397; *Works*, **VI**, p. 288. Letter dated 3 September 1659. J. J. O'Brien. 'Commonwealth Schemes for the Advancement of Learning.' *Brit. J. Educ. Studies*, (1968) **16**, 30.

|| *Works Fol.*, **V**, p. 273; *Works*, **VI**, p. 102: Turnbull, *H. D. & C.*, p. 289–291.

¶ (1587–1663). Regius professor of divinity, Oxford: ejected by the parliamentary visitors, 1648; after the Restoration became Bishop of Lincoln. *D.N.B.*

cases of conscience.* Thomas Barlow, who was then librarian of the Bodleian Library, acted as intermediary in this matter, and wrote to Boyle to say "with what thankfulness the good old man accepts of this your great charity (rare in any age, and unheard of in this . . . I doubt not but the present age and posterity may have just cause to bless God in this particular, and thank you for being the only occasion, and sole encourager of so good a work."† A little later Boyle tells us that "Mr. Pocock‡ is at my request printing a translation of Grotius's Book of the Truth of the Christian Religion§ into Arabick, and I need not tell you, how fit he is for such a work. The book is partly printed off already: we would gladly be advised, how it may be disposed into several parts of the East, to the greatest advantage of the design wch he and I persue in it."‖ The cost of printing this book amounted to £75 0*s*. 6*d*., and was paid by Boyle. His zeal for propagating the Christian religion extended, moreover, to providing ways and means to convert the natives of New England.¶

In satisfaction of Hartlib's insatiable thirst for news Boyle sent all manner of writings, both manuscript and printed, from Oxford; among these was one on coffee, translated from the Arabic by Pococke.**

The publication in 1659 of Boyle's *Seraphick Love*,†† a work which had been written many years before (*cf.* chapter II, p. 64), gave Evelyn the occasion to write a long letter on this "incomparable book," and to discourse on numerous famous manifestations of this

---

* *De Obligatione Conscientiae Praelectiones Decem: Oxonii in Scholâ Theologicâ Habitae. Anno Dom. MDCXLVII.* London, 1660. English translation by Robert Codrington, London, 1660. Madan., No. 1967*B.

The translator, Robert Codrington (?–1665) of Magdalen College, Oxford, was a prolific writer of complimentary poems to persons from whom he was able to derive some tangible benefit. (J. Keevil. *The Stranger's Son.* London, 1953, pp. 125, 126, 151.) His Latin odes to RB are in B.L., **II,** 21 and B.P. **XXXV,** and **XXXVI.**

† *Works Fol.* **V,** p. 406; *Works* **VI,** p. 301. Letter dated 13 September 1659.

‡ Edward Pococke, (1604–1691). Orientalist. *D.N.B.*

§ Hugo Grotius, (1583–1645). Jurist. *De Veritate Religionis Christianae.* Editio Nova cum Annotationibus, Cui accessit versio Arabica. Oxford, 1660. Madan No. 2498.

This work carries a dedication by Pococke to RB. The translation was made for the purpose of advancing missionary efforts among the Arabs, and received high commendation from Hartlib, Worthington and Richard Baxter.

‖ Worthington **I,** p. 161. Letter of RB to S. Hartlib dated 3 November 1659.

¶ B.L., **I,** 31. Letter of R. Baxter to RB dated 20 October 1660. See also W. Kellaway. *The New England Company 1649–1776.* London, 1961. p. 42.

** [David Antiochenus.] *The Nature of the drink Kauhi, or Coffe, and the Berry of which it is made.* Oxford, 1659. Madan, No. 2438.

†† Fulton, A.I, p. 2.

state of feeling.* John Beale likened the book to "a deep river, carrying more strength than noise."†

In common with other virtuosi of the period there was hardly a branch of knowledge that did not claim Boyle's attention. He acquired a new telescope,‡ but he made no astronomical discoveries of note, though his observations on a sunspot in 1660 were published at a much later date.§ He was interested in trades, manufacturing processes and all conceivable information obtainable from foreign parts, — as was the Royal Society after its formation, when Boyle was a member of committees set up to consider these matters. To satisfy such lively curiosity and genuine desire for knowledge he sought to enlarge his sources of foreign information: his published works witness his success in this respect.

Boyle's interest in anatomy prompted him to commission a translation of Louis de Bils' anatomy tract,|| which he dedicated to Samuel Hartlib, who distributed many copies of it, and told Boyle, "Really you have done an excellent work, for spreading this anatomical magnale upon the honest learned world."¶ Boyle had a particular interest in this tract, which dealt with the anatomizing of human bodies so as to "shew the inward state of a man's body. For all the parts of the body are in the body, except the guts and brains, which lye by." In his dedication to Hartlib, Boyle recalls that "in some papers I formerly told you of, I had mentioned divers things I had intended to try . . . in order to the preservation of animal substances, and the making some of them more durable subjects for the anatomist to deal with." The method subsequently adopted by Boyle for the preservation of specimens of this type was immersion in *spiritus vini*, or in turpentine.**

---

* *Works Fol.*, V, p. 399; *Works*, **VI**, p. 291. Letter of J. Evelyn to RB dated 29 September 1659.

† *Works Fol.*, V, p. 292; *Works*, **VI**, p. 130. Letter of S. Hartlib to RB dated 1 November 1659.

‡ *Works Fol.* V, p. 290; *Works*, VI, p. 127. Letter of S. Hartlib to RB dated 31 May 1659.

§ *Phil. Trans.* (1671), **6**, No. 74, p. 2216: *Works Fol.*, **III**, p. 213; *Works*, **III**, p. 511.

|| Louis de Bils. *A Tract touching Skill of a better Way of Anatomy of Man's Body*, 12mo. London, 1659. Fulton, p. 138. *Works Fol.*, **I**, *Life*, p. 140; *Works*, **I**, p. ccxix. RB's preface is dated 13 October 1659. Worthington, **I**, p. 343, note 2.

¶ *Works Fol.* V, p. 292; *Works*, **VI**, p. 130. Letter of S. Hartlib to RB dated 1 November 1659. Worthington, **I**, p. 159.

** *Phil. Trans.* (1666), **I**, No. 12, p. 199; *Works Fol.*, **II**, p. 543; *Works*, **III**, p. 138, B.L. **VII**, 54: Birch *History*, **I**, p. 374, 378, 393.

Robert Sharrock* was engaged about this time in giving Boyle editorial assistance in seeing his books through the press, as well as translating the *New Experiments Physico-Mechanicall* into Latin,† in which capacity Henry Oldenburg helped also after his return to England 1600. It is not surprizing to find that Sharrock was interested in chemistry, and that discussions on chemical experiments are dispersed through his correspondence with Boyle. In several of these letters there is mention of Peter Sthael,‡ who, on the authority of Anthony Wood, was brought to Oxford by Boyle in 1659, though it would seem not unlikely that Hartlib had something to do with the negotiations, if only as an intermediary, § leading to Sthael's settlement in England.

After the death of the Protector, Oliver Cromwell, on 3 September 1658, the country's republican government gradually weakened, and so conditions became favourable to a restoration of the monarchy. Charles II landed in England, 25 May 1660, and triumphantly entered London four days later. So soon as 5 June following, Robert Boyle publicly declared his loyalty to the King in pursuance of the Declaration of Breda by subscribing to the required document before Sir Harbottle Grimston, Speaker of the Convention parliament. || Extensions of this general pardon to Ireland in respect of Robert Boyle and other members of his family occurred in later years.¶ In November 1660, Boyle petitioned the King for a 31 year lease of various named properties, impropriations, tithes, etc., in co. Mayo.**

The Beginning of The Royal Society

The year 1660 witnessed not only the Restoration of Charles II to the throne of England, but also the formal establishment of an assembly that was later incorporated as The Royal Society. The general development of scientific knowledge in Western Europe during the first half of the seventeenth century produced a number of informal

* (1630–1684), divine: *D.N.B.* He wrote *The History of the Propagation & Improvement of Vegetables*, Oxford, 1660, and dedicated it to RB, saying that it had been accomplished entirely through his encouragement. Another work by the same author carried a joint dedication to RB: Madan Nos. 2528, 2529.

† Madan, No. 2547.

‡ Wood, *Life*, **I**, pp. 290, 472–3.

§ *Works Fol.* **V**, p. 286; *Works*, **VI**, p. 122. Letter of S. Hartlib to RB week ending [30 April 1659].

|| The document is printed *in extenso* in Hutchins, **III**, p. 674.

¶ *C. S. P. Ireland* (*1660–1662*), p. 316, 25 April 1661: R.C.H.M., 10th Rept. Part V, p. 15, November 1666.

** B.M., Egerton MSS 2549, fol. 96.

local assemblies of persons interested in experimental philosophy. England was no exception, and about 1645 such informal meetings were taking place weekly at Gresham College in the City of London, or in its vicinity.* It should be emphasized that the Royal Society did not develop as an off-shoot of the Gresham College establishment, but evolved from the gatherings of certain individuals who found it convenient to meet there (or elsewhere): and the first home of the Royal Society was at Gresham College because that was the most convenient place. Because of the disturbed state of the country about 1648–9 several of this company decided to remove to Oxford, where, being joined by others, they constituted the counterpart of the London assembly meeting at Gresham College, with which they retained close contact. The Oxford group was joined, as we have already seen, by Robert Boyle in 1655–6. The activities of this group suffered a great loss when John Wilkins left Oxford in 1659 to go to Cambridge; and the meetings which had taken place in Wilkins' rooms in Wadham College then transferred to Boyle's lodgings. Though the activities of the London group had been disturbed in 1659 by the quartering of soldiers in Gresham College, they were renewed more vigorously during the months preceding the Restoration. Some few months afterwards His Majesty spent an evening at Gresham College, " where he was entertained with the admirable long tube, wth which he viewed the heavens, to his very great satisfaction . . . . His Maj. hath also threatned to bestow a visit upon Mr. Boyle."† A momentous decision was taken at Gresham College on Wednesday, 28 November 1660, after the close of Christopher Wren's astronomy lecture when various persons, amongst whom was Robert Boyle, withdrew

> " for mutual conversation, into Mr. Rooke's apartment, where, amongst other matters discoursed of, something was offered about a design of founding a college for the promoting of physico-mathematical experimental learning. And because they had these frequent occasions of meeting with one another, it was proposed, that some course might be thought of to improve this meeting to a more regular way of debating things; and that, according to the manner in other countries, where there were voluntary associations of men into academies for the advancement

---

* Ornstein: Birch. *History:* J. Ward. *The Lives of the Professors of Gresham College.* London, 1790: H. Hartley and C. Hinshelwood. *Gresham College and the Royal Society.* [London, 1960]: D. McKie. *The Origins and Foundations of the Royal Society of London*, in *The Royal Society, Its Origins and Founders*, edited by Sir Harold Hartley, F.R.S. London, 1960. M. Purver. *The Royal Society: Concept and Creation.* London, 1967.

† Worthington, I, p. 215. Letter of S. Hartlib to J. Worthington dated 15 October 1660.

of various parts of learning, they might do something answerable here for the promoting of experimental philosophy."*

A week later, on 5 December 1659, Sir Robert Moray "brought word from the court, that the king had been acquainted with the design of the meeting, and well approved of it, and would be ready to give an encouragement to it."† It was one of Moray's great services to the newly formed Society that through his influence the King was pleased not only to offer himself to be entered one of the Society, but also to grant a Charter of Incorporation. In this connection Moray wrote to Christiaan Huygens, saying, "Dans quelques iours nous esperons que nostre societé sera establie de la bonne sorte;"‡ and to Boyle, saying, "The King hath been most graciously pleased to grant the desires in our petition, and you should not be away from the finishing what you helped to contrive and promote. We could allow those at Oxford any thing with less envy, than the fruition of those persons, whose presence with us can only turn the scales on our side: but if justice alone prevail not to make you haste to us, let your affection to the promoting of our business work."§ Robert Boyle was in London for much of the time from January to October 1661, so he probably added his influence to Moray's efforts for the promotion of the Royal Society. Boyle's standing with his fellow members in the Society is well shown by a remark in Moray's letter written only four days after the one just quoted. In reference to a proposal to change the hour of the Society's assembly from 3 p.m., to 9 a.m., he adds, "but it is lyke there will be no change without your Concurrence."||

The Royal Society's First Charter did not pass the Great Seal until 13 August 1662; it was found to be defective, and was followed by a Second Charter which passed the Great Seal 22 April 1663. A Third Charter, by which the Society is still governed, was granted in 1669.¶ In both the First and Second Charters Robert Boyle was named as one of the members of the Society's Council.

* Birch, *History*, p. 3.

† *Ibid*, p. 4.

‡ Huygens, III, p. 285.

§ *Works Fol.*, **V**, p. 510; *Works*, **VI**, p. 452. Letter from Sir Robert Moray to RB dated 7 October 1661.

|| B.L., **VII**, No. xviii. Letter from Sir Robert Moray to RB dated 11 October 1661.

¶ On the early history of The Royal Society consult Birch, *History:* A. Robertson. *The Life of Sir Robert Moray*. London, 1922. Chapter VIII: *The Record of The Royal Society of London*. 4th edition. London, 1940: *The Royal Society, Its Origins and Founders*, edited by Sir Harold Hartley, F.R.S. London, 1960: R. E. W. Maddison. 'The Accompt of William Balle from 28 November 1660 to 11 September 1663.' *N. & R.* (1960), **14**, 174.

The Wardship of Jane Itchingham

In November 1660 Boyle was granted the wardship of Jane Itchingham.* About February 166^{0}_{1} he petitioned the King because "After the patent was passed the ward was stolen away and secretly married without the petitioner's consent, contrary to law, by Sir Arthur Chichester of Dublin. His and her mother and divers other persons were privy thereto." He therefore prayed "for a letter to the Lords Justices of Ireland commanding them to order the prosecution of those guilty persons, as a mark of your displeasure and that petitioner may not be deprived of the benefit of his grant." The King acceded to this petition and ordered the Lords Justices of Ireland to "prosecute these persons according to the severity of the law," and also ordered that a grant be made to Boyle "of the benefit of all the mortgages, encumbrances, and other claims upon the estate of Jane Itchingham . . . which claims are forfeited to the King."† I have not traced the final outcome of this case, which was still being pursued in 1662, when Lady Chichester complained to friends that Robert Boyle "is now molesting her father and mother in the Exchequer," and solicited "the King's order in Council concerning the remitting of the wardship of Ireland for the time past."‡

The rapid entry of Robert Boyle into the favour of official circles is shown from the Minutes of the Council for Foreign Plantations, for on 7 January 166^{0}_{1} he was appointed one of a committee "to meet at Grocer's Hall, and inform themselves of the true state of the Plantations in Jamaica and New England, and to prepare such overtures and propositions as may be most fit for the King's service and the advantage of those Plantations."§ Then in May the same Council appointed him one of a committee to consider the "representations of the Quakers of their sufferings in New England", which had been referred from the Privy Council.‖ About the same time, by order of the King in Council, the Attorney-General was directed to prepare a Charter of Incorporation for the Company for Propagation of the Gospel in New

* Patent Rolls. 12 Car II. 30 November 1660. For the family ramifications of the persons involved, see Appendix p. 313. On wardships, see H. E. Bell. *An Introduction to the History and Records of the Courts of Wards and Liveries*. Cambridge, 1953.

† *C. S. P. Ireland (1669–1670)*. p. 399, about February 166^{0}_{1}; *ibid (1660–1662)*. p. 278, 26 March 1661; p. 284, 30 March 1661.

‡ R. C. H. M. Egmont II, p. 3. Letter of George Perceval to his brother Sir John Perceval dated June 1662.

§ *C. S. P. Colonial (1661–1668)*, p. 1.

‖ *Ibid*, p. 32. 20 May 1661. Other matters, *ibid.*, p. 41, 24 June 1661; p. 58, 11 November 1661.

England, and to include therein the name of Robert Boyle,* who very probably was concerned in drawing up the petition for continuing the existence of the Society for Propagation of the Gospel in New England, which had been established in 1649 during the Interregnum, and swept away at the Restoration by the Act of Oblivion and Indemnity.† The Company's Patent of Incorporation was granted 7 February 166½, "And his Majesty appoints the aforesaid Robert Boyle to be the first Governor of the Company during good behaviour with power to summon courts or meetings."‡ This position be held until 1689, when he felt obliged to resign on account of his failing health. Boyle's value to the Company "was his outstanding ability in managing its affairs and his remarkable conscientiousness . . . [He] hardly ever missed a meeting . . . [and] was so genuinely concerned in the Company's affairs that he had every question referred to him for consideration. The charge for horse hire to take one of the Company's officers to Pall Mall in order to consult him was a normal sundry in the Company's accounts."§

In addition to his duties with official organizations, Boyle, whilst in London during 1661, was very active with the Society meeting at Gresham College, of which Hartlib said he "is one of the chief". Boyle was elected to various committees of investigation set up by the Society, and was constantly being put in mind of providing some experiment or other. We find him concerned in such matters as the mercury experiment; enquiries to be made throughout the world; Glauber's method of discovering minerals and the method of making oil of vitriol; a report on Brazil wood; insects; correspondence with Prince Leopold of Tuscany; the erection of a library; a history of vipers and poisons; the sowing of seeds rained down from the sky; a query by the King on sensitive plants.||

---

* *Acts of the Privy Council, Colonial,* London, (1908), **I,** p. 309; *C. S. P. Colonial (1661–1668)*, p. 30, 17 May 1661; p. 51, 5 August 1661.

† B.L., **I,** 31. Letter of R. Baxter to RB dated 20 October 1660: B.P. **XL.** Item No. 7. Petition relating to Gospel in New England. For a detailed account of the Company see W. Kellaway. *The New England Company 1649–1776*. London, 1961.

‡ *C. S. P. Colonial (1661–1668)*, p. 71: *Acts of the Privy Council, Colonial,* **I,** p. 316: *Patent Rolls* 14 Car. II, Pars 11, No. 17. See also RB's will, Appendix, p. 260, note §.

§ W. Kellaway, *The New England Company 1649–1776*, London, 1961, p. 58.

|| Birch, *History*, **I,** pp. 12–34, *passim*. B. L. **VII,** 54. The sensitive plant, *Mimosa pudica*, was introduced into England in the seventeenth century (P. C. Ritterbush. 'John Lindsay and the Sensitive Plant.' *Annals of Science* (1962, published 1964,) **18,** 237).

Three works were published by Boyle in 1661. *Some Considerations Touching the Style of the Holy Scriptures**, dedicated to his brother, Roger, Earl of Orrery, was the first of several of theological tenor in which he was concerned to show that there are no contradictions in the Bible, nor any real conflict between science and Scripture. Richard Baxter said, ' I read your theology as the life of your philosophy, and your philosophy as animated and dignified by your theology, yea indeed as its first part."† *Certain Physiological Essays*‡ were mostly addressed to his ' hopeful ' nephew, Richard Jones, and may be regarded as paving the way for *The Sceptical Chymist*§, which is certainly the best known of Boyle's works. It is always mentioned in connection with the history of chemistry, though probably few of those who do so have ever read the book, which on the whole provides rather dull reading, and is not so refreshing, interesting, informative, or significant as some of his other writings.

Boyle's interest in the constitution of matter went back to his Stalbridge period when he wrote a non-surviving essay, " Of Atoms". In a letter written 1 March 1660 to Oldenburg, then at Paris, Robert Southwell says,

> " I am extreame glad to understand that Mr. Boyle is ingaged in soe advantageous a designe as ye collation of Chymicall experiments with Atomicall notions. He is a person of soe much ability, and soe much caution withall, that I doe not think he would have publisht unto his designe without haveing made some successefull and happy essay therein allready. Soe yt I feare not but he will most effectually compasse his designe, if the fickle changes of England, and ye blanketting of every new power does not divert him, and persuade him to bestow some time in laughing at ye strange kind of Poleticks afoot." ||

The Sceptical Chymist is directed mainly to disposing of the Paracelsian doctrine of the three hypostatical principles, (salt, sulphur and mercury), and the peripatetic doctrine of the four elements, (earth, water, fire and air).

> " The Chymists do generally affirm *Mercury*, *Salt* and *Sulphur* to be the compounding principles of all compounded things; which Doctrine is learnedly and solidly confuted by the English

* Fulton, p. 32.

† *Works Fol.* V, p. 553; *Works*, **VI**, p. 516. Letter of R. Baxter to RB dated 14 June 1665.

‡ *Fulton*, p. 20.

§ *Ibid*, p. 25: Madan, Nos. 3261, 3260: Marie Boas, ' An Early Version of Boyle's *Sceptical Chymist*'. *Isis* (1954), **45**, Part 2, No. 140, p. 153.

|| B.L. **VII**, No. xvii.

> Philosopher, I mean the Famous *Robert Boyle* in his *Sceptical Chymist*."*

It contains his oft-quoted definition of an element, which with certain reservations is still valid:

> " I now mean by elements . . . . certain primitive and simple, or perfectly unmingled bodies; which not being made of any other bodies, or of one another, are the ingredients of which all those called perfectly mixt bodies are immediately compounded, and into which they are ultimately resolved: now whether there be any one such body to be constantly met with in all, and each, of those that are said to be elemented bodies, is the thing I now question."†

He also put forward the notion of groups or clusters of particles, which may be regarded as equivalent to our molecules:

> " . . . . of these minute particles divers of the smallest and neighbouring ones were here and there associated into minute masses or clusters, and did by their coalitions constitute great store of such little primary concretions or masses as were not easily dissipable into such particles as composed them."‡

Boyle was a corpuscularian, and his explication of things by corpuscles he styled " the Phoenician Philosophy ", or sometimes also the " Mechanical Hypothesis or Philosophy." These views were more fully expounded later in *The Origine of Formes and Qualities.* § He regarded matter as being compounded of particles which by coalition produce corpuscles; according to the manner of their coalition various kinds of bodies having different properties are produced. On the view that all bodies are " made of one Catholick Matter common to them all, and . . . differ but in the shape, size, motion or rest, and texture of the small parts they consist of " ‖, it is a logical conclusion that transmutation of one into another is possible. As will be seen later, Boyle was always interested in this possibility, but not in the sense of alchemical projections.

His views on ultimate particles had great influence on the Russian natural philosopher, Lomonsov,¶ who notes in one of his memoranda,

---

* D. Abercromby, *Academia Scientiarum* . . . London, 1687, p. 54.

† *Works Fol.*, **I**, p. 356; *Works*, **I**, p. 562.

‡ *Works Fol.*, **I**, p. 300; *Works*, **I**, p. 475.

§ Oxford, 1666: Madan, No. 2739: Fulton, p. 55: Marie Boas, 'The Establishment of the Mechanical Philosophy.' *Osiris* (1952), **10**, 412: R. H. Kargon. *Atomism in England from Hariot to Newton.* Oxford, 1966, Chap. **IX.**

‖ *Works Fol.*, **II**, p. 513; *Works*, **III**, p. 93.

¶ Mikhail Vasil'evich Lomonosov, (1711–1765).

" After I had read through Boyle, a passionate desire to investigate the minute particles of substances took possession of me ".*

Boyle presented copies of his published books to Hartlib, who sent them to John Beale for " his best perusal ". In his acknowledgement, Beale said,

> " if an oracle had afforded me my choice of what questions to be informed, I am well assured, that those that are handled by Mr. Boyle had been the very first matter of my election. He wants a spirit that takes not pneumaticals to be the spring of life and weightiest of naturals, and now most considerable, since Jordanus Brunus and others have inclined our modern philosophers to approximate air to the chief of the ambient heaven. And for my own part, I have long conceived this ambient air to consist of as many & different ingredients, as our earth and seas, but I never expected in this life to see the diversities so handled & traced out, & the properties so fully discovered. Yet this I am enforced to confess before I have seen half of the book. In the other Treatises I am informed of the cause of such frequent misconduct in the best kind of medicines, and the strange Protean force of salt peter, and that niter w[ch] is highly styled by some, as if it were the spirit of the world. For the Corpuscularian Philosophy, I had long ago complained to you, that Sir K. Digby had said enough of it to make me giddy in their *Pro et Contra*, & my refuge was L[d] Bacon's " *Circa Ultimates rerum frustranea est Inquisitio.*' But now I see a stay for the light of reason and experience. Neither had I anything more in chase, then the cause of firmness, w[ch] in L[d] Bacon's language (who first awaken'd my attention to it) is frequently called consistency. Being now in my devotions before the oracle, you can expect but few lines, and in these my advice, that you procure them for yourself, before they are all gone out of the shops, for I supplicate, that these may rather return to you in cash, than in kind. ' Tis my great joy that Mr. B[oyle] is so far engaged to give us the rest of his notes and following experiments. In these he hath obliged all the intelligent inhabitants of this world, and hath given us hope, that we shall shortly complete humane sciences. Some families amongst us have answered all L[d] Bacon's votes for advancement of learning. And this honourable family deserves to be

* Quoted from B. N. Menshutkin, *Russia's Lomonosov*, Princeton, N. J., 1952. p. 30: H. M. Leicester. ' Boyle, Lomonosov, Lavoisier, and the Corpuscular Theory of Matter.' *Isis*, (1967), **58**, 240.

reputed the first college in this university or oecumenical academy."*

In a following letter Beale continued, saying:

"I was attentive in my devotions before the noble oracle, wch I may now, upon the reviewed solemnities of holy oath, style the most satisfying oracle in my apprehensions, that ever appeared in the converse of mortals, on this habitable globe, for discovery of all the works of nature . . . For philosophical satisfactions, I did chiefly address to philosophical experiments, in wch I seemed to have the best overtures of aid from Ld Bacon, but of this I complained, that in the progress of late years we had not brought his experiments or added our own, to any degree of ripeness. And this was indeed my discouragement. Now I confess I am surprised with wonder at the present advancement, & I dare promise our posterity, that knowledge shall in this following age abound in very great perfection, & to the best of noble operations. I can now no longer forbear to enquire for the return of Mr. Oldenburg, wch was promised some weeks ago. From him I hope to be informed when Mr. Boyl will be pleased to oblige the world with the publication of his other works . . . And this we may well discern that nothing can drop from his pen, that doth not oblige the future ages universally, neither can present envy appear against such bright lustre."†

Notwithstanding the vast extent of Boyle's writings it is most noticeable that he does not participate in, or deliberately provoke, controversy. However, there was one instance in which he was constrained to reply to criticism. His *New Experiments Physico-Mechanicall* (1660) had excited much interest, as well as objections by Thomas Hobbes‡ and Francis Line§. To the second edition (1662) of his work Boyle added *A Defence of the Authors Explication of the Experiments, Against the Objections of Franciscus Linus, And, Thomas Hobbes*||, which is of historical importance, for it contains the enuncia-

* Worthington, I, p. 366–369. Letter of J. Beale to S. Hartlib quoted by the latter in his letter to J. Worthington dated 26 August 1661.

† *Ibid.*, pp. 369–375.

‡ (1588–1679) philosopher: *D.N.B.* The book in question is *Dialogus physicus, sive De natura aeris conjectura sumpta ab experimentis nuper Londini habitis in Collegio Greshamensi*, London, 1661.

§ Known also as Franciscus Linus, *alias* Hall (1595–1675): *D.N.B.:* Conor Reilly, S. J., 'Francis Line, Peripatetic'. *Osiris* (1962), **14**, 222. Line's book is entitled *Tractatus de corporum inseparabilitate*, London, 1661. See also Conor Reilly, S. J., 'Robert Boyle and the Jesuits.' *The Dublin Review* (1955) No. 469, p. 288.

|| Fulton, p. 13: Madan, No. 2586.

tion of what is known as " Boyle's Law " (see p. 224). A copy of this book was sent by Sir Robert Moray to Christiaan Huygens, who commented as follows :

> " Je fus estonnè d'abord de veoir qu'il [*i.e.*, Boyle] avoit pris la peine d'escrire un si gros livre contre les objections si frivoles que celles de ses deux adversaires, mais ayant commencè a le parcourir et voyant que parmy ses refutations il a inserè quantitè de nouuelles decouuertes et observations non encore vues, j'ay souhaitè qu'il fut encore beaucoup plus long. J'ay estè sur tout bien aise d'y trouuer les deux experiences touchant la condensation et rarefaction de l'air, qui prouvent assez clairement cette proprietè remarquable a sçavoir que la force de son ressort suit la proportion contraire des espaces ou il est reduit . . . Pour n'avoir encore que parcouru le livre de Monsieur Boile je n'ay pu remarquer toutes les belles choses, qu'il contient, mais aux endroits que j'ay leu je voy paroistre beaucoup d'esprit et de modestie, avec cette retenue ordinaire qui l'empesche de parler definitivement, ainsi que font la pluspart des philosophes d'aujourdhuy."*

Whilst on a visit to England in 1661 Huygens had taken part in the meetings at Gresham College, and so had become known to many of the London and Oxford virtuosi. He paid visits to Boyle with whom he remained in friendly relationship.† The exchange of civilities between them almost invariably took place through a third party, for example Moray or Oldenburg. It was a rather curious characteristic of Boyle's that communications and comments on experiments, as well as gifts of his books, were most frequently made in this way. Boyle's reputation was steadily increasing, not only at home but also abroad, with the result that great demands were made on his time by visitors. " He had so universal an esteeme in Foraine parts; that not any Stranger of note or quality; Learn'd, or Curious coming into England, but us'd to Visite him with the greatest respect & satisfaction . . . In the Afternoones he was seldom w[ith]out company, which was sometimes so incom̄odious, that he now & then repair'd to a private Lodging in another quarter of the Towne.‡ . . . Nor did any Ambassador, Person of Quality or Learning, Travellers come over to see this Kingdom, think they had seen any-thing, til they had visited

* Huygens, IV, No. 1032. Letter of C. Huygens to R. Moray dated 14 July 1662 (N. S.)

† H. L. Brugmans. *Le Séjour de Christian Huygens à Paris*, Paris, 1935: R. E. W. Maddison, ' Robert Boyle and some of his foreign Visitors.' *N. & R.* (1952), 9, 3–14.

‡ B.M. Add. MSS 4229, fol. 58–59: Evelyn (Bray), III, p. 346. Letter of J. Evelyn to W. Wotton dated 29 March 1696.

Mr. Boyle."* Dr. Thomas Dent, Prebendary of Westminster Abbey, said,

> " I have known him severall mornings . . . entertain persons of severall nations viz—french, spaniards, German & english & yt in different dialects, & most readily with a most pleasing aire & genious; wch was in discourse alwaies agreable & inclinable to y[e] whole company."†
>
> " . . . His easiness of access, and the desires of many, all Strangers in particular, to be much with him, made great wasts(*sic*) on his Time; yet as he was severe in that, not to be denied when he was at home, so he said he knew the Heart of a Stranger, and how much eased his own had been, while travelling, if admitted to the Conversation of those he desired to see; therefore he thought his obligations to Strangers, was more than bare Civility, it was a piece of Religious Charity in him."‡
>
> " When some of his Friends have seemed to blame him for suffering himself to be so frequently interrupted by the Visits of Strangers, and condescending to answer all their Queries, he has replied, ' That what he did was but *Gratitude*, since he could not forget with how much Humanity he himself had been received by learned Strangers in foreign Parts, and how much he should have been grieved, had they refused to satisfy his Curiosity.' His Laboratory was constantly open to the Curious, whom he permitted to see most of his Processes."§

Severe criticism of certain of Thomas Hobbes' works, including the *Dialogus Physicus* mentioned above was made by John Wallis in *Hobbius Heauton-timorumenos . . . In an Epistolary Discourse . . . To the Honourable Robert Boyle, Esq.*|| Hooke has left an account of a meeting he had with Hobbes at the shop of the instrument-maker, Richard Reeves,¶ which reveals Hobbes in an unpleasant light. He says,

> " I was, I confess, a little surprised at first to see an old man so view me and survey me every way, without saying any thing to me . . . I soon found by staying that little while he was there, that the character I had formerly received of him was very significant. I

* *Ibid.* fol. 57. For another copy see B.M. Add. MSS. 28104, fol. 21. Letter of J. Evelyn to W. Wotton dated 12 September 1703.

† B.M. Add. MSS. 4229, fol. 51[v]. Letter to W. Wotton dated 20 May 1699.

‡ G. Burnet. *A Sermon Preached at the Funeral of the Honourable Robert Boyle*, London, 1692, p. 30.

§ E. Budgell. *Memoirs of the Lives and Characters of the Illustrious Family of the Boyles*. Third edition. London, 1737, p. 144.

|| Madan, No. 2618.

¶ E. G. R. Taylor, *The Mathematical Practitioners of Tudor & Stuart England*, Cambridge, 1954, p. 223.

found him to lard and seal every asseveration with a round oath, to undervalue all other men's opinions and judgments, to defend to the utmost what he asserted though never so absurd, to have a high conceit of his own abilities and performances, though never so absurd and pitiful . . . He would not be persuaded, but that a common spectacle-glass was as good an eye-glass for a thirty six foot glass as the best in the world, and pretended to see better than all the rest, by holding his spectacle in his hand, which shook as fast one way as his head did the other; which I confess made me bite my tongue. But indeed Mr. Pell's description of his deportment, when discoursed with about mathematical demonstrations . . . surpasses all the rest."*

Other works dedicated to Boyle in this year, 1662, were Christopher Merrett's translation of Antonio Neri, *The Art of Glass;*† John Gauden, (Bishop of Exeter), *A Discourse concerning Publick Oaths* . . ;‡ John Evelyn, *Sculptura.*§ This last, " To which is annexed a new manner of Engraving, or *Mezzo Tinto*, communicated to the Authour of the Treatise ", was prepared, according to the author, upon the " reiterated instances " of Robert Boyle.

His scientific interests were pursued without intermission. Richard Lower,‖ at Oxford, sent Boyle a long account on anatomical matters including Thomas Willis' dissection of the brain,¶ and injection of liquors into the jugular vein of dogs.** Later communications from Lower dealt again with these subjects, as well as blood transfusion, the use of Boyle's laudanum and millipedes, sheep rot, and much else besides.†† John Wallis, also at Oxford, sent a long letter dealing with his teaching a deaf and dumb person to speak and understand a language.‡‡ Boyle was still in touch with Hartlib for whose benefit he promised to find Helmont's Ludus, from which an oil could be procured to cure the stone. Hartlib's hopes of being cured of his affliction were doomed to disappointment, for the Ludus was never forthcoming.§§

* *Works Fol.*, V, p. 533; *Works*, VI, p. 486. Letter of R. Hooke to RB assigned to 3 July 1663.

† Fulton, p. 160.

‡ *Ibid.*

§ *Ibid.*, p. 159.

‖ (1631–1691), physician and physiologist, F.R.S.: *D.N.B.*

¶ (1621–1675), physician, F.R.S.: *D.N.B.* He wrote *Cerebri anatome* . . . London, 1664, which was not superseded for over a century.

** *Works Fol.*, V, p. 517; *Works* VI, p. 462. Letter dated 18 January 166$\frac{1}{2}$.

†† Letters of R. Lower to RB dated 27 April, 4 June 1663, 8 June 1664, 24 June 1664, 3 September 1666.

‡‡ *Phil. Trans.* (1670), No. 61, p. 1087; Boulton, I, p. 292.

§§ Worthington, II, p. 91. Letter of S. Hartlib to J. Worthington dated 16 December 1661.

At Gresham College Boyle was appointed to various committees, such as that for the pendulum experiment; the examination of Mr. Towgood's waterworks; a new method of preparing musk; the manuscript on natural philosophy written by Mr. Sinclair of Scotland. He was also concerned with, or produced, experiments relating to sound; weighing of air; an epilatory; dissolving boiled mutton in a fluid; the ascent of water in sand in a tube; the coagulation of two liquids on mixing; the behaviour of oil of turpentine on spirit of wine; the bursting of a pewter vessel full of water when frozen.*

Boyle conferred a great benefit on the Royal Society by foregoing his claim to the services of Robert Hooke, and supporting the latter's appointment as Curator of Experiments to the Society in which position he was subsequently confirmed 'for perpetuity.'† His skill and felicity in devising experiments were in great demand by the Royal Society for many years to come.

King Charles II having "been moved, out of the earnest desire we have to promote solid learning, and the good not only of our subjects but of mankind in general, to found a society of fit persons for improving all useful sciences, arts and inventions by experiments, erecting the same into a corporation by a charter; we have thought well to bestow on it half of the fractions of debentures vested in us by the late Act of Parliament for the settlement of that our kingdom", then instructed the Lord Lieutenant of Ireland to make grants to Robert Boyle and Sir Robert Moray in trust for the Royal Society.‡ This was only one of several abortive attempts under Royal patronage to provide the Royal Society with adequate financial support.

Further evidence of Boyle's good standing in court circles is shown by the fact that in this same year, 1662, a grant of forfeited impropriations in Ireland was secured unbeknown to him from the King. His intentions and hopes for their application to the advance of learning and dissemination of the Christian religion were the subject of correspondence between him and his kinsman, Michael Boyle, Bishop of Cork.§ In connection with family affairs he was appointed a trustee on behalf of the Earl of Kildare.||

---

* Birch, *History*, I. 1662 *passim*.

† Birch, *History*. 5 November, 12 November 1662, 11 January 166$\frac{4}{5}$.

‡ *C. S. P. Ireland* (*1660–1662*). p. 602; 668; *ibid* (*1669–1670*), p. 249; *ibid* (*1663–1665*), p. 497.

§ *Works Fol.*, **V**, p. 243, 632; *Works*, **VI**, p. 56, 638. Letter of RB to M. Boyle dated 27 May 1662; letter of M. Boyle to RB dated 13 August 1662: *Orrery Papers*, p. 17–19. Letter of Sir James Shaen to Earl of Orrery dated 22 January 166$\frac{1}{2}$: *C. S. P. Ireland*, (*1666–1669*), p. 255, 19 December 1666.

|| *C. S. P. Ireland* (*1660–1662*), p. 536: *Calendar of Treasury Books*, I, p. 417.

The Turkish New Testament and the Lithuanian Bible

In view of Boyle's known enthusiasm for all schemes directed to propagating the gospel throughout the world we are not surprised to find that he was encouraging William Seaman, probably the foremost Turkish scholar in England at the time, in his translations of the catechism and the New Testament.* The latter was published in 1666, partly at the expense of Boyle.†

We find that Samuel Boguslaus Chylinski‡ approached Boyle in connection with his difficulties in bringing to a conclusion his translation and publication of the Bible in Lithuanian. Chylinski had come to Oxford in 1657 bearing letters of recommendation from the University of Franeker, whither he had been sent some years earlier in order to study divinity at the expense of some churches in Lithuania. At Oxford he was mainly dependent on the benevolence of well-wishers. He conceived the plan of translating the Bible into Lithuanian as a suitable service and duty to the churches of his homeland for having sent him abroad to advance his education. The translation of the New Testament was finished in 1658, and the Old Testament about the end of 1659. Printing of this Bible was started on the strength of some subscriptions received for this purpose. Insufficient money was forthcoming to settle the account of the printer, who stopped the impression, when it was about one half completed. Chylinski's application to the Lord Mayor and Court of Aldermen of London resulted in a recommendation to the King for a public collection. At this stage (1660–1) a certain John de Kraino Krainski came to this country to solicit help for the Protestant churches of Lithuania. Krainski, whilst appearing to be favourably disposed and helpful to Chylinski's efforts to raise money in order to complete the printing of his translation of the Bible, did, in fact, what he could to hinder that work. By the end of 1661 Chylinski was in a very impoverished condition, and seems to have appealed for help to Boyle, who sent all the relevant papers to John Wallis for comment. The latter, having set out a full statement of the business for Boyle's benefit, concluded by saying:

> " Mean while, the Person who upon his own charge, with ye supply of his private friends, & his own Study & Credit, hath

---

* *Works Fol.*, V, p. 422; *Works* VI, p. 324. Letter of E. Pococke to RB dated 13 March 166$\frac{0}{1}$. Madan, No. 2480.

† Madan, No. 2727: *Works Fol.*, V, p. 549, 550; *Works*, VI. p. 511. Letters of W. Seaman to RB dated 5 and 19 October 1664 respectively.

‡ Born (before? 1616, died in England, 1668). Consult Stanislaw Kot, ' Chylinski's Lithuanian Bible; Origin and Historical Background.' in *Biblia Litewska Chylińskiego. Nowy Testament*. Tom II-Tekst. Poznań, 1958, p. xliii: B. L. II, 13. Letter of S. B. Chylinski to RB dated 1 June 1660.

> hitherto carried on that work, wants subsistence to live on (having been forced to pawn what he hath) & ye work of Printing ye Bible in that Language . . . is actually hindred . . . I know not well . . . what is more advisable for perfecting ye Intended Edition of ye Bible in that Language . . . than that by Order of his Matie & Counsel, or by such ways as they may think fit, so much out of ye monies collected for that purpose, be payed to Mr. Chilinski or ye Printer . . . by ye Treasurers im̄ediately . . . as shall be necessary for that work."*

We may hazard the guess that Robert Boyle had Royalist sympathies which were largely responsible for his absence from London, the seat of Cromwellian authority, during most of the time after his return from the continent in 1645. His life of partial retirement, involving retreat to Stalbridge, frequent visits to relatives, absence in Ireland for nearly two years, and settlement at Oxford, was modified at the Restoration. Though, no doubt, during the Interregnum he gratefully and philosophically accepted that protection of his interests which those members of his family and circle adherent to the Cromwellian regime were able to secure for him, he was not then the recipient of such marks of favour as were accorded later by Charles II. Even though such favours were not solicited personally by Boyle, but were procured by relatives and friends, possibly by way of recognition of Stuart obligations to the Boyle family in general, he was obviously *persona grata* with Charles II. Otherwise, it is inconceivable that he would have been named one of the Council of the Royal Society in its First and Second Charters;† named Governor of the Company for Propagation of the Gospel in New England;‡ asked to enquire of the King about the disposal of Chelsea College on behalf of the Royal Society; § given access to the King to show him John Eliot's Indian Bible; || elected into the Company of Mines Royal;¶ or allowed to make a belated claim in respect of his Irish affairs.** These instances do not exhaust the list. It is perhaps significant that, at the Restoration, Boyle came

* B. L. **VI**, 11. Letter of J. Wallis to RB dated 14 March $166\frac{1}{2}$: B.P. **IV**, 114, 115. Petition of Chylinski: Madan, No. 2423. S. B. Chylinski, *An Account of the Translation of the Bible into the Lithuanian tongue* . . . Oxford, 1659: Worthington **I**, p. 180; **II**, p. 74, 86.

† *v. supra*, p. 100.

‡ *v. supra*, p. 102.

§ Birch. *History*, **I**, p. 408, 6 April 1664.

|| J. C. Pilling. *Bibliography of the Algonquian Languages*. Smithsonian Institution, Bulletin, No. 13, Washington, 1891, p. 141.

¶ *Works Fol.*, **V**, p. 328; *Works*, **VI**, p. 184. Letter of H. Oldenburg to RB dated 10 December 1664.

** *C. S. P. Ireland (1663–1665)*, p. 188. 27 July 1663.

to London, where he spent much of the time during the years 1660 to 1664, and where he could be in touch with powerful friends at court, such as Sir Robert Moray, even though he himself was not a courtier.

His continued stay in London for most of 1663 and 1664 enabled him to give frequent attendance at meetings of the Royal Society, where he exhibited specimens of various materials; performed experiments of rarefaction, or compression, with his air-pump; gave observations on vipers, swallows, the volatile extract of sponges, rock samples submitted by Sir Robert Moray, tides, salt water, shining bodies. He was appointed to committees to consider John Beale's proposal to propagate cider fruit in England; Mr. Buckland's proposal to distribute potatoes to prevent famine; to report on the Statutes of the Royal Society; to draw up enquiries for Egypt and Ethiopia; the propagation of saffron; the Bolonian stone; Clayton's diamond. He was appointed also to the Society's permanent ' chymical ' committee, which included Nicaise Le Febure (Nicolas Le Fèvre), Sir Kenelm Digby, and the Duke of Buckingham.

Prominent foreign visitors who met Boyle during 1663 and 1664 were Balthasar de Monconys,* noted for the observations he recorded during his travels; Samuel Sorbière,† Historiographe du Roi and permanent secretary of the Montmor Academy, a person of greater reputation than merit; Christiaan Huygens for a second time.

In August 1664 Boyle returned to Oxford, where in association with others he was ordered by the Royal Society to make preparations for observing the conjunction of Mercury with the Sun.‡ Evelyn records on 25 October 1664, " I went to visite Mr. Boyle now here, whom I found with Dr. Wallis & Dr. Chr: Wren in the Tower at the Scholes, with an inverted Tube or Telescope observing the Discus of the Sunn for the passing of ☿ that day before the Sunn; but the Latitude was so greate, that nothing appeared."§. The observations of the Fellows of the Royal Society at London were equally fruitless.

Boyle's return to Oxford in 1664 opens a period which is replete with letters from John Beale, Henry Oldenburg and Robert Hooke. The latter two were punctilious in keeping Boyle supplied with public news both domestic and foreign, as well as with detailed accounts of the

* (1611–1665.)

† (1615–1670.)

These visits have been described in detail by R. E. W. Maddison. " Studies in the Life of Robert Boyle, F.R.S., Part I. Robert Boyle and some of his Foreign Visitors.' *N. & R.* (1951), **9**, 1.

‡ Birch. *History*, **I**, p. 466.

§ Evelyn (De Beer), **III**, p. 384.

activities at Gresham College. Their letters provide a useful supplement to what is found in Thomas Birch, *The History of the Royal Society of London*, but they ceased when Boyle came to settle permanently in London in 1668.

John Beale, a neglected character in the history of seventeenth science, was made known to Boyle by Hartlib, who said, " There is not the like man in the whole island, nor in the continent beyond the seas, so far as I know it; I mean, that could be made more universally use of, to do good to all, as I in some measure know, and could direct."* Beale's letters are lengthy, rather tedious to read, and show his wide range of reading by the numerous quotations from classical and contemporary authors. He writes to Boyle on religious matters, husbandry, animal husbandry, cider making, arboriculture, mnemonics,† and discusses Boyle's experiments and books, the gift of which always brings forth praise and profuse thanks. In one of his letters he proffers advice which posterity can only regret was not acted upon by Boyle.

> " And I may put you in mind to collect and preserve such of your letters, as do direct and solicit philosophy; for upon such records they will do the same work every where over and over, and in following ages. It is indeed fit to be a posthume work, but it requires our serious care in your life-time, as we see by the sad examples of the great wrong done to excellent persons, by after handling; as Scaliger and lord Bacon, amongst many others, may justly complain. Your own amanuensis, and your own directions, are the best conservatories of these; and you may then expunge and enlarge, as you see it of public conducement."‡

In view of Boyle's self-confessed haphazard way of dealing with his papers it is not unexpected that the plea went unheeded.

One of the very rare occasions on which Boyle took active interest in a political matter concerns the Irish Cattle Bill of 1664. The purpose of this bill, which was provoked by jealousy of Irish farmers, was to prohibit the importation of Irish cattle into England. In this connection, Boyle wrote to his brother, the Earl of Burlington, to point out the injurious effect the bill would have on the economy of

---

* *Works Fol.*, **V**, p. 275: *Works*, **VI**, p. 102. Letter of S. Hartlib to RB dated 27 April 1658: Birch, *History*, **IV**, p. 235.

† Stimulated by Caleb Morley's charts, which had been sent to him by Hartlib. *Works Fol.*, **V**, p. 423; *Works*, **VI**, p. 326. Letter of J. Beale to RB dated 25 February 166$\frac{2}{3}$: Worthington, *Diary*, **II**, p. 92, 102, 107.

‡ *Works Fol.*, **V**, p. 487; *Works*. **VI**, p. 417. Letter dated 10 August 1666.

Ireland.* Despite vigorous protests from influential quarters, the bill was passed and supplemented by another in 1667.

During 1664 William Faithorne† drew the likeness of Robert Boyle (Plate I), which was the basis for his engraving (Plate XXI) where the original *Machina Boyleana*, introduced at the suggestion of Hooke, has replaced the landscape of the drawing,‡ Boyle was then thirty-seven years of age.

Between 1663 and 1666 Boyle published more books. There appeared *Some Considerations Touching the Usefulness of Experimental Naturall Philosophy* §, considerable portions of Part I of which had been written when he was 21 to 22 years old. This part is designed to show the utility of experimental science for improving man's knowledge; the second part deals with its usefulness in medicine; and the appendix contains recipes. The work must have been in demand for a second edition appeared in the following year, 1664, when the important *Experiments and Considerations Touching Colours* || was published. Like other of his works this one contains a bewildering number of experimental facts, which even now afford stimulating reading. It also paved the way for the important optical researches of Hooke and Newton, who advanced the theory of light.¶ Boyle discovered and described here the ability of certain classes of substances to affect the opalescence of a solution of *Lignum nephriticum*, and to change the colour of extracts from certain plants, e.g., syrup of violets, litmus.** His recognition that, by this means, substances of 'acid' and 'alkaline' nature could be differentiated played an important part in combating the mistaken acid-alkali theory of matter, which had considerable

* *C. S. P. Ireland (1663–1665)*, p. 652; R.C.H.M. 2nd Report, p. 19: R. S. Correspondence. Early Letters, Guardbook B. I. 95, 96, 97: *The Diary of John Milward*, Edited by Caroline Robins. Cambridge, 1938, p. xxviii: E. Fitzmaurice, *The Life of Sir William Petty*. London, 1895, p. 141-4: B. Fitz-Gerald. *The Anglo-Irish*. London, 1952, p. 191–193: E. Strauss, *Sir William Petty*. London, 1954, p. 123–125.

† The elder (1616–1691). Portrait painter, engraver, and print-seller. *D.N.B.*: L. Fagan, *A Descriptive Catalogue of the Engraved Works of William Faithorne*. London, 1888.

‡ For a full account of this and derivatives, see R. E. W. Maddison. 'The Portraiture of The Honourable Robert Boyle.' *Annals of Science*. (1959, published 1962), 15, 154–159.

§ Fulton, p. 37: Madan No. 2634.

|| Fulton, p. 43.

¶ A. R. Hall. 'Further Optical Experiments of Isaac Newton.' *Annals of Science* (1955), 11, 27.

** E. N. Harvey. *A History of Luminescence*. (*Memoirs of the American Philosophical Society*, Vol. 44), Philadelphia, 1957, p. 391 *et seq:* J. R. Partington. 'Lignum Nephriticum.' *Annals of Science*. (1955), 11, 1: M. B. Hall. 'Introduction.' to RB's *Experiments and Considerations Touching Colours*, Reprint, New York, 1964.

vogue in the seventeenth century. Boyle was, therefore, the originato of the use of indicators for analytical purposes, though his work in thi respect was overlooked, or forgotten, in the next century, when w find James Watt advocating the use of an extract of red cabbag (previously used by Boyle) as the best indicator.* Since that time indicators have been extensively developed for titrimetric analysis.†

This *History of Colours* called forth interesting comments fron Nathaniel Highmore‡ and Christiaan Huygens,§ both of whom ha received a presentation copy from Boyle. In his letter of thanks t Moray, Huygens says:

> " Comme je l'avois attendu avec impatience, je n'ay pu differer d le lire aussi tost d'un bout a l'autre. Tout est excellent, plein d nouuelles descouuertes et de subtiles reflexions, toutefois devan que de pouuoir obtenir la veritable hypothese des couleurs, i faudra avoir celle de la lumiere et des refractions qui me semblen de mesme qu'a Monsieur Boile extremement mal aisées a penetre . . . Au reste j'ay estè bien aise de veoir a la fin du livre l'histoir du diament, augmentée, et je me sens honnorè de la façon qu'i y est parlè de moy."

Newton, too, read the book.|| So did Samuel Pepys, who commente that it " is so chymical, that I can understand but little of it, bu enough to see that he is a most excellent man."¶

The *History of Colours* concludes with " Observations Made thi 27th of October 1663 about Mr. Clayton's Diamond.", which is a early account of the luminescent property of a certain type o diamond.**

The *Occasional Reflections upon Several Subjects*†† dedicated to hi sister, Katherine, is a youthful work which was composed mainl during the Stalbridge period, and consists of short moralistic medita tions suggested by simple events, or casual observations, in the dail life of a country gentleman, such as, the sight of clouds; the behaviou of his spaniel; making a fire; fishing; listening to a lute; trapping of

* J. Watt. 'On a New Method of Preparing a Test Liquor to show th Presence of Acids and Alkalies in Chemical Mixtures.' *Phil. Trans.* (1784), **74** 419.

† E. Rancke Madsen. *The Development of Titrimetric Analysis till* 1805 Copenhagen, 1958.

‡ B. M. Sloane MSS. 548, fol. 21. Draft letter assigned to April 1664.

§ Huygens, **V**, No. 1253, p. 107. Letter of C. Huygens to [R. Moray] date $\frac{19}{29}$ August 1664.

|| Newton. *Correspondence*, **III**, p. 154, note 4.

¶ *Diary*, 2 June 1667.

** J. M. Sweet and A. A. Moss. 'Mr. Clayton's Diamond.' *J. of Gemmology* (1955), **5**, No. 3: M. Bauer, *Edelsteinkunde*, 3rd edition, Leipzig, 1932, p. 345.

†† Fulton, p. 47.

ırk; and so on. Manuscript versions in Boyle's hand of some of the ıeditations (both printed and unprinted) are extant, and bear dates Iarch to May 1647.* It was, perhaps, his attack of ague lasting early a month in 1649 that prompted the several meditations on that ıdisposition.

The meditations are written in delightful vein, and in a distinctive tyle, which prompted Jonathan Swift to write *A Meditation upon a Broom-Stick*,† With regard to this satire, Birch said that its author has certainly not shewn in that piece a just regard to the interests of eligion, any more than to the character of Mr. Boyle, by allowing imself to treat such subjects, and so excellent a person, with the most centious buffoonry‡": and John Weyland, Jun[r], wrote:

> "This is one of the numerous exhibitions of unprovoked sarcasm, and of the spirit of misanthropy, for the frequent indulgence of which, its author richly deserved a degree of chastisement from the other component part of the broom. It will remain one among many instances of the extent, to which wit and genius can immortalize the frailty and perversity of man, and of the ease with which, in support of the worst arguments, he can get the laugh on his side, by gratifying the indiscriminate thirst of the vulgar for entertainment, and by becoming the pander to their malicious propensity for personal detraction. It is by no means improbable that this witty but vulgar piece of buffoonery had a considerable tendency to put Mr. Boyle's Reflections and his other moral works out of fashion, as no detached editions of them have been published since that time. Yet a greater superiority in all the qualities worthy the estimation of a great and good man, and in the benefits their writings were calculated to confer upon mankind, can scarcely be imagined, than that which Mr. Boyle enjoyed over Dean Swift."§

If, indeed, it be true that the reflection on "The Eating of Oysters" ıggested to Swift the writing of *Gulliver's Travels*, then we should be rateful.

---

* B.P. **XIV**.

† J. Swift. *Gulliver's Travels and Selected Writings in Prose and Verse*. onesuch edition, London, 1942, p. 456 *et seq;* and notes p. 848–9. Another tire is *An Occasional Reflection on Dr. Charlton's Feeling a Dog's Pulse at resham-College* in *The Genuine Remains in Prose and Verse of Mr. Samuel utler*. 2 vols., London, 1759, Vol. **I**, p. 405.

‡ *Works Fol.*, **I**, *Life*, p. 44; *Works*, **I**, p. lxxii.

§ J. Weyland, Jun. Preface to his edition of RB's *Occasional Reflections*, ondon, 1808, p. xviii.

The work *Occasional Reflections*, when first published, was on the whole well received, but in a letter to Richard Baxter, the divine, Boyle says:

> " I find that these harmless papers have not escaped the censure of three or four learned men, who yet seem not to dislike them for their own sake, but mine, pretending that composures of that nature might well have been spared by a gentleman whom they are pleased also on this occasion to look upon as a philosopher, and not unfit to write books in that capacity."*

Katherine, his sister, commented thus:

> " Though you tell me not who objected against your writing *Occasional Meditations* . . . I think, the very objections . . . shew its made out of fear that such things written by a philosopher may do that good in the world, that the objector . . . would have done, and therefore decries it for its meanness whilst he dares not look at the light it breaks out with, least that should make him out of love with himself. This makes me hope you will not give such persons as prevailing a power over your thoughts, as . . . to cut them off from being published for the further good of those, who own themselves to be benefited by those you have already published . . . "†

Reflections, or meditations, on things, occasions, or events were a rather popular form of literary activity in the seventeenth century. Robert Boyle's sister, Mary, Countess of Warwick, wrote them too; and very pious, religious productions they are. An interesting variation was provided by William Spurstow,‡ who published *The Spiritual Chymist: or, Six Decads of Divine Meditations*,§ (obviously an echo of *The Sceptical Chymist*), thereby initiating a form of writing which spiritualizes common objects and events, and which found subsequent imitators, particularly John Flavel.||

In 1665 Boyle published *New Experiments and Observations Touching Cold*.¶ This was a subject on which he had been working,

---

* *Works Fol.*, **V**, p. 556; *Works*, **VI**, p. 520. Letter of RB to R. Baxter assigned to June 1665.

† *Works Fol.*, **V**, p. 559; *Works*, **VI**, p. 525. Letter of Katherine Jones to RB dated 29 June [1665].

‡ (1605?–1666). Puritan divine; intruded as master of Catharine Hall Cambridge, 1645; ejected at the Restoration. Vicar of Hackney, and friend of Richard Baxter. *D.N.B.*

§ London, 1666.

|| (1630?–1691). Presbyterian divine. *D.N.B.*

¶ Fulton, p. 50: M. B. Hall. 'What happened to the Latin Edition of Boyle's *History of Cold?*' *N. & R.* (1962), **17**, 32.

and collecting information, for some years: and the subject was pursued thereafter. He became (1675) an adventurer in the Hudson's Bay Company in order to be able to collect information on the effects of cold from the Company's servants, such as sea-captains, in that remote area. He conversed with those who had travelled in cold climates, and engaged with Samuel Collins,* physician to the Czar of Russia, to supply him with observations from that inhospitable region.†

The book opens with an account of weather-glasses, and describes Boyle's improvement in constructing an hermetically sealed weather-scope. It is of interest to note that he was aware of the necessity for a standard of cold; and that he advocated the use of spirit of wine coloured by cochineal as the liquid for his thermometers because it was free from the inconvenience of freezing in the course of his experiments. The main part of the book is devoted to numerous experiments on cold, many of them having been performed during the very frosty winter of 1662. There are various important observations and discoveries, such as Boyle's account of his discovery of freezing-mixtures; the preservation of food by cold storage; the expansion of water on freezing; and observations connected with physiology.

In 1663 Blaise Pascal's *Traités de l'équilibre des liqueurs et de la pesanteur de la masse de l'air* appeared posthumously. The work, which was reported on to the Royal Society by Boyle,‡ suggested to him certain investigations which resulted in publication of the *Hydrostatical Paradoxes* in 1666.§ Samuel Pepys recorded his 'extraordinary content' in reading this work, "which the more I read and understand, the more I admire, as a most excellent piece of philosophy" . . . and "of infinite delight".||

Another book, *The Origine of Formes and Qualities*.¶ was published by him in the same year, though it had been completed several years previously. In this work the author has espoused the atomical philosophy, and taken his stand as a corpuscularian. There is one universal matter common to all bodies, and it must have local motion

---

* (1619–1670). M.D., Padua; physician to the Czar 1660–9. *D.N.B.*: L. Loewenson. 'The Works of Robert Boyle and "The Present State of Russia" by Samuel Collins (1671).' *The Slavonic and East European Review*. (1955), **33**, 470.

† *Works Fol.* **V**, 633, 635; **II**, 400; *Works*, **VI**, 639, 641; **II**, 729. Letters of S. Collins to RB dated 1 September 1663, 20 November 1636, 29 August 1664 respectively.

‡ Birch. *History*, **I**, p. 401, 431, 434. 23 March, 25 May, 1 June 1664.

§ Fulton, p. 52: Madan, No. 2738.

|| S. Pepys, *Diary*, 24 July 1667; 25 August 1667.

¶ Fulton, p. 55: Madan, Mo. 2739.

to produce the variety of existing natural bodies. The association of atoms gives corpuscles, whose " accidents " are responsible for texture and other sensible qualities of substances. The notion that certain qualities, such as colour, odour, have a relation to our senses, as well as being independent of us, paved the way for the Newtonian concept of light. Boyle's views on a universal matter led him to believe in the possibility of a true transmutation of one metal into another. He believed he had achieved an artificial transmutation of water into earth by repeated distillation.* Daniel Coxe was asked by the Royal Society to undertake a trial of this experiment in both glass and metal vessels,† but seems not have made any report. This notion of the transmutation of water into earth persisted for a very long time: it was accepted by Joseph Priestley;‡ it was rejected by Bryan Higgins§ and was finally proved to be wrong by Antoine Lavoisier.||

Several notable events occurred in 1665. The Second Dutch War was declared, in which the English defeated the Dutch Navy at the Battle of Solebay, 3 June 1665, when Richard Boyle, second son of the Earl of Burlington, and nephew of Robert Boyle, was killed.¶ The plague which had been prevalent since the end of the previous year broke out with such severity that not only the meetings of the Royal Society were suspended, but also the Court, Parliament, and Law Courts migrated to Oxford, which town remained free from the pestilence. Boyle went to Oxford in June 1665. In August it was rumoured that he had been nominated to the Provostship of Eton become vacant by the death of Dr. John Meredith, but in the event Dr. Richard Allestree was appointed.** John Beale had imparted the rumour to Henry Oldenburg, who hastened to urge Boyle's acceptance of the position.†† Boyle promptly replied, thanking Oldenburg " for your resentment of an information, wherein you supposed me concerned, though indeed I were not at all so. For neither do I believe, that any such thing, as our worthy friend Dr. Beale was told, was at

---

* *Works Fol.*, **II**, p. 523; *Works*, **III**, p. 102: Ambrose Godfrey and John Godfrey. *A Curious Research into the Elements of Water* . . . London, 1747.

† Birch. *History*, **II**, 86 (25 April 1666); he was given a reminder on 15 August 1666.

‡ J. Priestley. *Philosophical Empiricism*, London. 1775, p. 57; *ibid.* The advertisement.

§ B. Higgins. *A Syllabus, of Chemical and Philosophical Enquiries* . . . London, [1775], p. 17.

|| A. L. Lavoisier. *Oeuvres*. 6 vols, Paris, 1862–1893, Vol. II, pp. 11–28.

¶ See Genealogical Chart, [F. d. 5].

** *C. S. P. Domestic* (*1664–1665*). p. 485, 21 July 1665.

†† *Works Fol.*, **V**, p. 333; *Works* **VI**, p. 192. Letter of H. Oldenburg to RB dated 29 August 1665.

all designed; neither if it had been proposed, would it have been much welcomed."*

On 8 September 1665 Robert Boyle was created Doctor of Physic of the University of Oxford, "at whose presentation Dr. [James] Hide, regius professor, made a speech in praise of him, etc., and then presented him."† Robert is silent on the conferment of this distinction.

This year, 1665, saw also the appearance of the first scientific journals in Europe, namely, the *Journal des Sçavans*, and the *Philosophical Transactions*.‡ The latter was commenced by Henry Oldenburg not only to disseminate scientific views, but also to provide a means of augmenting his income. Boyle contributed numerous papers to the *Philosophical Transactions*.§

Though many others fled the town on account of the plague, Oldenburg remained at his home in Pall Mall, having taken care to dispose the books and papers belonging to himself, the Royal Society, and Robert Boyle into separate parcels.‖ The large influx of persons into Oxford caused Boyle to complain of "ye great Concourse of strangers yt is now made here fro[m] all ye parts of England [which] has brought mee not only this day but all this Weeke such a multitude of Visits, yt I am resolved, if God permit yt in case I doe not goe wthin this weeke to Lees I will remove for a while to a place 4 or 5 miles hence yt I may have some time for those necessary things calld Eating, & Sleeping, & Reading & Writing."¶

This intention was undoubtedly acted upon soon after, for Daniel Coxe** in one of his letters says:

> "I understand by my Correspondents att Oxford that you are removed to Stanto[n] St John.†† The vulgar Spiritts are not so

---

* *Works Fol.*, **V**, p. 249; *Works*, **VI**, p. 66. Letter of RB to H. Oldenburg dated 8 September 1665.

† Wood, *Fasti*, **II**, p. 286: Wood, *Life*, **II**, p. 57.

‡ D. McKie, 'The Scientific Periodical from 1665 to 1798.' *Phil. Mag.* Commemoration Number, July 1948., p. 122: W. P. D. Wightman, 'Philosophical Transactions of the Royal Society.' *Nature* (1961), **192**, 23: [Anon]. 'Scientific Century.' *Times Literary Supplement*, 22 July 1965, No. 3,308, p. 620: E. N. de C. Andrade. 'The Birth and Early Days of the *Philosophical Transactions*.' *N. & R.* (1965), **20**, 9.

§ Fulton, p. 138: a list is given here.

‖ *Works Fol.*, V, p. 320, 332; *Works*, **VI**, p. 31, 32. Letters of H. Oldenburg to RB dated 4 July and 10 August 1665 respectively.

¶ Royal Society. Early Letters, Guardbook B. I. 98. Letter of RB to H. Oldenburg dated 11 November 1665.

** (?–1730). Proposed candidate for F.R.S., by Dr. Wilkins, and elected 22 March 166$\frac{4}{5}$. He was an active member of the Society. For contemporary references to him see F. H. Ellis. "The Author of Wing C 6727." *N & R.* (1963), **18**, 36: Venn.

†† N. E., of Oxford.

> much Concerned about the Court as ye more Refined are for you, all your moc[i]ons are attended as if the Destiny of ye learned world were included in yours, the ingenious spiritts in The University greivously resent your Retirement as being thereby deprived of their greatest glory & cheifest Ornament . . . I do allmost Envy that happy & truly ingenious Phisitian Dr. Rugely* who accompanies you in your retirement."†

We do not know exactly, when, or how, Boyle came to be acquainted with Daniel Coxe, who must have been a quite young man at the time. Coxe evidently knew several of the *virtuosi* at Oxford, and he was delighted to be favoured with a correspondence with Boyle, whose part is unfortunately missing. In the same letter from which we have just quoted, Coxe says: " You were pleased to Command mee to lay aside Courtship (a crime I am not Conscious of) in my Epistolar Commerce, & to Converse with the Liberty Philosophy affords mee." Coxe was greatly interested in chemistry, and treated Boyle as his master. He reports to Boyle his trials and views on menstrua, Helmont's alcahest, distilling mercury, reducing alkalis into earths, the Bow dye, factitious gems, and medical cases.‡ About the beginning of February 166^{5}_{6} Boyle was ill, or at least indisposed, again; and he asked Coxe to seek out another lodging for him. Coxe did so with alacrity. He found a suitable place at Stoke Newington, then a pleasant village about two miles from London, and solicited Oldenburg to pay a visit there. The latter reported to Boyle, " tho I was not wthin the house, yt is to be taken up for you, yet I lookt upon ye places about it; and must needs say, yt it seems to me very convenient for you: there being a large orchard, a walk for solitary medita[ti]on, a dry ground round about, and in all appearance, a good Air; all wch Mr. Coxe affirms to be accompanied wth a civill Landlord, and faire Landlady."§

The plague not only brought unwelcome visitors to Boyle, but also a request for assistance in combating the pestilence. Hugh Chamberlen‖ sent Boyle " An Essay of ye Plague " asking for his judgement thereon " or a better way of proceeding " in connection with his propositions for freeing the City of London of the plague in three months, or to lose

* Not identified.

† B. L. **II**, 54. Letter of D. Coxe to RB dated 19 January 166^{5}_{6}.

‡ B.L. **II**, 54; 58; 78 and 62; Letters of D. Coxe to RB dated 19 January, 5 February, 19 February 166^{5}_{6} respectively. There are other undated letters.

§ *Works Fol.*, **V**, p. 353; Works. **VI**, p. 221. Letter of H. Oldenburg to RB dated 17 March 166^{5}_{6}.

‖ Undoubtedly a member of the famous family of physicians; probably the eldest son of Peter Chamberlen, (1601–1683).

his reward. This communication was accompanied by an enclosure for Mr. [Secretary] Williamson, which presumably was forwarded by Boyle, for it is not with the other correspondence.*

Whatever distractions and inconveniences the plague may have caused Boyle, he found time to write *The Excellency of Theology, Compar'd with Natural Philosophy*, though it was not published till 1674.† The author's preface shows his concern for the materialistic outlook of the age, and he reproves those friends of his who deplored the time he devoted to theological pursuits, saying that it was not reasonable to expect him to confine himself to *one* subject, namely, natural philosophy.

Two works were dedicated to Boyle in this plague year; Ralph Austen, *A Treatise of Fruit-trees*,‡ and Richard Lower, *Diatribae Thomae Willisii . . . de Febribus vindicatio . . .*§ In 1666, John Eliot's *Indian Grammar*,‖ and in 1667 Walter Needham's *Disquisitio anatomica formato foetu*¶ were dedicated to him.

*Valentine Greatraks,*** the stroker*

Of English descent, Valentine Greatraks was born in Ireland, where he served in Cromwell's army. Thereafter he passed some uneventful years in the improvement of his estate until he became convinced that he was possessed of unusual powers of healing. About the year 1662, he says, " I had an Impulse, or a strange perswasion in my own mind . . . that there was bestowed on me the gift of curing the King's Evil." For some three years he treated those afflicted with this malady by laying on of hands and stroking the affected parts. He then had another ' Impulse ' which suggested that " there was bestowed upon me the gift of curing the Ague." The fame of his ability spread far and wide; his ministrations were sought by the low as

* B.L. **II.** 5, 7, 8. Communications of H. Chamberlen to RB dated 17 January 166$\frac{5}{6}$.

† Fulton, p. 81: *Works Fol.* **III,** p. 404; *Works.* **IV,** p. 1.

‡ Oxford, 1665. Madan, 2692: Fulton, 160.

§ London, 1665. Fulton, 160.

‖ Cambridge, Massachusetts. W. Kellaway. *The New England Company 1649–1776*. London, 1961, p. 137–138.

¶ London.

** (1629–1683.) *D.N.B:* V. Greatrakes, *A Brief Account . . . in a Letter addressed to the Hon. Robert Boyle . . .* London, 1666: Henry Stubbe, *The Miraculous Conformist . . .* Oxford, 1666: Madan No. 2758: *Works Fol.* **I,** Life. p. 45, *et seq; Works.* **I,** p. lxxiii *et seq:* Various correspondents in *N & Q.* (1953), **198,** 268, 315, 453: *ihid* (1954) **199,** 408, 545: R. A. Hunter and I. Macalpine. " Valentine Greatraks." *St. Bartholomew's Hospital Journal.* (1956), **60,** 361: G. Horne, *Letters on Infidelity*. Oxford, 1806, p. 88 *et seq.*

well as the high born, including Anne, Viscountess Conway,* the invalid for whom Robert Boyle unavailing prepared his *ens veneris* as a curative medicine. Greatraks came to England to treat her, but failed to cure her.

Boyle became more deeply involved in this Greatraks sensation than he might otherwise have been as a result of Stubbe's *Miraculous Conformist*,† This piece was not at all to Boyle's liking, for after receiving a copy of it, he immediately wrote a long letter‡ expressing his displeasure and disagreement with certain matters contained therein. The printed version of this letter does not append Boyle's notes and thoughts on Greatraks and his cures. As these remarks are of considerable interest for showing Boyle's care in approaching and examining a problem, they are reproduced here.

Robert Boyle's enquiries about the stroker

" What Complexion Mr Gr. is of; whether Melancholy, Sanguine, Phlegmatick &c. What Deseases he is, or has bin subject to; wr he be lean or fat; vigorous, or weak, & of wt Age.

How long since, & at wt time of his[a] life he first tooke notice of, & tryd ye Impulse he speaks of to cure Deseases? and whether yt were preceded by any fit of Sicknesse, or of Melancholy, or any Accident yt might have any extraordinary Influence upon his mind, & in wt maner he received his first Impulses?

Whether before he applys[b] himself to doe a cure, he uses any Ceremony, or other words, or prayer; & if a Prayer whether he uses arbitrary words or some set forme? and if ye later wt it is; As also whether whilst he is stroaking he imploy's any peculiar Rites, or words, or doe, or doe not, use to give thanks when he has done. And whethr he require yt Patient to doe, or say any thing, before & after he has stroakd.

Wr he p[er]form's any wonders by barely applying his hand to ye part affected, or must also rub it, & in case whether Friction be necessary, whether he make it light or strong, & how long he usually continues it?

Wr he can p[er]forme Cures, & remove pain's when ye Patients Cloths are on, or whether ye im[m]ediate Contact of his hand be requisite.

---

[a] Followed by *are* deleted.

[b] Followed by *his hand* deleted.

---

* (?–1679) *D.N.B: Conway Letters*, edited by M. H. Nicolson. London, 1930.

† It is written as a letter to RB, and dated 18 February 166$\frac{5}{6}$.

‡ *Works Fol.*, **I**, *Life*, p. 47; *Works.* **I**, p. lxxxvi. Letter dated 9 March 166$\frac{5}{6}$.

Wr his left hand, & his right hand doe stroake wth aequll efficacy; & whether ye Contact of his Cheek, breast &c aequall that of his hand?

Wr his Gloves, Shirt, & Stockens yt have bin worne im[m]ediately upon his flesh, be of aequall efficacy wth his hand in ye Cure of Deseases, or at least in ye removeing of Paine; & whether, in case they be effectuall, how long their efficacy will last after his ceaseing to ware them; & whether ye degrees or dura[ti]on of their vertue will be encreasd if they be taken fro[m] his Body when he is sweating, or at least more then ordinary warme.

Wt will be ye efficacy of his spittle, & his urine, & if they have any, how long it will last.

Whether his Urine be more effectuall when warme or cold. Wr its Sanative Vertue will be preservd longer if it kept close stopt, or otherwise; & whethr a little spt of Salt, or Nitre &c. dropt into it, whilst it is yet fresh & sanative, will by changeing the Texture destroy or impair the Vertue; & whether if his Unaltered Urine be wearily distil'd *in Balneo*, ye spirit, or ye phlegme, or residence, or all or none of it, will retain a sanative Vertue.

Wt Deseases they are yt Mr. G. cannot at all cure, & among those wch he some times cures, wch are they that he succeeds best, & wch he succeeds worst wth. And among ye former sort wt Complexions, ages, sexes &c. doe ye most favor, or disfavor his Cures.

Wr upon[c] Mr. Gr. his Touch, or stroaking there ensue ordinarily, or at any time, any quick & manifest evacua[ti]on of peccant matter by Vomit, Seige, Sweat, Urine, or Saliva[ti]on.

Whether Mr. Gr: his sanative power will react to ye cureing[d] or removeing pains in Horses, Doggs, or other Bruits.

Wr it be true yt is affirm'd yt ye Effluvia of Mr. Gr. his Body are well sented, when he has imploy'd no propper meanes to make it soe.

Wr Mr. Gr. is able to cure any possess'd Person, or any Desease (if there be any such) produc'd by witchcraft.

Wr Mr. Gr. be able to cure Men of differing Religions, as Roman Catholicks, Socinians, Jews &c. as also Infants, Naturalls, or destracted Persons to whose recovery ye faith of ye Patient cannot concurre.

How many times it is usually necessary for the Patients to be touchd according to their respective infirmitys, & wr wt Intervalls betwixt those times.

---

[c] Interlined.
[d] Followed by *of* deleted.

Wr ye efficacy, or inefficacy of his Touch be alwayes proportionall to ye greater or lesser danger, or obstinacy of ye Malady, consider'd as Phisitians are wont to aestimate it."*

In one of his weekly letters to Boyle, Henry Oldenburg reported the success of Greatraks in curing Monsieur de Son† of great back-aches with which he had been troubled for many years,‡ Daniel Coxe reported other cures.§

Widely divergent opinions were held concerning Greatraks; some regarded him as an imposter, or a trickster; others, certainly those who benefited from his ministrations, saw him as a saviour. Not unexpectedly, he incurred the hostility of the clergy as well as of the medical profession; and he had been attacked in print by a pamphlet *Wonders no Miracles*, which called forth Greatraks's own vindication of himself in *A Brief Account* addressed to Robert Boyle to whom he was not then known personally. They did become acquainted, and Boyle was " a spectatour of at least 60 performances " by Greatraks‖.

Greatraks had returned to Ireland by early summer 1666, where he continued to treat any who came to his house. He did not pay another visit to England till 1668, in which year he met Boyle again. His letter¶ dated 12 September of that year, written on his way back to Ireland again, gives an account of fresh cures he had wrought. He visited some of those he had treated and cured on his previous visit to England two years before, and found them still well. He was evidently still subjected to vilification from certain quarters, for in this letter to Boyle, he asks him to write to the various persons he has named, " & yt if yu finde every thinge I have herein writ to be truth, yt yu will put yor testimony to it, if not publish me to ye world to be a Lyar, & Imposter." That he was convinced of his divinely inspired power, from which he declined to derive mercenary advantage, is shown in the post-script to the letter, where he asks Boyle to tell Dr Coxe that

> " might I gaine a 1000^{lb} a yeare duringe my life I would neith[er] live in a Citty (to loose ye happines of my inocent Countery life)

* B.L. **III,** 33.

† *recte* d'Esson, seigneur d'Aigmont. (1604–?) French engineer and engraver. Frequently mentioned in the Boyle correspondence, and in Huygens, vol. **V** *et seq.*

‡ *Works Fol.* **V,** p. 352; *Works.* **VI,** p. 219. Letter from H. Oldenburg to [RB] dated 13 March 166$\frac{5}{6}$.

§ B.L. **II,** 64. Letter of D. Coxe to [RB] dated 5 March 166$\frac{5}{6}$.

‖ *Conway Letters*, edited by M. H. Nicolson. London, 1930, p. 273. Letter of Henry More to Anne, Viscountess Conway, dated 29 April [1666].

¶ B.L. **III,** 21.

nor take money fro[m] others for yt wch God has freely bestowed on me. In London I was resolved to doe no thinge, for severall reasons, but since I left yt place I have improved my Talent to Gods glory, & ye good of many poore afflicted persons, & to speake ye truth to yu I have found ye power of God wonderfully assisting of me."

At the time of the Great Fire of London, which raged from the 2nd to the 6th of September 1666, Boyle was somewhere near the town, perhaps at Chelsea, though he had departed for Oxford before the 10th of that month.* Gresham College was taken over by the Lord Mayor and the Court of Aldermen, so the Royal Society had perforce to meet elsewhere. The destruction round about St. Paul's Churchyard caused great loss to the stationers and printers, and caused great difficulties in producing the *Philosophical Transactions*. Because of the precarious state of his finances Oldenburg asked Boyle to solicit a recommendation from Lord Brouncker and Sir Robert Moray for his employment as Latin Secretary in the Government service.† He said, "I have good hopes ye generous Inclina[ti]ons of those two worthy Persons, ye Ld Brounker and Sr R. Moray, thus excited by ye ready favor of yr recommenda[ti]on, will prove effectuall for my relief. I am eternally obliged to you for yr so carefull a concern for me."‡ He was, however, destined to be disappointed.

Arising from the disastrous losses of the booksellers in the fire,§ there is an interesting outcome mentioned in a letter of Boyle's sister, Katherine, which rather implies that Boyle had an option on some of Benjamin Worsley's books. Evidently, he had not severed all contact with Boyle. Katherine says,

"... I have since taken to myself the mortification of seeing the desolations, that God, in his just and dreadful judgment, has made in the poor city, which is thereby now turned indeed into a ruinous heap, and gave me the most amazing spectacle, that ever I beheld in my progress about and into this ruin. I dispensed your charity amongst some poor families and persons, that I found yet in the fields unhoused. Since then, Mr. Worsley has been with me, and given me very considerable particulars of

---

* *Works Fol.* V, p. 357; *Works.* **VI**, p. 228. Letter of [H. Oldenburg] to RB dated 10 September 1666.

† *Ibid; Works Fol.* V, p. 361; *Works.* **VI**, p. 233. Letter of H. Oldenburg to RB dated 16 October 1666.

‡ *Works Fol.* V. p. 359; *Works.* **VI**, p. 231. Letter of H. Oldenburg to RB dated 25 September 1666.

§ Evelyn (de Beer). **III**, p. 459.

> providence, that assisted to his preservation, and that of his goods, which he probably enough thinks raised in value, as to that part of them, wherein you have an interest, by the great consumption, that has been of that sort of commodity, both at Sion College, and also in St. Faith's church, where all thereof was destroyed. He now very ingenuously told me, he now desired to perfect to you that bill of sale for them that he at first promised to give you, not seeing, how your title to them can otherwise be legally good and sure; and he proposes then their being in your name put into Dr. Sydenham's hand, as his security for 250*l.* for six months; which, if you have no exceptions to, will fall in well for my occasions, one of the most pressing of which is the paying of 100*l.* to the doctor, which I owe him, and for which your bond was given, by way of collateral security, to confirm the engagement of this house to him, therefore, so as he need lay out but 150*l.* and pay himself 100*l.* upon a security worth 5 or 600*l.* and that but for six months."*

Boyle agreed to this arrangement.† Can we surmise that Boyle was buying Worsley's unwanted chemical books, now that the latter was an official in the Council of Trade?

In 1662 Boyle's sister, Katherine, occupied a house in Pall Mall which was the sixth from the Palace end; and she was next-door neighbour to Thomas Sydenham,‡ the physician, who dedicated his book on fevers to Boyle.§ About 1664 The Earl of Warwick assigned two houses in Pall Mall to Lady Ranelagh, and the Rate Books show that from 1665 onwards she occupied premises further east in that

---

* *Works Fol.*, **V**, p. 561; *Works*, **VI**, p. 528. Letter of Katherine, Viscountess Ranelagh to RB dated 12 September [1666].

† *Works Fol.*, **V**, p. 562; *Works*, **VI**, p. 529. Letter of Katherine, Viscountess Ranelagh to RB dated 18 September [1666].

‡ (1624–1689) M.B., M.A., Oxon; M.D., *h.c.*, Cantab; established himself as a physician in Westminster: *D.N.B.*: K. Dewhurst. *Dr. Thomas Sydenham; His Life and Original Writings*. London, 1966.

§ T. Sydenham, *Methodus curandi febris* . . . London, 1666. The dedication contains the following interesting remarks which are taken from the translation of R. G. Latham, *Works of Thomas Sydenham*, London, 1848–50, Vol. **I**, p. 10.

> " . . . I profess to have placed my work under your patronage for the two following reasons: It was on your persuasion and recommendation that I undertook the subject; and it is by your own experience that the truth and efficacy of some of the matters delivered in the Treatise, have occasionally been tested. Hence you are made a sufficient witness; since in the fulness of your humanity you have gone so far as to accompany me in the visiting of the sick. Herein you have exhibited a kind spirit, descending to offices, which, however honorable, are little recognised by the spirit of the age we live in."

thoroughfare.* The above letter seems to bear on this business of change of residence, and is perhaps connected with Robert Boyle's plan to live with his sister. In a post-script to a later letter she says; 'I have ordered Thomas to look out for charcoal, and should gladly receive your order to put my back-house in posture to be employed by you, against your coming, that you might lose no time after."† The provision of some kind of laboratory facilities was clearly envisaged.

In these two years which were so disastrous for London, Boyle was as active as ever in the affairs of the Royal Society. He sent communica-ions on a monstrous calf; a monstrous head of a colt; oil of Florence; he lodestone. Various matters were referred to him for considera-ion, such as; Kuffler's new oven; do animals die *in vacuo* for lack of air, or because their lungs are compressed. He was concerned in various experiments with the air-pump, the transfusion of blood, the efrigeration of bodies, and music. He was also elected to committees for improving artillery, and for astronomical experiments.

During the first half of 1667 Boyle was for the most part at London, o there was more opportunity for 'good discourse' with his sisters Katherine and Mary. He was, as ever, active at the Royal Society's meetings, being concerned with further experiments on animals, such as bleeding them to death, and injecting air into them; the drying of air; magnetical experiments; breathing under water; shining phosphorescent) wood and fish. With Robert Hooke he was en-rusted with the experimental demonstrations for the entertainment of the Duchess of Newcastle‡, when she visited the Royal Society 30 May 667.§ It is very noticeable that nearly all the experiments derived from the work of Boyle.

During this same period Boyle was in correspondence with Samuel Colepresse about mines and minerals. A letter from the latter gives replies to ninety queries that Boyle had raised concerning these matters in Devonshire and Cornwall.|| Sundry boxes of ores were sent

---

* London County Council. *Survey of London*, Vol. **XXIX**. *The Parish of St. James Westminster*. Part One. South of Piccadilly. London, 1960, p. 67: City of Westminster Public Library. Rate Books. St. Martin-in-the-ields.

† *Works Fol.*, **V**. p. 562; *Works*, **VI**, p. 530. Letter from Katherine, Viscountess Ranelagh, to RB dated 13 November [1666].

‡ Margaret Cavendish, (?–1673); described by Evelyn as "a mighty pre-ender to Learning, Poetry and Philosophy."

§ See the Diaries of Pepys and Evelyn under this date: Birch *History*, Vol. I, p. 175, 177.

|| *Works Fol.*, **V**, p. 573; *Works*, **VI**, p. 546.

to Boyle for his examination. John Locke, too, was corresponding with Boyle on chemico-medico-matters.*

In the second half of 1667 Boyle was at Oxford, so Hooke's and Oldenburg's frequent letters reporting the work of the Royal Society, as well as general news, were resumed. In June of this year, when there was much alarm in the country on account of the disastrous turn of events in the war with Holland, Henry Oldenburg incurred the suspicion of the Government, with the result that he was incarcerated in the Tower of London on warrants signed by Lord Arlington. It was undoubtedly his extensive foreign correspondence which caused him to be suspected, especially because he had been critical of the conduct of the war. Commenting thereon, Pepys wrote in his diary under date 25 June 1667, " I was told, yesterday, that Mr. Oldenburg, our Secretary at Gresham College, is put into the Tower, for writing newes to a virtuoso in France, with whom he constantly corresponds, in philosophical matters; which makes it very unsafe at this time to write, or almost do any thing."

Under this misfortune, Oldenburg was not the first, nor the last, to discover who were, and who were not, his true friends. The warrant for his release bears date 26 August 1667. Shortly afterwards he wrote to Boyle saying,

> " . . . My late misfortune, I feare, will much prejudice me; many persons, unacquainted wth me, and hearing me to be a stranger, being apt to derive a suspicion upon me. Not a few came to the Tower, meerly to inquire after my crime, and to see the warrant; in wch when they found, that it was for dangerous desseins and practices, they spred it over London, and made others have no good opinion of me. *Incarcera audacter, semper aliquid haeret* . . I have learned, during this commitment, to know my real friends."†
>
> " Though it be above a forthnight since I gave notice to some of my corresponding friends, that I was return'd to my former station, yet have I received as yet no answer from any but yrself and Dr. Beale."‡

Boyle remained at Oxford until April 1668. In February Oldenburg informed Boyle of the arrival of ' two of the Florentine virtuosi.' They

* K. Dewhurst. *John Locke* (1632–1704). *Physician and Philosopher* London, 1963.

† *Works Fol.*, V, p. 364; *Works*, VI, p. 237. Letter of H. Oldenburg to RB dated 3 September 1667.

‡ *Works Fol.*, V, p. 364; *Works*, VI, p. 238. Letter of H. Oldenburg to RB dated 12 September 1667: D. McKie. 'The Arrest and Imprisonment of Henry Oldenbourg'. *N. & R.* (1949), 6, 28.

were Lorenzo Magalotti* and Paolo Falconieri,† who were received and treated by the Royal Society with all the consideration due to their connection with the court of the Grand Duke Ferdinand II of Tuscany. Magalotti visited Oxford, and saw Boyle some time between 1 and 8 March $166\frac{7}{8}$.

Magalotti wrote to Prince Leopold de' Medici, who with his brother the Grand Duke founded the famous Accademia del Cimento, saying:

> "... trovai il Boile così garbato uomo, che il far sole 50. miglia per andarlo a trovare mi parve troppo poco; non si può mai dire quant' egli sia cortese, discreto, a obbligante, e quanto sia amabile e cara la sua conversazione. Io in due volte ne godei intorno a 10. ore, e se non avessi avuto paura ch'ei mi cacciasse, non gli sarei uscito di casa ne' due giorni e mezzo, che mi trattenni in quella città. Mi fece vedere diverse esperienze altre appartenenti alla pression dell'aria, altre alla mutazione di diversi colori prodotta dal mescolamento di diversi fluidi. Parlò di V. A. con quei sentimenti di venerazione e d'ossequio, che eran confacevoli alla sua virtù e al mio desiderio, ed io gli corrisposi con quelle espressioni che credetti più proprie alla stima, che l'A. V. fa di questo degno soggetto; procurai di dargli animo a scriverle, immaginandomi che non l'avesse mai fatto per l'addietro, nè seppi ben rinvenirmi dalla sua risposta s'io m'ingannassi. Mi disse che il Cav. Finkio‡ l'aveva invitato a quest'onore, ma che egli se n'era astenuto, non bastandogli l'animo di farlo di proprio pugno a cagione della sua vista inferma, e non avendo a chi dettare in altra lingua che nell'Inglese. Cercai di sopire queste difficoltà fino a quel segno, che giudicai di poter fare senza ingerirli sospetto di aver commissione di stimolarvelo; per quello che tocca a me gli ho promesso la communicazione di tutto quello che crederò atto a lusingare la sua nobile curiosità; e perchè mi fece dono d'un de' suoi libri tradotto in latino che alla mia partenza non era arrivato in Italia, gli promessi di servirlo d'un

---

* (1627–1712) Courtier; occupied various positions of state in Tuscany; elected F.R.S., 1709.

† (1638–1704) Courtier and sonneteer. Concerning the first visit of Magalotti to England, see A. M. Crinò. *Fatti e Figure*. Florence, 1957, p. 123; 127; 151: R. E. W. Maddison, "Studies in the Life of Robert Boyle, F.R.S., Part I. Robert Boyle and some of his Foreign Visitors." *N. & R.*, (1951), **9**, 22-24: R. D. Waller, 'Lorenzo Magalotti in England, 1668–9'. *Italian Studies*, (1937), **1**, 49.

‡ Sir John Finch, (1626–1682); physician; original F.R.S.; minister to the Grand Duke of Tuscany. Finch was then at Florence, whence he wrote to RB to tell of Boyle's high reputation in Italy. (*Works Fol.*, V, p. 601; *Works*, VI, p. 589. Letter dated 6 February 1668 (N.S.).

esemplare di quello di V. A., la quale supplico umilmento ad aggiungere a quelli, che per donare o per vendere verranno in Inghilterra."*

In the notes made by Magalotti for his *Relazione d'Inghilterra* we find the following about Robert Boyle:

" . . . Di questo savio e virtuoso Gentilhuomo non si può mai parla tanto in sua lode, che ei non meriti malto più. Pieno di religione verso Dio di magnanima carità verso il prossimo, di generosità di affabilità, di cortesia, di gentilezza verso tutti. Egli è assai ancora giovane, ma d'una complessione cosi inferma che non gli permette interi i suoi giorni. Parla benissimo in Franzese, e l'Italiano ha però qualche impedimento nella favella, la quale gli è spesso interrotta da una spezie d'imputamento, che pare che sia costretto da una forza interna di ringoiarsi le parole, e con le parole anco il fiato, onde par talmente vicino a scoppiare che fa compassione in chi lo sente.†"

From April to June 1668 Boyle was in London, and attended the Royal Society's meetings, where he gave an account of the experiments he had made during his absence from London; he presented the Society with a gauge for measuring a vacuum; he demonstrated his portable baroscope. The experiment on sounding organ pipes *in vacuo* described in *I Saggi di Naturali Esperienze* of the Accademia del Cimento was referred to Boyle for consideration.

He was appointed to the committee to report on John Wilkin's *An Essay towards Real Character and a Philosophical Language*, which had been recently published.

In connection with the expected grant of Chelsea College with its lands to the Royal Society,‡ Boyle was solicited for a contribution to the building fund;§ he gave £50.||

About the end of June 1668 Boyle left London for Leese, the home of his sister Mary, Countess of Warwick, where he stayed till 4 August. From then onwards he lived with his sister Katherine, Viscountess Ranelagh, in Pall Mall, London.

---

* A. Fabroni. *Lettere Inedite d'Uomini Illustri*. 2 vols. Firenze, 1773–5. Vol. I, p. 301: A. M. Crinò. 'Robert Boyle visto da due Contemporanei.' *Physis*, (1960), **II**, 318.

† A. M. Crinò. *Fatti e Figure*. Florence, 1957. p. 158.

‡ *The Record of The Royal Society of London*. 4th edition. London, 1940., p. 24.

§ *Works Fol.*, **V**, p. 382; *Works*, **VI**, p. 262. Letter of H. Oldenburg to RB dated 14 January 166$\frac{7}{8}$.

|| C. R. Weld. *A History of the Royal Society*. 2 vols. London, 1848. Vol. I, p. 211.

IV

The London Period

(1668–1691)

During the reign of King Charles II the district of Westminster saw considerable development in the area between Whitehall and St. James's Palace, as a result of which St. James's Street, St. James's Square, Pall Mall, the Haymarket and other smaller streets were created to form an exclusive, aristocratic quarter close to the Royal Court and the seat of government.* When Robert Boyle went to live with his sister, Katherine, she had already been in Pall Mall since 1660. "One of those four pretty handsome [houses] in Pall Mall by my Lady Ranelagh's" was bought for £1,400 in 1667.† During the years that Robert was there, many well-known people lived in Pall Mall, including Dr. Thomas Sydenham, the physician, who had been one of the earliest to take up residence in that street; the Countess of Portland; the Countess of Southesk; Sir William Temple, author and diplomat, whose wife was Dorothy Osborne; Sir William Coventry, politician; Mrs. Mary Knight, singer; Mrs. Eleanor Gwyn, actress; Mary Beale, portrait painter; Abraham Storey, speculative builder. In 1668–69 Boyle's immediate neighbours on the south side of the road were Sir William Temple on the east, and the Countess of Portland on the west. What he thought of having Nell Gwyn next door to the Countess of Portland, we do not know: he never mentions her.

Pall Mall has the distinction of being one of the first streets in London to have been illuminated by gas-light. This technical achievement occurred in 1807 through the efforts of Friedrich Albrecht Winsor (or Winzler), an energetic promotor of gas-lighting, who, too,

---

* N. G. Brett-James. *The Growth of Stuart London*. London, 1935: City of Westminster Public Library. Rate Books. Parish of St. Martin-in-the-Fields.

† J. Cartwright. *Sacharissa*. London, 1901., p. 188.

resided in that street.* We may be sure that Boyle would have been surprised, yet gratified, could he have known that such a development would result from the simple experiment of distilling coal which had been performed by the Rev. John Clayton†, who had reported the experiment to Boyle in answer to one of the queries raised by the latter. The particular query was the following: " Whether Air be ingenerable, or incorruptible, & whether it be producible by Art, or only extricable out of other bodys? "‡

With Robert Boyle's settlement in London there was no longer need for Henry Oldenburg and Robert Hooke to send those regular news-letters which provide so much information on persons and daily events, scientific or otherwise, both at home and abroad. John Beale continued to write, but his letters were shorter than formerly. It was now the turn of Boyle's Oxford associates to write. Occasionally, they did, but no sustained correspondence having the importance of Oldenburg's ever developed.

Residence in London enabled Boyle to devote more personal attention to the Royal Society, as well as to the affairs of the New England Company. He joined the Hudson Bay Company in March 1675, but took no active part in its business, though he took advantage of his connection to obtain information from the Company's ships' captains on all possible matters of philosophical interest. He was appointed to a committee of the East India Company, 27 April 1677; and, at his desire, made free of the Turkey Company, 16 May 1678. In both cases the motive was probably to improve his opportunities for acquiring local information from distant parts.

Hooke's *Diary* reveals that he frequently visited and dined with Robert Boyle; that many topics were discussed between them; that he was also concerned in an advisory capacity with alterations or additions to the residence in Pall Mall§; that he " promised to design him [Boyle] a house "||; and that he " contrived his laboratory "¶.

---

* Dean Chandler and A. Douglas Lacey. *The Rise of the Gas Industry in Britain.* London, British Gas Council, 1949. Chapter 4.

† John Clayton, 1657–1725; Rector of Crofton; F.R.S.

‡ B. L., **VI**, 62. Letter of [John Clayton] to [Robert Boyle], undated, but assigned to 1687–88.

Walter T. Layton. *The Discoverer of Gas Lighting.* London, 1926.

Dean Chandler and A. Douglas Lacey. *The Rise of the Gas Industry in Britain.* London, British Gas Council, 1949. Chapter 1.

§ *The Diary of Robert Hooke*, Edited by H. W. Robinson and W. Adams, London, 1935., pp. 250, 279, 290.

|| *Ibid.*, pp. 257, 260.

¶ *Ibid.*, pp. 307, 308, 320: Evelyn (de Beer), **IV**, p. 124. 20 November 1677

Relationships between Boyle and Hooke, who had the reputation of being difficult and crabbed, must have been amicable on the whole during the many years of their acquaintance, though there were times when Hooke felt that he had been slighted by Boyle and his sister, Lady Ranelagh. On one occasion " she scolded ", and Hooke confided to his *Diary* " I will never goe neer her againe nor Boyle ".* However, his ill humour did not last long, for in less than a week he was once more on visiting terms.

Henry More†, the Platonist, visited Boyle in 1671, when some discussion took place on the philosophy of Descartes‡. More's book, *Enchiridion Metaphysicum*, appeared later in the same year with some references to Boyle, which " had the ill hap to give offence to that incomparable person "§. Apparently, Boyle had no great opinion of More's excursions into experimental philosophy. In connection with More's *Remarks upon two late ingenious Discourses* . . . ||, Lady Conway wrote: " I hear Mr. Boyle sayes you had better never have printed it, for you are mistaken in all your experiments."¶ To which More replied: " If I be mistaken in my Experiments, I suppose Mr. Boyle will show me my mistakes, and all the world besides, which makes me conclude it is better that I have printed them that I and such as I may be undeceived by him or some he may employ against my Lord Hailes** and me. For that pinches Mr. Boyle most in those Remarks which my Lord Hailes and myself agree in which is the exploding that monstrous spring of the Ayre."††

It was convenient for Boyle that Henry Oldenburg was one of the residents in Pall Mall. This energetic secretary of the Royal Society derived part of his income by making translations, including some works of Boyle;‡‡ and his house provided facilities for some of Boyle's experiments. Oldenburg died, 5 September 1677; and his wife died twelve days later, leaving two very young children, Rupert and Sophia. The lack of other relatives in England, and the absence of wills caused a

---

* *The Diary of Robert Hooke*, p. 364.

† (1614–1687), writer of theological and philosophical works.

‡ M. H. Nicolson. *Conway Letters*. London, 1930, p. 333; and next reference.

§ *Works Fol.*, V, p. 552; *Works*, VI, p. 515. Letter of Henry More to RB dated 4 December [1671].

|| London, 1676.

¶ M. H. Nicholson. *Conway Letters*. London, 1930, p. 420. Letter of Lady Conway to H. More dated 4 February $167\frac{5}{6}$.

** Sir Matthew Hale (1609–1676); lord chief-justice.

†† *Ibid.*, p. 423. Letter of H. More to Lady Conway dated 9 February $167\frac{5}{6}$.

‡‡ A. R. Hall and M. B. Hall. " Some Hitherto Unknown Facts about the Private Career of Henry Oldenburg." *N. & R.*, (1963), **18**, 94.

difficult and confused state of affairs.* Robert Boyle provided money to bury the deceased, and undertook oversight of the children to supply them with money. Many of the papers both of the Royal Society and of Boyle had been deposited with Oldenburg; their recovery from the administratrix was effected not with difficulty.

Benjamin Worsley, Boyle's intimate in earlier years, died about the same time as the Oldenburgs. Lady Ranelagh then wrote to her brother, Robert, as follows:

> " I cannot, my brother, but condole with you the remove of our true, honest, and ingenious friends in their several ways, Dr. Worsley and Mr. Oldenburg, since it has pleased God to call them hence so soon one after another . . . They each of them in their way diligently served their generation, and were friends to us. They have left no blot upon their memories (unless their not having died rich may go for one) and I hope they have carried consciences of uprightness with them, and have made their great change to their everlasting advantage . . . "†

The laboratory facilities at Pall Mall enabled Boyle to engage in much experimental work with the aid of the assistants he was fortunate to have either as collaborators or as paid servants. A pneumatical engine was set up, and over the years numerous experiments were carried out with it, much of the work being done with the help of Denis Papin‡. There was a long series of experiments involving the effect of diminished, or increased, pressure on shining flesh, all kinds of vegetables, flowers, fruits, small animals and insects.§ Other helpers in this last period of Robert Boyle's life include Daniel Bilger, Hugh Greg, Ambrose Godfrey Hanckwitz, Frederick Slare, Christopher White, to name only the better known. It is of interest to note that Hanckwitz, Slare and Papin became Fellows of the Royal Society; White became operator of the *Officina Chimica* of the Ashmolean Museum, Oxford.

The activities of the Royal Society, devoted as they were to the advancement of the New Philosophy, excited some ridicule and much opposition in certain quarters. For example, the University of

---

* B. P. **XL.** The two documents in question have been printed by A. R. Hall and M. B. Hall, *N. & R.* (1963), **18,** 99: Birch, *History.* **III,** p. 343, 352.

† *Works Fol.*, V, p. 563; *Works*, **VI,** p. 531. Letter of Lady Ranelagh to RB dated 11 September [1677].

‡ (1647–1712), F.R.S.: Huygens. **VIII.** No. 2174, p. 172. Letter of D. Papin to C. Huygens dated 25 May 1679: References in Hooke's *Diary* show that Papin was often with RB.

§ *Works Fol.*, **III,** p. 305; **IV,** pp. 115, 133, 140, 144, etc.; *Works.* **III,** p. 653; **IV,** pp. 532, 556, 567, 572, etc.

Cambridge would not license the publication of some verses by Peter du Moulin in praise of the Royal Society together with other of his poems*. There was also antagonism on the part of the College of Physicians, which resented the Royal Society's incursion into the affairs of medicine and anatomy. *The History of the Royal Society* by Thomas Sprat, published in 1667, and *Plus Ultra* by Joseph Glanville, published the year after, were designed to combat this hostility; and the latter work has a chapter devoted to " what hath been done by the illustrious Mr. Boyle for the promotion of useful knowledge ", which asserts that he had " done enough to oblige all mankind, and to erect an eternal monument to his memory ", and that his writings contain the " greatest strength and the genteelest smoothness, the most generous knowledge and the sweetest modesty, the noblest discoveries and the sincerest relations, the greatest self-denial and the greatest love of men, the profoundest insight into philosophy and nature, and the most devout, affectionate sense of God and of religion."†

These two works provoked Baldwin Hamey‡, a prominent member of the College of Physicians, who disapproved of a

> "Rival Society treading close upon the heels of the Aesculapians, whose vortex would be so great as to comprehend everything . . . [so he] therefore found out a person of his own profession but a country Practiser, one Dr. Henry Stubbs, a Man of as much Acrimony as Wit with as knowing an head, as he had an able hand, and that wanted no ill nature to compleat the Satyrist in him, and this Man he generously retaynd for his Champion against the Royal Society: Stubbs then drew his Pen with great virulence and layd about him most furiously indeed, and was well gratified by Dr. Hamey for it . . . "§

The attack launched by Stubbe developed into a pamphlet war which lasted for over a year.||

" Amidst these outrages against that Society ", (as Birch has it), Stubbe maintained a correspondence with Robert Boyle, who seems to have been very tolerant of Stubbe's trenchant remarks on various matters as well as his candour and freedom in criticising some of

* *Works Fol.*, V, p. 594; *Works*, VI, p. 579. Letter of P. du Moulin to RB dated 28 December 1669.

† p. 92.

‡ The younger, (1600–1676), M.D.

§ R. Palmer. " Life of the most Eminent Dr. Baldwin Hamey." 1733. MS at Royal College of Physicians.

|| A list of the pamphlets will be found in Jackson I. Cope. *Joseph Glanvill, Anglican Apologist*. St. Louis, 1956, p. 27.

Boyle's own work. Stubbe's pleaded with Boyle to " abandon that constitution ", the Royal Society, saying, " It is too much, that you contributed to its advancement and repute; the only reparation you can make for that fatal error, is to desert it betimes."* Boyle did nothing of the kind. The Royal Society survived all these attacks. Stubbe's pen became engaged in other polemics some few years later; and, after his unfortunate death by drowning, his funeral sermon was preached by his former literary antagonist, Joseph Glanvill.

It would be tedious to list all the matters dealt with by Robert Boyle in connection with the Royal Society. It will suffice to say that he was invariably solicited to provide entertainment for distinguished visitors, such as the Venetian ambassador, Piero Mocenigo, (1 July 1669), and two Florentine nobles, (9 March 1671). His own demonstrations before the Society covered a wide range, including such as related to the water hammer; unannealed glass; swing of the pendulum *in vacuo*; freezing experiments; electrostatic experiments with a ball of sulphur; production of light on mixing two liquors; his hydrostatical device for detecting counterfeit guineas; his method of embedding insects in colophony so as to resemble amber. At the request of the Society he reported on such matters as oil and water sent from Palestine; the Bolognian stone; *resina nigra*; Leibniz's book *Hypothesis physica nova* and Newton's *Theory of Colours*; and much else besides.

At the Royal Society's anniversary meeting, 30 November 1680, Robert Boyle was elected President of the Society. In declining this mark of distinction he sent the following letter to his respected friend Robert Hooke.

> " Though since I last saw you, I met with a lawyer, who has been a member of several parliaments, and found him of the same opinion with my council in reference to the obligation to take the test and oaths you and I discoursed of; yet not content with this, and hearing, that an acquaintance of mine was come to town, whose eminent skill in the law had made him a judge, if he himself had not declined to be one, I desired his advice, (which because he would not send me till he had perused the Society's charter, I received not till late last night,) and by it found, that he concurred in opinion with the two lawyers already mentioned, and would not have me venture upon the supposition of my being unconcerned in an act of parliament, to whose breach such heavy penalties are annexed. His reasons I have not now time to tell

* *Works Fol.*, **I**, *Life*, p. 59; *Works*, **I**, p. xcv. Letter of H. Stubbe to RB dated 4 June 1670.

you, but they are of such weight with me, who have a great (and perhaps peculiar) tenderness in point of oaths, that I must humbly beg the Royal Society to proceed to a new election, and do so easy a thing, as among so many worthy persons, that compose that illustrious company, to choose a president, that may be better qualified than I for so weighty employment. You will oblige me also to assure them, that though I cannot now receive the great honour they were pleased to design me, yet I have as much sense of it, as if I actually enjoyed all the advantages belonging to it. And accordingly though I must not serve them in the honourable capacity they were pleased to think of for me, yet I hope, that, God assisting, I shall not be an useless Member of that learned Body, but shall manifest in that capacity both my zeal for their work, and my sense of their favours. This you will oblige me to represent in such a way, as may persuade the Virtuosi, that you will discourse with, how concerned I am to retain the favourable opinion of persons, who shall reckon your good offices on so important an occasion among the welcomest favours you can ever do."*

At first sight, Boyle's refusal to become President of the Royal Society seems merely to imply unwillingness to take the presidential oath as laid down in the Society's Second Charter of 1663; but it is difficult to see why, on the score of that oath alone, he should have gone to the trouble of getting three legal opinions, especially as he had, at an earlier date, been sworn a member of Council of the Royal Society in the same form†. In the above letter to Hooke he refers to his "great (and perhaps peculiar) tenderness in point of oaths", which expression in a seventeenth century context often suggests someone of the dissenting persuasion. In the same letter he refers to the "obligation to take the test and the oaths", which obviously relates to the Test Acts of 1673 and 1678. It would seem that Robert Boyle's legal advisers were probably of the opinion that the Presidency of the Royal Society could be regarded as "a place of trust from or under his Majesty . . .", and that acceptance of that office (though it is not specifically mentioned) would necessitate his complying with the requirements of those Acts in respect of taking the Oaths of Supremacy and Allegiance, as well as providing a certificate of having received the "sacrament of the Lord's Supper according to the usage of the Church of England".

---

* *Works Fol.*, I, *Life*, p. 74; *Works*, I, p. cxix. Letter of RB to R. Hooke dated 18 December 1680.

† Birch. *History*, III, 114. 4 December 1673.

If Robert Boyle approved of, and conformed to, the Church of England, as his contemporaries assure us he did, why should he have had a tenderness of conscience about taking the oaths required by the Test Acts? He had aleady declared his loyalty to the king at the Restoration in connection with the Declaration of Breda;* but the taking of oaths was a matter that greatly concerned him. It may be noted that John Gauden, Bishop of Exeter, wrote a dedicatory epistle to Robert Boyle in his "*A Discourse Concerning Publick Oaths, and The Lawfulness of Swearing in Judicial Proceedings* (1662)," which was an answer to the scruples of Quakers. Boyle was very tolerant in his religious outlook;† he gave financial assistance to necessitous non-conformist divines and to many protestant refugee ministers from the continent; and he was on good terms with dissenters, such as Richard Baxter, "the fittest man of the age for a casuist, because he feared no man's displeasure, nor hoped for any man's preferment."‡ One wonders if Boyle were a dissenter at heart, and fearful of incurring the harsh penalties of the Test Acts should he reveal his private beliefs by refusing to "take the test". However, Gilbert Burnet says "He was constant to the Church; and went to no separated Assemblies, how charitably soever he might think of their Persons, and how plentifully soever he might have reli[e]ved their Necessities".§

It may be recalled that in the early days of the Royal Society, before the First Charter had been granted, Robert Boyle had presided, or as we should say acted as chairman, at some of the Society's meetings, and in this sense had been "President".

During this last period of Boyle's life there was no falling off in the number of books dedicated to him, either out of genuine regard, or from a desire on the part of their authors to benefit from his patronage‖. Nor was there any decrease in the number of his visitors. Some came to make his acquaintance on account of his high reputation, or because of common interest, (*e.g.*, Daniel Georg Morhof, Martin Kempe, Gottfried Wilhelm von Leibniz, Holger Jacobaeus and many others, including a Chinaman, a convert brought to this country by le Père

* 5 June 1660. See p. 98.

† See the letter of Thomas Dent to William Wotton dated 20 May 1699. *Works Fol.*, **I**, pp. 89 and 90; *Works*, **I**, pp. cxli and cxlii. Also G. Burnet. *A Sermon Preached at the Funeral of the Honourable Robert Boyle*. London, 1692.

‡ Samuel Palmer. *The Nonconformist's Memorial*. London, 1775. Vol. **II**, p. 173.

§ G. Burnet, *loc. cit.*, p. 29.

‖ Fulton. p. 162 *et seq.*

Couplet, and recommended to Boyle by Thomas Hyde).* Many came to offer information, much of which is recorded and acknowledged by Boyle in his numerous writings; others sought his advice about their illnesses;† his help was often requested by refugees, especially French ministers of religion, victims of the revocation of the Edict of Nantes.‡ Not unexpectedly, we find that attempts were made to sell him some secret process,§ or to interest him in some commercial venture;|| there were letters begging the loan of money;¶ and attentions from the occasional crank.**

His disbursements for charitable and other purposes continued. He contributed towards the expense of publishing *The History of the Reformation of the Church of England* by Gilbert Burnet††; he contributed to John Ray's *De Historia Piscium Libri Quatuor* the plate for the Plaice and Dab;‡‡ and he contributed to Robert Morison's *Historia Plantarum Oxoniensis* a plate on which he is described as *rerum naturalium indagatorum Coryphaei*§§. He paid for the translation and printing of *The Four Gospels . . . and the Acts of the Holy Apostles*;|||| he helped William Seaman to produce the *Grammatica Linguae Turcicae*¶¶; and was at the expense of paying for the translation and printing of the *New Testament* in Irish, which necessitated production of a fresh font of type.*** The set of Small Pica Irish characters cast by Joseph Moxon for that work remained the sole font of Irish type in this country until the beginning of the nineteenth

* R. E. W. Maddison. "Studies in the Life of Robert Boyle, F.R.S. Part IV. Robert Boyle and Some of his Foreign Visitors." *N. & R.* (1954), **11**, 38: B.P. **XXI**. Item 69, 26 July 1687.

† In particular RB was regarded by his contemporaries (*e.g.*, Samuel Pepys) as an authority on the eye; even Dr. Daubeney Turberville, when in doubt, would send a patient to him for examination. R. A. Hunter and F. C. Rose. "Robert Boyle's 'Uncommon Observations about Vitiated Sight', (London, 1688)". *British Journal of Ophthalmology* (1958), **42**, 726.

‡ Letter of Renaudot requesting assistance in settling a dispute with ecclesiastical authorities; undated: B.L., **V**, 38.

§ G. à Sonnenberg offered to sell the secret of the philosopher's stone: B. L. **V**, 103, 104.

|| H. Usher offered an interest in a mining proposition: B. L. **VII**, 36.

¶ Le Comte d'Ostanne wanted to borrow ten guineas: B. L. **IV**, 91.

** Doggerel from Anselm: B. L. **I**, 7.

†† London, 1679.

‡‡ London, 1686. C. E. Raven, *John Ray, Naturalist*, Cambridge, 1950. p. 357: G. Keynes, *John Ray, A Bibliography*, London, 1951, p. 70.

§§ Oxford, 1680. Madan, No 3271.

|||| Oxford, 1677. Madan, No: 3164.

¶¶ Oxford, 1670. Madan, No: 2863. RB had subsidized Seaman's translation of the *New Testament* into Turkish at an earlier date; *ibid*, No: 2727.

*** R. E. W. Maddison. "Robert Boyle and the Irish Bible." *Bulletin of the John Rylands Library*, (1958), **41**, 81.

century; some of the original punches of Moxon's Irish type still exist.

Considerable sums of money were devoted by Robert Boyle every year to the relief of distress. " He took care to do this so secretly, that even those who knew all his other Concerns, could never find out what he did that way; . . . no body ever knew how that great share of his Estate, which went away invisibly, was distributed; even he himself kept no Account of it . . . "* On one occasion, we know that he placed £250 in the hands of Sir Robert Southwell for the relief of distress among preachers in Ireland;† and on another, £300 to relieve necessitous converted Indians.‡ George Vertue§ in his notes for a history of art in England records Boyle's charity towards Philip Duval‖: " the great Mr. Boyle show'd so much affection to this painter (who was not so successfull in the rewards of his Art finding) he had a good deal of knowledge in Experimental chimistry. which practice both wasted his substance & Time. therefore towards his support as long as Mr. Boyl livd he allowed him 50 pounds yearly."¶

The absence of comment on contemporary events is still noticeable in Bolye's correspondence. There is no mention of the great fire which devastated the Middle Temple, London, on the night of 26 January 167$\frac{8}{9}$;** nor of the murder of Thomas Thynne, which took place in Pall Mall, 12 February 168$\frac{1}{2}$; nor is there comment on Titus Oates and the Popish plot. Matters of domestic interest can be gleaned only from sources such as the diaries of his sister Mary, Countess of Warwick, and of Robert Hooke, or from the *Orrery Papers* and an occasional letter. From these sources we learn of Boyle's meetings with other Fellows of the Royal Society, his visits to Leese, and his instructions regarding management of his Irish estates. The *London Gazette* has an advertisement offering a reward for information on a boy who absconded from the house in Pall Mall taking 28 guineas, as well as some gold and silver.††

---

* G. Burnet. *A Sermon Preached at the Funeral of the Honourable Robert Boyle*, London, 1692, p. 32.

† Royal Commission on Historical Manuscripts. *Egmont*, **II**, p. 189.

‡ W. Kellaway. *The New England Company*, 1649–1776. London, 1961, p. 58.

§ (1684–1756). Engraver and antiquary. *D.N.B.*

‖ (?–1709?). French painter who settled in England.

¶ B.M., Add. MSS. 23069, fol. 32^{r}.

** *The Diary of William Lawrence*, Edited by G. E. Aylmer. Beaminster, 1961, p. 41: *Elias Ashmole*, Edited by C. H. Josten. 5 vols, Oxford, 1966. Vol. **IV**, p. 1635.

†† No: 2258 for 7/11 July 1687.

Visit of Cosmo III to England

The Hereditary Prince of Tuscany, who became Cosmo III,* Grand Duke of Tuscany, visited England in 1669, where he arrived at Plymouth, 22 March,† and thence progressed to London. Under date 5 April 1669 Samuel Pepys wrote in his diary, " . . . took coach again, and went five or six miles towards Branford [Brentford], the Prince of Tuscany, who comes into England only to spend money and see our country, comes into the town to-day, and is much expected; and we met him, but the coach passing by apace, we could not see much of him, but he seems a very jolly and good comely man." The list of over 145 persons who went to greet Cosmo at Brentford does not include the name of Pepys, but we do find in it the names of " M:r Oldenburgh " and " M:r Boyle brother to the Earls of Orrery and Cork."‡

On arriving in London, Cosmo was lodged at the house of the Earl of St. Albans in St. James's Square, and subsequent days were passed in receiving and visiting persons of distinction. Ten days after his arrival he paid a visit to the Royal Society. During the month of May he visited Cambridge, Oxford, and other places nearby. He was back again in London on 6 May, and on the morning of 8 May 1669 Cosmo

> " went in his carriage to the house of Mr. Robert Boyle, whose works have procured him the reputation of being one of the brightest geniuses in England. This gentleman not only reduced to practice his observations on natural philosophy, in the clearest and most methodical manner, rejecting the assistance of scholastic disputations and controversies, and satisfying the curiosity with physical experiments, but, prompted by his natural goodness, and his anxiety to communicate to nations the most remote and idolatrous the information necessary to the knowledge of God, caused translations of the Bible into the Oriental languages to be printed and circulated, in order to make them acquainted with the Scriptures; and has endeavoured still further to lead the most rude and vicious to moral perfection, by various works, which he has himself composed. Indeed, if in his person the true belief had been united with the correctness of a moral life, nothing would have remained to be desired; but this

---

* Cosmo de' Medici (1642–1723), son of Ferdinand II.

† *Travels of Cosmo III, Grand Duke of Tuscany, through England, during the Reign of King Charles the Second (1669) Translated from the Italian Manuscript in the Laurentian Library at Florence. To which is prefixed, A Memoir of his Life*. London, 1821.

Wood, *Life*, Vol. **II**, p. 156–162.

R. E. W. Maddison. " Studies in the Life of Robert Boyle, F.R.S. Part I. Robert Boyle and some of his Foreign Visitors." *N. & R.* (1951), **9**, 26–29.

‡ A. M. Crinò. *Fatti e Figure*. Firenze, 1957. p. 209–211.

philosopher, having been born and brought up in heresy, is necessarily ignorant of the principles of the true religion, knowing the Roman Catholic church only by the controversial books of the Anglican sect, of which he is a most strenuous defender, and a most constant follower; his blindness, therefore, on this subject, is no way compatible with his great erudition. He shewed his highness, with an ingenious pneumatic instrument invented by himself, and brought to perfection by Christiaan Huygens of Zuylichem, many beautiful experiments to discover the effect of the rarefaction and compression of the air upon bodies, by observing what took place with animals when exposed to it: and hence may be learned the cause of rheumatism, catarrhs, and other contagious disorders produced by air, and of various natural indispositions. It was curious to see an experiment on the change of colours: two clear waters, on being poured into one another, becoming red, and by the addition of another red, becoming clear again; and the experiment of an animal shut up in a vacuum, and the whole exposed to the pressure of the air. There was an instrument which shews of itself the changes of the air which take place in twenty-four hours, of wind, rain, cold, and heat, by means of a watch, a thermometer, a mariner's compass, and a small sail like that of a windmill, which sets an hand in motion, that makes marks with a pencil as it goes round; there was also another instrument of a most curious construction, by means of which a person who has never learned may draw any object whatever. He shewed also to his highness, amongst other curiosities, certain lenses of a single glass, worked facet-wise, which multiplied objects; a globe of the moon of a peculiar construction, and several other things worthy of attention. Having gratified his curiosity in the most agreeable manner, and it being now near noon, his highness returned home."*

"To satisfy his curiosity, his highness went again, on the morning of 15 May, to the house of Mr. Robert Boyle, who had prepared other experiments in natural philosophy for his amusement. He there found Mr. Henry Oldenburg, secretary of the Royal Society. At dinner-time, he returned home."†

Cosmo was accompanied by a large suite of attendants, amongst whom was his physician, Giovanni Battista Gornia, It was, no doubt the latter who gave Boyle a receipt for Apoplectic Balsam.‡ Lorenzo

* *Travels of Cosmo the Third* . . . London, 1821, p. 291–293.
† *Idem*, p. 317.
‡ B.P. **XXXVI**. Index of Experiments No.: 1748.

Magalotti was also in attendance, and we learn from a letter which he wrote from Paris in the autumn of 1669 to the mathematician Vincenzio Viviani, that he himself was taken ill of the fever during the early days of May in London.* Boyle was so anxious to demonstrate his friendship for Magalotti, that he visited his bedside for two or three hours daily.† Both of them seem to have developed the same distrust of medical practitioners and their physic. On his return to Italy Magalotti felt constrained to write Boyle a long letter *Sopra controversie religiose*, confuting the errors, which he, as a Roman Catholic, considered Boyle to hold. Magalotti is said to have translated most of Boyle's *Seraphick Love* into Tuscan, but it appears not to have been published.

About the end of June 1670 Robert Boyle had a stroke, the seriousness of which is revealed by his sister, Mary, through her diary.

> 13 *July* 1670 . . . " before my going to bed had a letter from Mr Cox yt repeated to me ye sad news I had ye week before heard of ye weak languishing condition my dear Brother Robin was in, & how much danger he found him to be in; I was much troubled at ye news, & went to God for him before going to Bed."‡

On the same day his eldest brother, Richard, wrote in his diary,

> " hearing yt my brother Boyle was very ill J sent up Lts by Graham who went next morning post to London to visit from mee my Brother."§

> 14 *July* 1670. Mary wrote, " I had not so much life in my meditations as usual, being very sad for fear of the loss of my dear brother Robin, who was very sick." ||

> 22 *July* 1670. " This night I had the news that my brother Robin was better."¶

> 3 *August* 1670. " Then took coach to go to London to visit my dear and sick brother Robin . . . I found my dear brother in a very weak and languishing condition, yet had great satisfaction to see him again, which was a mercy I feared I should not have enjoyed."¶

---

* A Fabroni. *Lettere Familiare del Conte Lorenzo Magalotti.* 2 Vols. Firenze, 1769. Vol. I, p. 37.

† *Idem*, Vol. I, p. xxvi.

‡ B.M., Add. MSS. 27,358. " Collections out of Lady Warwick's Papers." fol. 132.

§ *Lismore Manuscripts*, XXIX. " Diary of the 2nd Earl of Cork, 1 September 1669–24 Nov. 1672."

|| *Memoir of Lady Warwick.* London, 1847.

¶ *Idem* and B.M., Add. MSS. 27,358.

> 4 *August* 1670. "Afterwards went to my sick brother, where I spent my whole day and had good discourse."*
>
> 5 *August* 1670. "Then spent all the day with my weak brother; had with him much good discourse."*
>
> 6 *August* 1670. "... when I had spent the morning with my brother, which I did in good discourse, I took coach to return home to Leez, having the satisfaction, by God's goodness to me, of leaving my brother in a more hopeful way of recovery, in the opinion of his doctors, than when I came to town."*

On 22 July 1671 Mary was able to write, "this day my deare Brother Robin came to see me [at Leese], and he haueing bene at deathes doore, and by Godes mercy reacouered, I fond my selfe much pleased to see him well againe, and my heart was carried out much to Blesse God for hearing all the prayares I had put upe for his life in the time of his siknes."†

In a letter to John Mallett, written for the benefit of a kinsman suffering from the like complaint, Robert Boyle gives his own account of the medical treatment he received.

> "... through the goodness of God, my paralytical distemper is much lessened, yet I am far from being fully cured of it; and during the space of an eleven months past since it first invaded me, I have taken so many medicines, and found the relief they afforded me so very slow, that it is not easy for me to tell you what I found most good by. The things, which to me seem the fittest to be mentioned on this occasion, are, that cordial medicines, especially such, as peculiarly befriend the *genus nervosum*, were very frequently and not unusefully administered. That I used during this sickness less purging physic, and that gentle, than in many years before; and found cause to think such evacuations very weakening, and, when they are not very necessary, dangerous. That the dried flesh of vipers seemed to be one of the usefullest cordials I took; but then I persevered in taking it daily for a great while. That I seldom missed a day without taking the air, at least once, and that even when I was at the weakest, and was fain to be carried in men's arms from my chair into the coach. That the best thing I found to strengthen my feet and legs, and which I still use, was sack turned to a brine with sea-salt, and well rubbed upon the parts every morning and night with a warm hand.

---

* *Idem* and B.M., Add. MSS. 27,358.

† B.M., Add. MSS. 27,352. "Diary of Mary Rich, Countess of Warwick. Nov. 1669–Mar 1672."

That for my hands I use several things, and particularly palm oil, which comes from Guinea, and a liquor somewhat like the *spiritus lavendulae compos.*, of the dispensatory, and also fomentations with cephalic flowers and herbs, one or other of which I yet daily continue. But yet I found nothing so available as frequent exercise of my hands and feet, in which I continued as far as my strength would possibly allow me, getting sometimes others to bend the joints of my arms and hands for me. And though this course makes me every day sore and weary, yet I continue to undergo it, because I think I find more benefit by it alone, than by all the outward applications of physicians and chirurgeons. These Sir, are in brief the chief things, that I have observed about my own distemper; but how far they will be applicable to that of your friend, I do not know half so well, as that in case he use them they will prove very effectual . . ."*

Salt Water Freshened†

In 1674 Robert Boyle published a tract entitled *Observations and Experiments about the Saltness of the Sea* in which, amongst other things, he dealt with the preparation of fresh water from sea-water by distillation. He expressed the opinion " that the sea derives its saltness from the salt that is dissolved in it "; and concluded, " we might the more hopefully attempt to obtain by distillation sweet water from sea-water; since, if this liquor be made by the bare dissolution of common salt in the other, it is probable that a separation may be made of them, by such a heat, as will easily raise the aqueous parts of sea-water, without raising the saline, whose distillation requires a vehement heat, as chymists well know to their cost. And such a method of separating fresh water from that which was salt, would make our doctrine of use, and be very beneficial to navigation, and consequently to mankind." The problem of ensuring an adequate supply of good drinking water during lengthy voyages at sea was one that had been of much concern through the ages; indeed, considerable attention is still devoted to this same problem.

* *Works Fol.* **I**, *Life*, p. 62; *Works.* **I**, p. xcix. Letter from RB to John Mallet dated 23 May 1671.

† R. E. W. Maddison. " Studies in the Life of Robert Boyle, F.R.S. Part II. Salt Water Freshened." *N. & R.* (1952), **9**, 196. A full account is given there. An account of water desalination covering a wider period than the above, which is concerned only with RB, is given by G. Nebbia and G. N. Menozzi. " A Short History of Water Desalination." *Acqua Dolce dal Mare.* IIa Inchiesta Internazionale, Milan, 18–19 April 1966: *Idem.* ' Una polemica brevettistica in Inghilterra alla fine del XVII secolo.' *Quaderni di Merceologia*, (1967), **6**, 23.

It had been long known that potable water could be prepared by distillation from sea-water. Balthasar de Monconys during his visit to England in 1663 was taken by Henry Oldenburg to see the engines for distilling water exhibited by Kuffler, the son-in-law of Drebbel, who had invented them*. Boyle was familiar with these Drebbel-Kuffler engines.

In 1675 a certain William Walcot† secured a 'patent' for his invention of *Purifying corrupted water, and making sea-water fresh, clear, and wholesome*‡, but his invention does not appear to have come into general use in succeeding years. In 1683 Robert Fitzgerald§ and partners‖ applied for Letters Patent relating to *Engines for purifying salt and brackish water, making it sweet and fit for drinking and purposes of cooking*, which as far as can be judged had the same scope as Walcot's patent, namely, the distillation of sea-water.¶ Fitzgerald was a nephew of Robert Boyle, and this fortunate relationship was undoubtedly of great advantage to him. In April 1683 the *London Gazette* carried the following announcement:

> "His Majesty having been humbly given to understand by the Honourable *Robert Fitzgerald* Esq. and some other Gentlemen, that they had found out the Art and Mystery of reducing the Salt Water of the Sea into good, perfect and wholesome Fresh Water,

* Cornelis Drebbel (1572–1633). Inventor, who finally settled in England. After his death his son-in-law, Dr. Johannes Sibertus Kuffler, endeavoured to exploit his various inventions.

J. S. Kuffler (1595–1677); studied at Padua. Married Drebbel's daughter, Catharina, in 1623.

† William Walcot (1633–1699); page of honour to Charles I whom he attended on the scaffold; said to have gone up to Oxford in 1650, and to have graduated B.A. in 1653, but not mentioned in *Alumni Oxonienses*; student of Gray's Inn 1654, and also of the Middle Temple 1657, but seems not to have practised at the bar; appears in 1696 to have been in straitened circumstances; died in the Province of Holland, unmarried and intestate.

‡ English Patent No: 184. 28 October 1675.

§ Hon. Robert Fitzgerald (died 1698, aged sixty); nephew of Robert Boyle, being the son of the 16th Earl of Kildare and Joan Boyle; in his youth active in the design to restore Charles II; Cornet in 1659; Lieut. of Horse Guards in 1664; created Doctor of Law of Oxford University 1677; Privy Councillor (Ireland) 1678; Capt. of Horse 1684; Commissioner of forfeited estates. Robert Boyle bequeathed him £30.

‖ Sir Theophilus Oglethorpe (1650–1702); Brigadier General; Principal Equerry to James II; from 1698 till his death M.P. for Haslemere, Surrey. William Bridgeman not identified; perhaps the F.R.S., elected 1679. Patrick Trant; not identified; he obtained a patent in 1677 for soap-making. Thomas Maule (died 1715, aged seventy); possibly Lieut. of the Yeomen of the Guard to Charles II.

¶ Fitzgerald's patent passed the Great Seal on 9 June 1683. Patent Rolls 33 Car II. Pars III. 14. English Patent No: 226.

in quantities sufficient to serve all the uses of any Ship at Sea, and that this Secret was to be performed without mixing any unwholsome Ingredient at all, whereby to endanger the Lives or Healths of any Person that should make use of it; which Proposal His Majesty received with great Grace and Favour, and was pleased to Command the Honourable *Robert Boyle* Esq, to attend Him within some few days after, who upon due Consideration so fully satisfied His Majesty of the Wholesomeness and Usefulness of the said Water, that His Majesty having received full Satisfaction therein, has been Graciously pleased for the publick Good (and to encourage so useful an Undertaking) to give the said Mr. *Fitzgerald* and his Partners, a Grant of the same; who do intend from time to time by themselves, or Persons to be appointed by them, to meet at times and places hereafter by them to be named, to receive such Proposals as may be reasonable from any Persons, who intend to have the Benefit and Use of their said Invention."*

About this time Fitzgerald published a pamphlet entitled *Salt-Water Sweetned.*† The dedication to Charles II commences:

"When Mr. Boyle, with my Partners, and my self attended your Majesty with the Experiment of Reducing *Salt Water* into *Fresh*; your Majesty seem'd so well pleas'd with an Invention of that Universal Benefit, . . . your Favour and Patronage was not fully obtain'd, till with the strictest Scrutiny you had first examin'd those Hopes and Probabilities, you vouchsaf'd to Encourage in Us . . . When we had obviated all Difficulties, and salv'd all Objections against this Undertaking . . . your Majesty was Graciously pleas'd to give us your Approbation, and to Order Us your *Letters Patents*."

The existence of two patents for a similar invention and granted to different groups of adventurers naturally caused trouble. Walcot protested at the grant of a patent to Fitzgerald, and took his case to the Privy Council, where his patent was declared void.‡

The dedication in Fitzgerald's pamphlet continues by saying:

"And if the Fruits of your Royal Grant has not hitherto been deriv'd to Us, 'tis partly by some Obstacles we met from the

---

* *London Gazette*, No: 1821. 30 April to 3 May 1683; *Ibid*, No: 1822. 3 to 7 May 1683.

† Robert Fitzgerald. *Salt-Water Sweetned; Or, a True Account of the Great Advantages Of this New Invention both by Sea and by Land: together with a Full and Satisfactory Answer to all Apparent Difficulties. Also the Approbation of the Colledge of Physicians. Likewise A letter of the Honourable Robert Boyle to a Friend upon the same Subject.* Sm. 4to. London, 1683.

‡ Walcot's case will not be pursued further here. See R. E. W. Maddison, *N. & R.* (1952), 9, 196.

suggestions of a private Person, but more especially by the Late *Horrid Conspiracy* . . .'

"This Experiment is in a great degree owing to the Eminent Mr. *Boyle*, and indeed well worthy so Ingenious a Promoter, being so much the more the Favourite of his happy *Genius*, as it is Universally useful to Mankind; But what ever Advantages this Country, or indeed the World may receive by it, his whole ambition is bounded in the publick profit, contenting himself with no other Benefit from it, than the satisfaction and pleasure of seeing it accomplished by his Friends . . . ."

The presentation of a copy of this pamphlet to the king was referred to in the *London Gazette:**

"Whitehall, Octob. 12. There was lately presented to His Majesty a small Treatise, entituled, *Salt Water Sweetned* (which is since Translated into several Languages) being an account of an Excellent Invention of rendring *Sea Water* Fresh and Potable in large quantities, with small Charge. In the said Treatise there are full Answers to all the Objections that have been made against it; with an Approbation from the most Eminent Physicians of the Colledge, and other Learned Persons, of the Wholsomeness of the said Water; together with a Letter from the Honorable Mr. *R. Boyle*, upon this Subject. All Merchants, Owners, and Masters of Ships, or others who suffer in their Estates or Dwellings for want of fresh Water, and are desirous to partake of the benefit of so useful an Invention, may repair to the *Carolina* Coffee-house in *Birchin Lane*, from Eleven a Clock to One; where they will find some of the Undertakers, or their Agents, to make them reasonable Proposals; who do publish this at the desire of several Merchants and others, as well at home as abroad."

The letter from Mr. Boyle mentioned in the above account was written to Dr. John Beale; it says:†

"To give you a short account (suitable to the little time I have to do it in) of the transaction, which I suppose must have given the rise to the mention made of my name in the publick gazette, I must inform you, that one of my nearest relatives (Captain *Fitzgerald*) and some other worthy gentlemen, having acquainted his majesty, that they had an invention for making sea water sweet and wholsome, in great quantity and with small charge, and that I had examined, and did approve the water so prepared; his

* *London Gazette*, No: 1868. 11-15 October 1683; and No: 1871. 22-25 October 1683.

† *Works Fol.* IV, p. 159; *Works*, IV, p. 593.

> majesty was pleased, with very gracious expressions, to command me to attend him with a further and more particular information. Having readily obeyed this order, and been made acquainted with the objections the king thought fit to make against the practicableness of the invention, which (though a private man had urged them) I should think the most judicious, that have been framed against it; I humbly represented to him, that I looked upon this invention as comprising two different things; a *mechanical* part, which related to the engine itself and the use of it a ship-board, and a *physical* part, which concerns the potableness, and wholesomeness of the liquor. About the former of these I did not pretend to clear the difficulties, especially such strong ones as his majesty had proposed, but left it to the patentees to give him satisfaction, which they were in a readiness to offer: but as to the wholsomeness of the prepared water, I had made some trials upon that liquor, which gave me no just grounds of suspecting it to be unwholsome, but several motives to believe it well conditioned, and of great use to navigators, and not to them only. And having hereupon briefly acquainted his majesty with the chief trials I had made to examine this sweetned water, he was pleased to look upon them as satisfactory; and vouchsafed, on that occasion, to discourse as a virtuoso of the sea, and brackish waters, and gave me some new, as well as instructive observations about them; and, in conclusion, dismissed the patentees with a gracious promise of his royal protection and peculiar favour."

He then gives the reasons that persuaded him " of the salubrity of this water ". He considered,

> " that almost all the rain water, that falls from the clouds on the main ocean, and which . . . is unquestionably received as wholsome, must be afforded by the sea, and consequently be but sea water, freed from its salt ". Next he found, " as his majesty himself had done, that the liquor was well tasted, and without any sensible brackishness; and so some of it continued for between four and five months in a large chrystal bottle . . . purposely kept unstopped, and for the most part in a south window, where it neither did, nor probably in a long time will, putrify, nor so much as appear troubled, or less transparent".

He found that it lavered very well: and that " this water will boil pease tender, which, amongst seamen, is one of the principal signs of good water ". Furthermore, he found the gravity to be only very slightly less than that of good potable natural waters. Lastly, he showed the ' patentees ' water to be more free from common salt, than waters, that are usually drank here in *London* ". Robert Boyle goes on to say,

> " and that, which more satisfied me myself, was a trial, that I carefully made by a way, which having mentioned, but not yet, for want of opportunity, disclosed to his majesty ". " By this way of trial, I found . . . that if there were in water so much as one grain of salt in above two ounces of water, I could readily discover it; and . . . by this critical examen, I could not detect so much as a thousandth part of salt in our prepared water."

He concludes his letter,

> " Thus, Sir, you have a short and artless account, such as my haste will permit, and the nature of the subject requires, of my part in promoting this profitable invention, to which I own myself a great well-wisher, not out of any private interest (though that were obligingly proffered me by the patentees) but as I think the bringing it into general use may prove a real service to mankind."

In the *London Gazette* during November 1683 the following announcement appeared:

> " Whitehall, Novemb. 2. His Majesty was pleas'd to Command the Honourable Mr. *Boyle*, to attend Him to give His Majesty an Ocular Proof of the Nicety of his way of examining the Freshness and Saltness of Water, and to apply it to the Sea Water, prepar'd according to the patentees Invention; which being done before His Majesty, his Royal Highness,* and the Duke of *Grafton*,† several Persons of Quality being also present, it was made apparent by a certain prepar'd Liquid which Mr. *Boyle* had brought with him, that a Discovery could be made if there were so much as a thousandth part of Salt in a propos'd Water; by which Trial His Majesty finding that the Prepar'd Sea Water, for which He has granted his Royal Patent, was at least as free from Salt as the best Waters used in this Town; received such satisfaction as to the wholsomeness of the said Water, that He was pleased to declare His Royal Intentions both to encourage the said Invention, and to have the said Water made use of in His several Maritime Garisons, which Nature has not furnished with wholsome Water."‡

---

* James, Duke of York (1633–1701), F.R.S. Afterwards King James II.

† Henry Fitzroy, 1st Duke of Grafton (1663–1690). Natural son of Charles II.

‡ *London Gazette*. No: 1874. 1 to 5 November 1683; *ibid*, No: 1876. 8 to 12 November 1683.

Robert Boyle's own account of this occasion and his method of examining waters did not appear till after his death. He writes:*

"Having since the publication of the foregoing letter, been commanded by the king, to show his majesty an experiment of the way herein mentioned, to examine the freshness and saltness of waters: I did in his presence (and that of his royal highness, his grace the duke of *Grafton*, and several persons of quality) make trial of it, both upon some water prepared according to the patentees way, and upon two or three natural waters, that were ordered to be brought. In all which trials (in some whereof his majesty, for greater certainty, was pleased to employ his own hands) the success was such, as moved him to vouchsafe the experiment the honour of his special approbation, and to give me an encouraging permission to communicate it, as a thing that may prove not unuseful to the publick. This I think fit to mention, not only to procure to my way of trying waters, the high advantage of a royal, and (on philosophical accounts) illustrious patronage; but that, if this method be found as beneficial as I wish it may, men may know, to whom they ought to acknowledge the early publication of it." . . . "It was no very difficult matter for me to think, that by a heedful application of the precipitating quality of common salt, one might discover, whether any particles of it (at least in a number any way considerable) lay concealed in a distilled water, or any other proposed to be examined." . . .

"I took some common distilled water in glass vessels . . . and put into a thousand grains of it one grain of dry common salt: into a convenient quantity, for example, two or three spoonfuls of this thus impregnated liquor, I let fall a fit proportion, for instance, four or five drops, of a very strong and well filtered solution of well refined silver, dissolved in clean aqua fortis; . . . upon which there immediately appeared a whitish cloud, which though but slowly, descended to the bottom, and settled there in a white precipitate." . . .

"I observed, that having let fall a few drops of our metalline solution into the liquor obtained from the sea-water by the patentees way of sweetning it, there did not presently ensue any white cloud or precipitate, much less such an one as had been newly afforded by the water, that was impregnated with less than a thousandth part of salt. And if after some time there happened to appear, (for it is not absolutely necessary it should) a little cloudiness in this fictitious liquor, it was both slowlier produced,

* *Works Fol.*, V, p. 199: *Works*, V, p. 744.

and much less than that, which appeared in the impregnated water."

The paper containing the above account was deposited with the Royal Society on 30 November 1683 " to remain sealed in the custody of the secretary."* The paper was opened after Boyle's death and read before the Society on 17 February 169½, when a demonstration made by Hans Sloane justified the claims of the author. The information it gives about the use of a silver nitrate solution to distinguish between good and bad drinking waters was considered at that time to be a very valuable secret. It is difficult to see why so much secrecy should have been attached to this, because Robert Boyle had in 1663 published the fact that a solution of silver in aqua fortis could be used as a qualitative test for the presence of common salt or spirit of salt in water.† That he did set great store by the test is proved by his own words quoted above. The remarks of Stephen Hales‡ on this point, too, are of interest. He says:§ " The solution of Silver in *Aquafortis* was at that time kept a great Secret, as to its Property of discovering the least quantity of Salt or Spirit of Salt in Water. Had either Mr. *Walcot* or Mr. *Fitz-gerald* had the free use of it, and known how to have used it, they might then probably have made a greater progress in what they were in pursuit of; but for want of it, they could not well distinguish, when the Distilled Water was free from Spirit of Salt, and when not, and therefore, failed in their Attempt." Hales ascribed his own success in preparing potable water from sea-water to the fact that this method of test was available to him.

Fitzgerald's pamphlet went through numerous editions, besides being translated into various languages. Many testimonials extolling the quality of the water produced by the process were obtained and widely advertised.

From these publications we learn that an apparatus costing £18, and having a diameter of 33 inches was capable of producing 90 gallons of water in 24 hours at a cost of less than a farthing per gallon, all charges considered. The following advantages were claimed for the apparatus. No chemist was required to operate it; the likelihood of failure was remote; it would function at sea except during the severest weather; there was no danger of fire, or inconvenience from smoke; the fuel consumption was small; and the cost of the ' ingredients ' was low. This mention of ' ingredients ' suggests that some substance may have

* Birch. *History*.
† *Works Fol.*, I, p. 496; *Works*, II, p. 112 113.
‡ (1677–1761), D.D., F.R.S. Physiologist and inventor.
§ S. Hales. *Philosophical Experiments*. London, 1739. See Preface.

been added to the sea-water prior to distillation. On this point Stephen Hales, who in the next century occupied himself with this very same problem, says:*

> "Mr. Boyle certified that the few Ingredients made use of by Mr. Fitz-gerald, are fixed in the Fire, and give no noxious Quality to the Water." . . . "Sir *Hans Sloane*, who saw the Cement which was used in Mr. *Fitz-gerald's* Method, tells me that it looked like common Brick Clay. But whatever it was, there was so small a quantity of it used; that what was sufficient for producing Sixty Tuns of good Water, might be contained in two Bushels." . . . "Which makes me very much suspect, that these Cements, as they were called, were only made use of as a Pretext." Also . . . "*General Oglethorpe* informed me, that his Father, Sir *Theophilus*, told him, that Lime was one of the Ingredients, of what he and the rest of the Patentees, in *Charles* the Second's time, called the Cement, with which they made distilled Sea-water wholsome."†

The patentees were very active during 1684 in their endeavours to establish the commercial success of their invention, and full use was made of the king's interest, as well as Robert Boyle's supporting testimony.

The importance attached officially to this invention may perhaps be gauged from the following most interesting announcement which appeared in the *London Gazette* during January 1685.‡

> "*London, January* 23. Yesterday His Majesty was pleased to send to the Lord Mayor closed in a Silver Box sealed up with His Majesty's Seal, the Receipts of the several Cements used by the Patentees for making Sea Water fresh; As also the Receipt of their Metalline Composition and Ingredients Certified under the Hand of the Honourable *Mr. Robert Boyle*, to be kept so sealed up by the present and succeeding Lord Mayors, lest a Secret of so great importance to the Publick might come to be lost, if lodged only in the knowledge of a few Persons, therein concerned."

It is very regrettable that this box with its contents is not now traceable.

In 1686 Edmund Arwaker§ published a "voluminous Song or Panegerick in Verse" entitled *Fons Perennis. A Poem On the Excellent and Useful Invention of Making Sea-Water Fresh*, which contains

* S. Hales. *Philosophical Experiments*. London, 1739. See Preface.

† *Idem. An Account of a Useful Discovery to Distill double the usual Quantity of Sea-Water, by blowing Showers of Air up through the Distilling Liquor*. Second edition. London, 1756. p. 26.

‡ *London Gazette*. No: 2002. 22 to 26 January 168$\frac{4}{5}$.

§ See page 242.

tributes to Robert Boyle and to King Charles II. In addition to being commemorated in verse, the invention was commemorated by " Medals engraven and cast in silver, Representing and Illustrating " the art. John Evelyn makes passing reference to " Dulcifying Sea-Water " as being " worthy the Memory and Honor of Medals ", but it would seem that he was unaware of their existence.*

In spite of all his influential advantages Fitzgerald's invention was not a commercial success. It was alleged that in eleven and a half years the Fitzgerald group had installed three ' engines ' only, viz., one at the garrison at Hull, which did not work satisfactorily; and one each in Jersey and Guernsey. According to Stephen Hales " . . . the Persons concerned with Mr. *Fitz-gerald*, finding themselves extreamly disappointed in their expectation, withdrew from any Partnership with him; Insomuch that his Instruments, which were dear enough before their Effect was known, were soon after sold for old Goods, for want of a vent for them at Sea ". In explanation of the commercial failure of both Fitzgerald's and Walcot's methods Hales says:† " I suspect the true Reason why both their Methods of preparing fresh Sea-Water were disused [was], *viz*, because when they had been used for a considerable time, they were found to disagree with those who drank them ". Hales also suspected that in some instances spirit of salt was present in the distilled water. It is most likely that the quality of the distilled water was variable, being sometimes good and at others bad, according to the nature of the sea-water that was employed. Both methods failed because they were not technically good enough to treat sea-water, no matter what its origin or condition. Undoubtedly both were put forward in all good faith, and as far as Robert Boyle is concerned, " it is not to be suspected that so worthy and good a Man as Mr. *Boyle* was, would impose a Falshood on the World, for the sake of any one whatsoever ".

That Robert Boyle did really consider himself to be the inventor of this method of purifying water is implied by some words written by John Collins‡ in his book *Salt and Fishery* published in 1682, the year before Fitzgerald secured his patent. In the section dealing with

---

* J. Evelyn. *Numismata*. London, 1697. Chapter IV, p. 167.

The medals in question have been identified quite recently. Edward Hawkins in *Medallic Illustrations of British History*, London, 1885, described two medals under Nos: 279 and 280 and ascribed them to the invention of a steam engine by Samuel Morland. The ascription is obviously wrong, as may be seen by reference to the illustration: they relate to Fitzgerald's invention. (I am indebted to Professor Giorgio Nebbia for informing me of this correct ascription.) See Plate XXVIII.

† S. Hales. *Philosophical Experiments*. London, 1739. See Preface.

‡ (1625–1682), F.R.S. Mathematician.

" New Inventions about making Salt of Sea Liquor, and separating good fresh water from Salt Water " he writes: " Discourse arising about another Patent " (undoubtedly Walcot's) . . . " Mr. Boyle affirmed, it had been before performed by himself, that he had presented his *Majesty* with some bottels of Water so made, and with the Secret, that it would be of great use at Sea in many Emergencies, yea and of Ordinary use in saving much Cask and Stowage; That all Pump-Waters had a little saltness in them, and that the Waters thus made, were wholsome, and no salter than Pump-Water."

From the evidence available it seems certain that Robert Boyle, knowing that sea-water could be made drinkable by distillation, was genuinely desirous of conferring a benefit on humanity, and therefore suggested to his nephew that here was a problem well worthy of solution on a commercial scale. Fitzgerald admits that Boyle was the prime mover in this affair.

However, the attitude adopted towards William Walcot throughout is rather puzzling. Even if Boyle were ignorant of him and his patent initially, he certainly could not have remained so for long. Walcot's strictures on Robert Boyle, contained in his various pamphlets,* seem to be justified, for the latter clearly gave his utmost support to Fitzgerald and furthermore the number of persons appearing in this matter who were related to Boyle, or who were his friends, suggests that he wanted them to benefit by whatever success might attend the commercial development of the invention, even if he himself did not wish to derive profit from it.

Phosphorus

During the autumn of 1677 there arrived in London a certain Johann Daniel Kraft† whose activities are important in connection with the history of phosphorus.

About 1676 Johann Kunckel von Löwenstein, the well-known chemist, was visiting Hamburg, where he heard of the remarkable substance which had been discovered by Hennig Brand of that town, and called by its discoverer ' *kaltes Feuer* ' or ' *mein Feuer* '. Kunckel wrote to Kraft about this extraordinary substance, whereupon the latter went to Hamburg. Both of them visited Brand, and seem to

* References in the article by R. E. W. Maddison. *N. & R.* (1952), 9, 196.

† Very little is known about him. Apparently, he had his origin in Miltenberg in Franconia. He studied medicine; travelled widely; held official positions in Germany; died in Holland. He was well known to Leibniz, who had a good opinion of him. On his visit to England, see R. E. W. Maddison. " Studies in the Life of Robert Boyle, F.R.S. Part I. Robert Boyle and Some of his Foreign Visitors ". *N. & R.* (1951), 9, 29.

have convinced him that they could sell the secret of this substance at a high price in appropriate quarters. Some underhand negotiations evidently took place, whereby Kraft obtained a quantity of this substance phosphorus. Kunckel alleges that Kraft paid Brand 200 thalers to keep the secret of preparing this substance from him. However that may be, the fact remains that Kraft having obtained possession of a quantity of this precious material, as well as certain, if incomplete, details of its method of preparation started on a demonstration tour from which, no doubt, he hoped to derive much profit. According to Boyle, Kraft showed the substance to Charles II.* In September 1677 Kraft visited Boyle, whose full account of the occasion was published by Hooke as follows :†

A SHORT MEMORIAL OF SOME OBSERVATIONS MADE UPON AN ARTIFICIAL SUBSTANCE, THAT SHINES WITHOUT ANY PRECEDENT ILLUSTRATION

September, 1677.

On *Saturday* the fifteenth of this month I was after supper visited by Mr. *Kraft*, a famous *German* Chymist, who was pleased to come and shew me a strange rarity he hath newly brought into *England*, to the sight whereof he allowed me to invite several members of the Royal Society, he being desirous, because the matter he imploys is very costly and of difficult preparation, to be a good Husband of it, and by shewing it to several curious persons at once, to exempt himself from the need of showing it often. The Company being met, the Artist took out of a pretty large box he had brought with him, divers Glass Vessels and laid them in order on the Table. The largest of them was a Sphere of Glass, which I guessed to be four or five inches in Diameter, being hollow and intire, save that in one place there was a little hole, at that time stopt with sealing wax, whereat to pour in the Liquor, which seemed to me to be about two Spoonfuls or somewhat more, and to look like muddy water made a little reddish with brick-dust or some other powder of that colour, he also took out of his Box three or four little pipes of Glass sealed, or otherwise stopt at both ends, being each of them somewhat bigger than a Swans quill, and about five or six Inches long, and having at one end a small fragment or two of that matter that was to shine in the dark.

---

* *Works Fol.*, **IV**, p. 20; *Works*, **IV**, 381.
† R. Hooke. *Lectures and Collections*. London, 1678.

He likewise laid upon the Table three or four Vials of several sizes, but none of them judged capable to hold above very few Ounces of water: in each of which Vials there was some Liquor or other, that was neither transparent nor well coloured, which Liquors I confess upon his making no particular mention of what they were to do, I was not curious to compare together, either as to quantity or as to colour. Besides all these substances which were fluid, he had in a small Crystalline button Bottle, a little lump of matter, of which he seemed to make much more account than of all the Liquors, and which he took out for a few moments to let us look upon it, whereby I saw that it was a consistent body, that appeared of a whitish colour, and seemed not to exceed a couple of ordinary Pease, or the kernel of a Hazel Nut in bigness, some other things 'tis possible Mr. *Kraft* took out of his Box, but neither I or (for ought I know) others of the Company took notice of them, partly because of his hast, and partly because the confused curiosity of many spectators in a narrow compass, kept me from being able to observe things as particularly and deliberately as I would gladly have done, and as the occasion deserved. Which Advertisement may I fear be but too applicable to a great part of the following Narrative.

The forementioned Glasses being laid in order upon the Table, the windows were closed with wooden-shuts, and the Candles were removed into another Room by that we were in; being left in the dark we were entertained with the ensuing Phænomena.

I. Though I noted above that the hollow Sphere of Glass had in it but about two Spoonfuls (or three at most) of matter, yet the whole Sphere was illuminated by it, so that it seemed to be not unlike a Cannon bullet taken red hot out of the fire, except that the light of our Sphere lookt somewhat more pale and faint. But when I took the liberty to hold this Glass in my hand and shake it a little, the contained Liquor appeared to shine more vividly, and sometimes as it were to flash.

II. I took one of the little pipes of Glass formerly mentioned, into my hand, and observed that though the shining matter had been lodged but at one end, yet the whole Glass was enlightened, so that it appeared a luminous Cylinder, whose light yet I did not judg to be always uniform, nor did it last like that which was included in the Vials.

III. In the largest of the Vials next the Spherical already mentioned, the Liquor that lay in the bottom being shaken, I observed a kind of smoke to ascend and almost to fill the cavity of the Vial, and near the same time there manifestly appeared as it were a flash of lightning that was considerably diffused, and pleasingly surprized me.

IV. After this I took up that small Crystaline Vial that I lately called (by a name familiar in our Glass-shops) a Button Bottle, wherein was contained the dry substance which the Artist chiefly valued, as that which had continued luminous about these two years, and having held that Vial long in my hand, in the same position in reference to my eye, and lookt attentively at it, I had the opportunity to observe (what I think none of the Company did) that not only this stuff did in proportion to its bulk, shine more vividly than the fluid substances, but that which was the Phænomenon I chiefly attended) though I could perceive no smoke or fumes ascend from the luminous matter, yet I could plainly perceive by a new and brisker light that appeared from time to time in a certain place near the top of the Glass, that there must be some kind of flashy motion in the matter that lay at the bottom, which was the cause of these little coruscations, if I may so call them.

V. The Artist having taken a very little of his consistent matter, and broken it into parts so minute, that I judged the fragments to be between twenty and thirty, he scattered them without any order about the Carpet, where it was very delightful to see how vividly they shined; and that which made the spectacle more taking, especially to me, was this, that not only in the darkness that invironed them, they seemed like fixt Stars of the sixth or least magnitude, but twinkled also like them, discovering such a scintillation as that whereby we distinguish the fixt Stars from most of the Planets. And these twinkling sparks without doing any harm (that we took notice of) to the Turky Carpet they lay on, continued to shine for a good while, some of them remaining yet vivid enough till the Candles being brought in again made them disappear.

VI. Mr. *Kraft* also calling for a sheet of Paper and taking some of his stuff upon the tip of his finger, writ in large Characters two or three words, whereof one being *DOMINI*, was made up of Capital Letters, which being large enough to reach from one side of the page to the other, and being (at least as I guessed) invigorated by the free contact of the external Air, shone so briskly and lookt so oddly, that the sight was extreamly pleasing, having in it a mixture of strangeness, beauty and frightfulness, wherein yet the last of those qualities was far from being predominant. And this Phænomenon did in more senses than one afford us the most of light, since not only the Characters shone very vividly upon the white Paper, but approaching it to my Eyes and Nostrils, I could discern that there ascended from them a fume, and could smell that fume to be strong enough, and (as it seemed to me) to participate of the odour of Sulphur and of that of Onions. And before I past from the mention of these resplendent Characters, I must not forget that either by their light, or that of the Globe, or both by the one and the other a man might discern those of his fingers that were nearest the shining stuff, and that this being held to the face though without touching it, some of the conspicuousest parts, especially the Nose, were discoverable.

VII. After we had seen with pleasure, and not without some wonder, the foregoing particulars, the Artist desired me to give him my hand, which when I had done, he rub'd partly upon the back of it, and partly on my cuff, some of his luminous matter, which as if it had been assisted by the warmth of my hand shone very vividly, and though I took not notice of any thing upon my skin, that was either unctuous or rough, yet I often times tried in vain by rubbing it with my other hand to take it off, or manifestly diminish its splendor, and when I divers times blow'd upon some of the smaller parts of it, though they seemed at the instant that my breath beat upon it, to be blown out, yet the tenacious parts were not really extinguisht, but presently after recovered their former splendor. And all this while this light that was so permanent, was yet so mild and innocent that in that part of my hand where it was largely enough spread, I felt no sensible heat produced by it.

By that time these things were done 'twas grown late, which made Mr. *Kraft*, who had a great way to go home, take leave of the Company after he had received our deserved thanks for the new and instructive Phænomena, wherewith he had so delightfully entertained us.

Because Mr. *Kraft* had twice attempted to fire heated Gun-powder with his Phosphorus, but without success; probably because the powder was not very good (as by some circumstances I conjected) and because it was not sufficiently heated before the matter that should set it on fire was put upon it, he promised me he would come another time to repair that unsuccesfulness: And accordingly, On the two and twentieth of *September* in the Afternoon I recived a visit from Mr. *Kraft*, who told me he came to make good his promise of letting me see that his shining matter was able to kindle heated Gun-powder, and because no strangers were present, I had the fairer opportunity to view it, which I was able to do better by day light, than I had done by its own light, for when he had taken it with a new Pen out of the liquor with which he kept it covered to preserve it, I perceived it to be somewhat less than the nail of one of my fingers and not much thicker than a shilling and I observed that when it had lain a little while upon a piece of clean Paper and discharged it self from its superfluous moisture, it began to emit whitish fumes which seemed to be very ponderous, since for the most part they did not ascend but surrounding the matter whence they issued, by their stagnation made as it were a little Pond or small atmosphere about it; so that lest it should wast too fast, he was obliged as soon as he had cut off a little corner less than half a pins head, to put the stuff nimbly back into the Vial out of which he had taken it; where I observed it for a very short time to send up exhalations into the liquor that covered it, and quickly after, as it were, quencht it. This done the Artist divided the little corner he had cut off into two parts, one of which he spread as far as it would reach upon a piece of white Paper, which he presently after held at a distance over a chafing-dish of burning Coals, by whose heat being excited it presently flasht and burnt away, and I having perceived that there was another part of the Paper which though not heeded by him, had been lightly besmeared by the same matter, I held it over the Coals, but at a considerable distance from them, and yet this little matter nimbly took fire and burnt a hole in the paper. And to satisfie my self that the heat did but excite the luminous matter, and that twas this its self that lighted the Paper, I held the rest of the same piece of Paper far nearer the fire and kept it there a pretty while without finding it at all scorched or discoloured. Lastly, the other part of the divided fragment of the hitherto mentioned matter, Mr. *Kraft* put upon the tip of a quil, and having at a distance from the fire, very well dryed and warmed some Gun powder upon another piece of Paper, he laid that Paper upon the ground, and then holding his quill upon it, as if it had been a match, within half a minute (by my guess) that powder took fire and blew up.

'Twill not perhaps be impertinent to add that on occasion of the operation I observed the Air to have on the shining substance when freely exposed to it. I took a rise to tell Mr. *Kraft* that I presumed it might be worth while to try whether his Phosphorus did shine by virtue of a kind of real or (if I may so call it) living flame, which like almost all other flames required the presence and concourse of the Air to maintain it, or whether it were of such a kind of nature as the Phosphorus of the learned *Baldwinus*, which I suspected to shine not like a flame or a truly kindled substance; but like a red hot Iron, or an ignited piece of Glass, wherein the shining parts are not repaired by fewel, as in other burning bodies, but are put by the action of the fire into so vehement an agitation as whilst it lasts suffices to make the body appear luminous. This conjecture Mr. *Kraft* seemed much to approve of when I told him that the way I proposed to examine his *noctiluca* by, was to put a little of it into our Pneumatick Engine, and Pump out the Air, whose absence, if it were of the nature of other flames, would probably extinguish, or very much impair its light, but yet since he offered not to have the trial made; probably because he had but very little of his shining substance left, I thought it not civil to press him. But to countenance what I said of the nature of *Baldwinus Phosphorus*, I shall recite an Experiment that I purposely made, to examin whether the presence of the Air were necessary to the shining of this Phosphorus, as I had long since found it to that of some pieces of shining wood.

We exposed for a competent time to the beams of a vigorous light, a portion of matter of about the breadth of the palm of ones hand, which we had prepared to be made luminous by them. And then causing the Candles to be removed (for we chose to make tryal by night) we nimbly conveyed the matter into a receiver that was kept in readiness for it, presuming (as the event shewed we might) that by using diligence the light would last as long as the experiment would need to do; making hast therefore to Pump out the Air, we heedfully watched whether the withdrawing of it would, contrary to my conjecture, notably diminish the light of the shining matter. And after we had thus withdrawn the Air gradually, we tryed whether by letting it return hastily, it would produce a more sensible change in the matter (which had been purposely put in without any thing to cover it, that it might be the more exposed to the Airs Action.) But neither upon the gradual recess of the Air, nor yet upon its rushing in when it was permitted to return, could we certainly observe any manifest alteration in the luminousness of the Phosphorus, other than that slow decrement that might well be imputed to the time during which the experiment was making. It being well known that this luminous substance requires no long time to make it decay, and by degrees to lose all its light; so that though once there seemed to one or two of the by-standers, upon the return of the Air, to be some recovery of part of the lost splendor, yet after repeated experiments it was concluded that the presence of the Air was not at all *necessary* to the shining of our matter, and it was judged most probable that the absence or presence of the Air, had *no manifest* operation on it. I might add to this that perhaps the presence of the Air is rather hurtful than advantagious to this sort of lights, since for having had a large Phosphorus that was much esteemed, and, whilst I kept it, exactly protected from the Air did very well; a part of the Glass that covered it, having by mis-chance been somewhat crackt, though none of the splinters appeared displaced, yet it seems some of the Corpuscles of the Air made a shift to insinuate themselves at these chinks (as narrow as they were) and in not many days made the matter cease to be capable of being made luminous as before. I cannot stay to inquire whether this unfitness or indisposition may be imputed to the bare moisture of the Air, or to some other substance or quality that alone or in conjunction with the moisture, may spoil that peculiar texture, or constitution that fits the matter of the Phosphorus assisted by the impressions of external light to become luminous. This, I say, I cannot stay to examine, though, That this Phosphorus is of a nice and tender constitution, and easily alterable, I was induced to think, by finding that the want of circumstances, seemingly slight enough, would keep it from being made; and I guess that a convention of circumstances did more contribute to the production than any peculiar and incommunicable nature of the matter: Because having had the curiosity to make some trial upon so obvious a material as quick Lime, though the success did not answer my designs, yet, neither was it so bad, but that some luminous quality was produced in the Lime by the action of the fire, and a saline Liquor; and I scarce question but other materials will be found capable of being made luminous by the same or the like operation, that is imploy'd by *Baldwinus*, when that learned man shall think fit to communicate his way to the Publick. But to return to what I was saying, that the contact of the Air might be rather hurtful than advantagious to the Phosphorus, I shall only add here as matter of fact, (for my conjectures about Light belong to my yet unpublisht Notes, *of the Origine of Qualities*) that whereas the contact of the Air, though it were not free, did in a few days destroy the luminousness of a good Phosphorus, yet having included another in a Receiver, whence we afterwards pumpt out the Air, this matter though inferior to the other in vividness was so little spoiled by lying open in our Vacuum, that at the end of not only some weeks, but some months, I found that the beams of a Candle passing to it through the Receiver, would notwithstanding the Vacuum it yet continues in, suffice to re-excite in it a manifest light.

In the course of his discussions with Kraft, Robert Boyle obtained a hint as to the nature of the material from which the phosphorus was derived, and thereafter made his own experiments directed to the preparation of this substance. He encountered difficulties, which were partly resolved by discussion with a visitor from abroad, who is designated by Boyle merely as "*A.G.*M.D.", and who is almost certainly Ambrose Godfrey Hanckwitz.* According to a pamphlet entitled *Historia Phosphori et Fama*† written by Hanckwitz, it was

* *Works Fol.* **IV**, pp. 21–2; *Works.* **IV**, pp. 382–3. On the life and work of Hanckwitz, see R. E. W. Maddison. "Studies in the Life of Robert Boyle, F.R.S. Part V. Boyle's Operator: Ambrose Godfrey Hanckwitz, F.R.S.". *N. & R.* (1955), **11**, 159.

† This pamphlet is mentioned by J. Ince, *Pharmaceutical Journal and Transactions* (1854), **13**, 280; and by H. G. Ince, "Memoranda of the Godfrey family . . . ". (MS in the library of the Pharmaceutical Society of Great Britain, London.) It was not published, and has now disappeared.

Boyle's intention that Bilger*, the chemist, should produce the phosphorus according to instructions, and then set Hanckwitz to work to prepare it; but, before Bilger could complete his part of the task and issue orders to Hanckwitz, the latter had set to work on his own initiative and produced a sample of phosphorus which he handed to his master.

On the subject of phosphorus, Robert Boyle published two works;† in addition he deposited with the secretaries of the Royal Society on 14 October 1680 a paper entitled "A Method of Preparing the Phosphorus of Humane Urine," and superscribed "...not to be open'd without y[e] consent of the writer Mr. Boyle."‡

Phosphorus as made by the original discoverer was described by Hanckwitz as a "dark, unctuous, dawbing mass", whereas his was made "into fine large white sticks, like tobacco-pipes." Hanckwitz eventually set up in business as a manufacturing chemist, and for very many years had virtually a European monopoly in making the substance, because of his mastery of the technical requirements of the process. He himself wrote:

> "I may without Vanity call my self, for near these Forty Years, the Sole Maker in Europe; for ... I could never hear of any body either at Home or Abroad ... that did prepare the Solid and Transparent *Phosphorus* besides myself. Some indeed have obtained a few Grains, but none could as yet produce the Quantity of 7 or 8 Ounces in one Distillation, except myself. And though this Preparation is intirely of my own finding out, yet I here confess with the utmost Sense of gratitude, That I am indebted for the first Hints of the Matter whence it is made, to that Ornament of the *English* Nation, the great Mr. *Boyle*, my kind Master, and the generous Promoter of my Fortune, who's Memory shall ever be dear to me."§

It is often stated that Boyle established a laboratory in Maiden Lane, Covent Garden, where he had Hanckwitz as his operator. This

---

* Evidently Daniel Bilger. See M. H. Nicolson. *Conway Letters*. Oxford, 1930, p. 409. Also, P. Shaw. *Chemical Lectures*, London, [1734], pp. 404–5.

† *The Aerial Noctiluca* (1680) and *New Experiments and Observations, Made upon the Icy Noctiluca* (1682); *Works Fol.* **IV**, pp. 19 and 70; *Works*, **IV**, pp. 379 and 469.

‡ Royal Society Classified Papers, 1606–1741. **XI** (1) 21. The paper was not published till after RB's death, when it appeared in the *Philosophical Transactions* (1693), **17**, No: 196, p. 583.

§ Ambrose Godfrey. *An Account of the New Method of Extinguishing Fires by Explosion and Suffocation.* [London], 1724. p. x.

statement is quite wrong. Hanckwitz did not enter his premises in Southampton Street (not Maiden Lane) until midsummer 1707.* Before that date (and during Boyle's lifetime) the area on which these buildings were erected (as well as the whole of Southampton Street and its connection with Maiden Lane), was part of the gardens of Bedford House, Strand. This is obvious when contemporary maps are consulted.

Robert Boyle and Georges Pierre

Mention has already been made of Boyle's life-long interest in the constitution of matter, and his views thereon, from which as a logical conclusion the conversion of one kind of metal into another should be possible. His interest in transmutation was linked also with his interest in supernatural phenomena as providing evidence to invalidate certain arguments of atheists.† Perhaps it was these interests, coupled with the possibility of advancing the Christian religion in Eastern parts, as well as performing acts of charity, that resulted in his being entangled with a certain Georges Pierre. The course of the relationship between the two of them can be judged only by the surviving letters of Pierre, for Boyle's part of the correspondence has not come to light; indeed, it may no longer exist. After reading these letters, one is left with the impression that the whole affair was a confidence trick.

The first mention of Pierre is found in an unsigned letter, written from Caen, 11 February 1677 (N.S.), to someone who is addressed as "Madame"; it says that le sieur Georges Pierre des Closets recommends the writer to her in order "de vous entretenir de loccupation dans la quelle ie suis depuis vn an a la lecture continuelle des Philosophes; à la quelle les experiences de la transmutation des metaus que luy mesme ma fait voir ont donné Lieu."‡ The writer asks for help and guidance, saying, "Mais iay besoin de vous pour scauoir ou ie dois prendre ce mercure vniuersel et indeterminé, que lon me dit estre par tout." After some speculations as to where it may be found, the letter closes with an offer of the writer's services.

---

* City of Westminster Public Library. Rate Book. Parish of St. Paul, Covent Garden. Accounts of the Overseers of the Poor. 1707.

† *Works Fol.*, V, p. 244; *Works*, VI, p. 57. Letter of RB to Joseph Glanville dated 18 September 1677.

‡ B.L. VI, 22. This is undoubtedly an incomplete copy of a missing original.

Among the Boyle Papers there is an item which is endorsed as follows:—

> "July.25.1677.
> Lre from George Patriarch of Antioch, to some Body to see Mr. Boyle, & to acquaint him wth some processes touching ye Philosophical stone."

The writer says to the unknown addressee . . .

> "vous nauez pas encore satisfait a la priere que je vous auois faicte de passer en Angleterre pour y seruir les grecs . . . Il me reste a vous dire, que sy vous faicte office pres de Sa Mte Britanique que vous ayez toustes les circonspections quil Vous conuiendra a fin que lon ne decouure pas ce que vous este[.] Il y a plusieurs amateurs de la *Verite* Phisolophique [*sic*] qui trauaillent bien vainement et Il ne pouroit que vous arriver de fascheuses suittes sy Vous manifestiez, nenuoyez donc aucun Hors Mr. Boille, au quel en particullier Vous temoignerez lestime particulliere que Jay pour sa personne, ensuitte Vous pouuez conferer auec luy et luy faire voir le projection, luy assurant que je vous ay ordonné de luy presenter Les oeuures de feu mes ayeules et que dorenauant Vous luy donnerez de mes nouuelles et quil ne sera pas long temps sans que dieu le fasse jouir du bonheur destre Veritable philosophe, Vous pouuez aussy lassurer que [ses] oeuures et ses experiences sont recues [?] de bon [one word not deciphered] par la Compagnie a lassemblée de laquelle Incesst Il remette Ses experiences Seulet de Medecine, ou de luy ou daultres et apres auoir satisfaict et lassemblée terminée vous passerez aupres de Bory* pour luy faire quelques remedes & retournez chez les[ieur] boille luy faire part de Ce que la compagnie Jugera apropos, Priez le aussy sil vous faictes laugmentation du poids dont nous luy expliquerons plus ampt la raison, de la faire publier dans le journal afin de voir ce que les Philosophes tenebreux diront . . . escrit au serail dans ma chambre de ste Sophie le 25 Juillet 1677 et de mon pontifficat le 2e.
>
> George Patriache dantioche."†

Although this letter is ostensibly signed by the Patriarch of Antioch, it is undoubtedly in the hand of Georges Pierre. With regard to the beginning, where mention is made of the Greeks, is it only a coincidence that at this time Joseph Georgirenes, Archbishop of Samos, was

* Could this be Giuseppe Francesco Borri, (1627–1695)?
† B.P. XXV.

in England in connection with the proposed establishment of an Orthodox Greek Church in London?*

In the autumn of this year Boyle was greatly interested by the account of Wenzel's transmutation of base metal into gold which was given to him by Graf Waldstein.† Boyle already knew of the Wenzel projection, for he had told Hooke about it in February 1675. In a letter which Boyle wrote to Joseph Glanvill in order to impress upon him the need for precise documentation when recording " stories of witchcrafts, or other magical feats ", he says,

> " . . . [it] has been particularly affirmed to me by the illustrious count of Wallestein, his imperial majesty's minister here, that the famous frier Wencel has several times actually made transmutations of baser metals into gold, in the presence of the emperor, and divers noblemen and good chemists; supposing the truth of this, I say, this one positive instance will better prove the reality of what they call the philosopher's stone, than all the cheats and fictions, wherewith pretending chemists have deluded the unskilful and the credulous, will prove, that there can be no such thing as the elixir, nor no such operation as they call projection . . . "‡

We come now to the series of letters over the signature of Georges Pierre. They are all somewhat rambling, hurriedly written, badly composed, and, at times, difficult to decipher. Various persons and places are mentioned, some of which are identifiable, whereas others are evidently in an arbitrary code. In the former category we have

du Moulins, presumably Peter, the younger;
le Moine, presumably Estienne;
Nointel, Charles François Olier, Marquis de, ambassador at Constantinople, 1670–1680;
Guilleragues, Gabriel Joseph de la Vergne, Vicomte de, ambassador at Constantinople, 1680–1685;

Among the names that are undoubtedly coded, we have Philalette, Doubabiel, Herigo. Finally, there are many other names that may be genuine, or concealed, but have not been identified.

---

* Wood. *Life*, **II**, p. 379.
E. Carpenter. *The Protestant Bishop*. London, 1956. Chapter **XIX**.

† Karl Ferdinand Waldstein, envoy of the Emperor to the English court, 5 June 1677–9 March 167$^{8}_{9}$: Bittner.

‡ *Works Fol.* **V**, p. 244; *Works*, **VI**, p. 58. Letter of RB to J. Glanvill dated 18 September 1677.

An account of Wenceslaus Seiler and his projection was " Published at the Request, and for the Satisfaction of several Curious and Ingenious, especially of Mr. Boyle, &c.," by Johann Joachim Becher, *Magnalia Naturae*, London, 1680.

Close relations between Boyle and Pierre had been establish[ed] sometime before November 1677, for the first available letter fro[m] Pierre is dated 23 December 1677 (N.S.).* He acknowledges receipt of " ce cristal, la grande lunette, des trebuchets, & le Globe," which are evidently intended for Georges du Mesnillet, Patriarch of Antioch,† who " mordonne aussy de vous dire que l'assemblée sera bien tost et que vous avez a trauailler achever promptemt ce que vous presenterez a Ces mrs." Throughout these letters there is the offer of gifts of fruit, wine, cider, roots of flowers, and other things to Boyle, who in turn is asked to send stockings, bottles of wine to Pierre, numerous costly presents for M. du Mesnillet, various gratifications for officials at the *Porte*, letters of authority for the consuls at Aleppo, Smyrna, and the Governor of Tangier. At one stage Boyle is informed that du Mesnillet

> " mescrit bien au large sur vostre charité & il desire auoir auec vous vn commerce ouuert comme le mien," and is grieved " de ne pouuoir vous enuoyer 4 ou 8 ballots dont 2 seulemt sont sortis du serail le 26 dec[bre] dernier et sont en alexandrie." Pierre says that these two bales contain " 20 pies brocard dor, 20 pies Satin de la chine en broderie dor & dargent, 20 pies satin de la chine de diuerses Couleurs 60 pies thoiles de perses auec or & argent, vn lit a la chinoise de bois, les pentes rideaux et autres ornements du lit en broderie chinoise, 3 paniers de porcelaine de la chine, 2 grand tapis de soye auec laine de perse charges dor & dargent 4 plus pettits, dont Il y en a 2 pour moy et 60 pies de diff[erente]s etoffes 4 phioles de baulme blanc, vne Lampe perpetuelle, dans vn Cristal de roche fait de sa main, 2 lingots dor. le tout est entre les mains dvn francois nommé Mr. Dupuy consul de france en symirne . . . Jay vne lettre a luy adresante pour vous enuuoyer & linserer dans Celle de vostre Consul qui aura soignt de vous enuoyer au plustost lesd[it]s ballosts."‡

Pierre acknowledged receipt of Boyle's letters for Aleppo and Smyrna at a later date.§

Much of the correspondence is concerned with attempts to persuade Boyle to be present at the " assemblée ", when he will be admitted as a member of a highly philosophical society, the exact nature of which is not clear; in fact, in the absence of confirmatory evidence, the society of which Pierre writes may never have existed. In pressing Boyle to

---

* B.L., IV, 106.

† I have not discovered any Patriarch of Antioch of this name at this date.

‡ B.L., IV, 127. Letter of G. Pierre to RB dated 28 February [1678] (N.S.).

§ B.L., IV, 117. Letter of G. Pierre to RB dated 17 May 1678 (N.S.).

n about attending the " assemblée ", Pierre
; or, if his health does not permit, then Pierre
. He tells Boyle that

e mr nostre patron est mort a hispane. Cest vous
ra sa place . . . Le Sr Philalette a fait disertation de
erite excelent a lassemblée generalle. Il dit que hermes
appelle et recognu par tous les sauens Le trois fois grand, en
mier lieu Il estoit grand prestre, grand Roy, & grand philo-
ophe. Il se seruit de cet exemple et dit que Lon deuoit vous cognoistre par cette mesme raison, grand par ses trauuaux Innombrables sa doctrine dont le public a tiré de merueilleuses lumieres, et non plus grand pour sa pieté et sa probité cognues de tous les gens de bien de langleterre et mesme cheri de son Roy & Enfin on concoit plus grand quhermes pour son Humilité dont Il a fait vne profession singuliere pendant tout le Course de sa vie, nous aurons Cette harangue latine et nous vous en ferons part."*

In this same letter Pierre again complains that correspondence to and from him is being opened either by the English or French postal services, and gives Boyle another accommodation address to use.

Apart from sending books, Boyle had also sent recipes to Pierre, who at one point complains to the former, " me distes sil vous plaist pour quoy ches vous auez vous quelqun qui ayant donné Coppie a Clomarez garcon appoti[cai]re qui demeure ches lefebure deuant ches vous de toustes les receptes que vous auez Enuoyée pour presenter a lassemblée . . . Je vous supplie de donner a Ceux qui sont a vostre seruice dans le Laboratoire quils nayent aucun commerce auec Cette sorte de Canaille . . . "†

The archives of the Royal Society contain an extraordinary document which concerns the election of Robert Boyle into a Sancta Sacraq[ue] Societas Cabalistica Philosophorum at Herigo In Comitatu Niceae; he was sponsored by Georgio Antiochae Patriarchâ, societatis Cambellario Cancellarioq[ue], a Petro Philalepta Cantuariae diocesis; vt & a Domingo Alvares Sarragocae Diocesis.‡

As we come nearer to the date proposed for the " assemblée ", so Pierre's enquiries as to whether or not Boyle will attend become more urgent. Apart from raising the question of financing the journey of

---

* B.L., IV, 110. Letter of [G. Pierre] to RB dated 24 March 1678 (N.S.).

† B.L., IV, 127. Letter of G. Pierre to RB dated 28 February 1678 (N.S.). Nicaise Le Febure came to England in 1660 and was appointed apothecary to the royal family. He died in 1669; and it would seem that an apothecary's establishment was carried on in Pall Mall under the Sign of Le Febure.

‡ B.P., XL. Dated 7 March 1678 (N.S.).

350 leagues from Caen (equivalent to 260 post stages),* he also raises the matter of tendering an oath, concerning which Pierre says, "vostre serment, qui ne vous obligera a rien qua seruir et ne vous induira dans aucunnes tentations mais vous sceruira comme deguillon pour aymer et seruir dieu plus parfaictemt & dvne aultre maniere que vous ne lauez fait cy deuant."†

Pierre's letter dated 13 May 1678 (N.S.)‡ opens with a recipe for "Banarum§ potable", and is followed by *Experience pour cognoistre lorsque les matieres quon prepare pour Le grand oeuure seront preparée fidellemt et auront acquis le tiltre desperance.* This letter acknowledges receipt of Boyle's letter of 22 April 1678 (O.S.) "et toustes Celles que vous mauez fait lhonneur de mescrire et la lettre de Croyance latine et Celle de nostre patron qui luy a esté despeschée et Coppie de la vostre du 22 . . . Je suis dans vne tristess Inconsolable de la perte de mr vostre Neueup et de mad vostre soeur." The condolence here offered by Pierre refers to the death of Boyle's sister, Mary, Countess of Warwick on 12 April 1678; the nephew has not been identified, but perhaps is Charles Boyle, second son of Francis Boyle, concerning whom no information has been found.

In a later letter Pierre acknowledges receipt of Boyle's letters for Aleppo and Smyrna, and tells him that the pendulum clock he sent has arrived damaged, and is being returned. With regard to the various items that Pierre had asked Boyle to supply, he goes on to say, "vous auez bien fait de ne risquer point les Canons [.] jen ay fait faire de 2 pieds que Jay enuoyez en la place des vostres". Further on he refers to the Pope "et les abus qui se commettoient dans les Indulgences quil octroye[.] mr Du mesnillet qui trauaille a la reunion de lEglise grecque auec la Latine et qui veut auusy faire vne reunion de la Caluiniste & luterienne ayant fait representer par le superieur de mont

* B.L., IV, 112. Letter of G. Pierre to RB dated 11 April 1678 (N.S.).

† B.L., IV, 114. Letter of G. Pierre to RB dated 21 April 1678 (N.S.).

‡ B.L., IV, 115.

§ Proof that arbitrary code words are used to conceal the true names of things and places is provided by two scraps of paper in B.P. **XXVIII**, which give the key to items bearing numbers 20–41 and 76–108. The word which is required to be concealed is given an arbitrary code word commencing with the next following letter in the alphabet. Thus

Butyrum antimonii	becomes	Climax
Calybs	"	Darisa
Digerere	"	Excitare
Sal tartari	"	Tasafa
Testa sicca	"	Vorago.

It is unfortunate that the complete key to these code words has not survived, for they occur not only in this correspondence but also elsewhere in RB's chemical notes and recipes, such as those to be found in B.P. **XXV**.

sinaj a enfin tant fait sur lesprit de Ce pape quil a reuocqué gener-[alemen]t toultes les Indulgences [.] Je vous en enuoye le bref afin que vous puissiez le faire voir a vos Euesques, nous voudrions bien que les gasettes publiassent cela car il est question de rapeller la plupart des esprits de lignorance ou Ils sont."*

In his next letter Pierre says, " il mest arrivé vn grand malheur a locasion des Canons de fontes que Jauois fait faire voyant limposibilité dauior les vostres, le fils de l'homme au quel vous madressez vos lettres appelle charles le mort voulant les tirer vn Canon a creue qui luy a Cassé la teste[.] jen suis fort blessé . . . Un officier de larchevesque de Cantorberi qui est a Caen qui est amy de mr la moine a esté tesmoing de ce malheur et vous dira que jay Couru grand danger de la vie."†

The " assemblée " which Pierre had tried his utmost to get Boyle to attend took place 25 June 1678 at the Château de Herigo, said to belong to Mr. Doubabiel, cousin of the Duke of Savoy. Pierre reported to Boyle that the chief of the Cosmopolites

> " ma fait vos Eloges dvne maniere a surprendre tous ceux qui laccompagnoient et a me donner la plus grande joye que Jaye resenty depuis longtemps. Il me tient ce langage. Cest a vous & pour le sr boille chers amy et confrere que larterisme‡ a ouuert aujourdhuy les scauants secrets de son Escolle philosophalle Transmutatoire, pour vous y faire voir a loeil, et toucher au doigt la veritable Interpretation de toultes nos Emblemes et de tous nos stilles, des quels les habitants de la montagne chymicque se sont seruis pour cacher leur terre feuillée aux Impies Ennemis Jurez de dieu, & des doctes [§] de la Nature, leurs allegories, parabales, problesmes, Types, Enigmes, dires naturels, fables pourtraicts & figures. Cest dans ce liure de vie que je te presente pour Nostre cher confrere afin quil se rasasira amplemt de la faim des sciences, cest dans Icelluy que tout y est parfaictemt explicqué et mis en son Jour, cest a vous den procurer au plustost de bons euen[men]ts tant pour la satisfaction particulliere de nostre confrere que pour la nostre en gen[er] &C. Le reste fut beaucoup de Complimens et vn grand Coffre fermé auec 3 Clefs qu'on maporte; pirodes|| en vostre nom sen est chargé

* B.L., IV, 117. Letter of G. Pierre to RB dated 17 May 1678 (N.S.).
† B.L., IV. Letter of G. Pierre to RB dated 30 May 1678 (N.S.).
‡ I have not been able to decide if this is *larterisme* or *lasterisme*. The word occurs frequently in the correspondence.
§ One word not deciphered.
|| An unidentified person.

aussy bien que de Ce liure qui nest aultre que vos lettres dacossiation."*

About July 1678 Boyle seems to have become disquieted in regard to his dealings with Pierre, for the latter in his letter 20 August 1678 (N.S.) opens by saying,

"Je vous considere dans destranges Inquietudes, et vous auez raison puis que Mr de la marche le plus meschant Homme du sciecle et auquel jauois fait donner 180li tous les 3 mois pour seruir de Copiste aux lettres estrangeres de nos maistres . . . a escrit au Roy [touchant lassemblée] qui en estant Informé a mis des personnes aux aguets, lon auoit surprins ses lettres et pour preuue de sa perfidie et du large desain de ce miserable que vous recognoistrez ou Il vouloit Jetter nos maistres dans le grand labirenthe je vous enuoye sa lettre au Roy qui a dissippé la plupart des Cosmopolites . . . par vn malheur extr[e]me le 25 de Juillet der[nier] le garde magazin du chasteau de herigo ayant esté gaigné par argent ou a[u]ltre part mit vne mesche qui a fait saulter les Murs les bastions et le superbe bastiment et plus de 30 de nos maistres[.] Jugez la douleur ou Je suis et lestat dans le quel Je me trouue Vous apprendrez Cette triste nouuelle par dautres que par moy, pirodes ny estoit pas et doit arriuer dans peu a londres dans la Conjoncture presente Je me veu[x] rendre ches vous et suiuré les ordres de Ceux qui sy doiuent rendre."†

By now, Boyle's suspicions seem to have been aroused, for he wrote to Descqueville, who finds passing reference in Pierre's earlier letters, for information. Descqueville in his reply says,

"Jaurois bien desyrey dauoir Veu mr pierre auant que de me donner le bien de Vous escryre mais il est party depuis huit Jours sans lauoir dit a personne mesme a son pere que Je Vis hier auquel il auoit laissey Vnne bouteille de verre dans laquelle il ia Vn peu deau quil dit par Vn billet estre de la mesme que Vous aues enuoyee laquelle il falloit Confronter auec celle quil ma rendue de Vostre part pour uoir si elle estoit pareille . . . il y eust lontemps quil nous eust dit a tous deux quil fayroit dans le mois de septembre Vn grand Voyage pour Vous et pour monsieur le cheuallier du mesnillet . . . Vous me distes monsieur que Je vous marque le temps que Je lay veu a paris Je ne pense pas uous auoir escrit que Je ly aie Jamais Veu Je ne Cognois que depuis le Voyage quil fist lan passey en angletterre et il na point

* B.L., IV, 125. Letter of G. Pierre to RB dated 25 June 1678 (N.S.).

† B.L., IV, 123. A contemporary attempt to transcribe this letter will be found at B.L., VI, 37.

party de Ceste Ville depuis cest donc Je Vous asseure et que toutes ses Cources ont abouty dans la Ville de baieux quy nest que six lieues dicy cest le lieu ou demeure ceste demoiselle a laquelle il a fait Vn enfant . . . "*

Georges Pierre had now disappeared. Boyle did not let the matter rest there; he instituted enquiries. A certain D[e] S[aintgermain] who visited Boyle in 1679 was asked to make enquiries on his return to the continent. He did so, and reported to Boyle by a letter in which Pierre's name is put in cipher, but he was unable to give any helpful information.† From some unrecorded source Boyle obtained the following extraordinary account of Pierre's subsequent career and end.

> " Exacte copie de cequi est contenu à L'égard de Mr. Pierre dans une lettre écrite de Caen à une personne à Londres, le 17 Juillet 1680.

Voicy ceque ie sçay du fameux Monsieur Pierre des Clozets. Il y a enuiron cinq à six mois quil éstoit icy de retour d'un voyage de prês d'un an. Je ne scay en quel endroit du monde mais enfin il contoit merueilles de toutes ses aduantures, et quoyqu'il en pust estre, nous le vismes reuenir et faire son entrée à Caën sur un cheual de soixante Loüis, auec un valet qui n'estoit pas moins bien monté, sans compter un autre cheual de bagage chargé de diuers habits fort riches, et autres meubles et curiositez tres rares et tres précieuses (il n'y a aucune éxageration dans tout ceque ie vous dis icy) il passa un bon espace de tems à L'Abaye de Fontenay, faisant des preparations chymiques, en attendant que la prise de corps, qui estoit contre luy à Bayeux fût leuée. Il vous pourra souuenir que ce fut pour auoir abusè d'une fille de qualitè de ceste ville apres donc qu'il n'y eut plus de danger pour luy, il vint à Caen, et se fit voir à touts ses amis dans un éclat merueilleux touts les iours changement d'habits, et touts fort riches un entrautres dont le drap estoit d'une grande beautè boutons et boutonnieres d'or; le revers des manches chamarrè de même, bas de Soye, veste de brocard à fleurs d'or, boucles de pierreries fines à la iarretiere et aux souliers, Castor fin, bref une petite Caleche fort propre, auec deux Cheuaux, un Cocher et un Laquais. Dans cette posture il n'estoit point de bonne maison à Caën, ou il n'allast debiter, tout cequ'il vouloit bien qu'on sçeust de luy, vray ou faux. enfin pour faire nargue a la médisance,

---

* B.L., II, 119. Letter of Descqueville to [RB] dated 22 September 1678 (N.S.).

† B.L., VI, 41. Letter of D.S. to RB dated 26 February 1680 (N.S.).

et faire voir que son estat n'estoit pas un faux brillant, il acheta à Breteuille une terre qui passoit par decret et qui luy fut adiugée sur le prix de 14000[lb] payez comptant, tout en or. Depuis cet achapt, toute son occupation fut a faire bastire sur ce lieu, à planter dresser des avenües, le meubler tres bien enfin d'en faire un seiour de plaisance, peutêtre pour s'y delasser le reste de ses iours, de toutes ses fatigues précedentes; mais à peine auoit il ébauchè de si beaux desseins, qu'il y deuint malade d'une inflammation de poumon dont il est mort il y a enuiron deux mois. Son corps fut apporté icy dans un carrosse tout couuert de deüil, et on luy fit une pompe funebre fort belle, par les ordres de son Frere le Curè. Son pere, Sa mere et touts ceux de sa maison, en ont portè, et portent encor un fort grand deüil, du moins à l'exterieur, ayans touts pris des habits de drap auec longue crespe, et long manteau; aussy sa succession le valoit elle bien, puisque l'on tient, qu'outre la terre qu'il auoit achetée de quatorze mil liures ils sont entrez en possession de beaucoup d'argent fait, et de bons meubles. telle fut la fin de cet homme, dont on a si peu connu le caractére, qu'apres sa mort on en sçait encor moins que lors qu'il estoit uiuant."*

There is no doubt but that Boyle greatly regretted his dealings with Pierre, for after the death of the latter he endeavoured to recover all he had written to him. In a letter dated 15 June 1682 and written to Monsieur La Marche at Caen, who is certainly the same individual complained of so bitterly by Pierre, and who had written to Boyle for assistance in finding employment in England, Boyle says,

"... I have some reason not to take it very kindly, that when I wrote to you three or four years ago, about Mr. Pierre des Closets, and seemed much concerned to receive an answer, (and therefore, if I mistake not, wrote more than once) you never vouchsafed me a line of answer: yet, if now you shall think fit to give monsieur [blank] who will be as soon as this letter in your parts, the best assistance you can, to recover all the papers of mine, that are to be found among those of your great friend monsieur Pierre, or wherever you know any of them are, (and which ought in justice to be restored to me) you will find, that you have not gratified an unthankful person, and, that this will be the readiest way you could take, to engage me to serve you, on those occasions, that may be offered ..."†

* B.L., VI, 43.
† *Works Fol.*, V, p. 245; *Works*, VI, p. 60.

It is not possible to say if Boyle did in fact succeed in recovering his papers. Probably he did not. Should his side of the correspondence with Pierre ever come to light, it will provide very interesting reading, for it might explain how Boyle came to be involved in this unfortunate episode.

Notwithstanding his unfortunate experience with Pierre, Boyle's interest, and, we must assume, his belief in transmutation persisted. Constantijn Huygens (the younger) wrote in his diary under date 20 June 1689 (N.S.): " Broer [i.e., Christiaan] was 'smergens bij m[rs] Boyle geweest, die hem onder anderen vertelt hadde, dat een man bij hem geweest hadde, die met een poeder dat root en claer was, een once goudt gemaaeckt hadded van loot; dat hij gehoort hadde, dat die man in Vranckrijck gearresteert was."* Boyle advocated,† and undoubtedly was mainly instrumental in securing, the repeal of the " Statute made in the Fifth Year of King Henry the Fourth, against the Multiplying Gold and Silver," so that experimental work connected with transmutation should not be impeded on legal grounds.‡

In May 1688 Robert Boyle published a two-page leaflet stating that he would hardly be able to publish any complete treatise in future, but only accounts of miscellaneous experiments, the reasons being that several Centuries of his experiments § together with other papers were lost. Furthermore, other manuscripts were irrevocably spoiled through a mishap with oil of vitriol over a chest of drawers in which they were stored. An additional note || alleges pilfering of his early bound notebooks, and complains strongly of plagiarism of his writings.

In a letter dated 30 May 1689 addressed to Jean Le Clerc¶, Boyle writes, " the want of health has, for this great while, so confined me to my lodgings, and so long exposed me to a multitude of visits, that rob

---

* *Historisch Genootschap te Utrecht*. Nieuwe Reeks. (1876), No: 23.

† *Works Fol.*, **V**, p. 246; *Works*, **VI**, p. 61. Letter of RB to Christopher Kirkby dated 29 April 1689.

‡ I Gul. & Mar. Ch. 30. The act is printed in *Works Fol.*, **I**, *Life*, p. 83; *Works*, **I**, p. cxxxii. This act passed the Royal Assent 20 August 1689; was partly repealed in 1869 by 30 & 31 Vict. Ch. 59; and finally repealed in 1948 by 11 & 12 Geo. 6. Ch. 62.

§ B.P. **XXXVI.** The first page of this leaflet is given in facsimile by Fulton, Fig. 22 facing p. 128. The leaflet was read at a meeting of the Royal Society on 11 July 1688. (Journal Book of the Royal Society, **VII**, 135). There has been much unprofitable speculation as to the identity of *J. W.* to whom RB addressed this *Advertisement about the Loss of his Writings*. The present writer has no useful suggestion to offer. One wonders if RB used ' Centuries of experiments ' in imitation of Francis Bacon in his *Sylva Sylvarum*.

|| B.P., **XXXVI.** The note is printed *in extenso* in *Works Fol.*, **I**, *Life*, p. 79–80; *Works*, **I**, p. cxxv-cxxviii.

¶ (1657–1736). Protestant minister and professor at Amsterdam.

me of all leisure, that I have not been able to send any thing of that kind [*i.e.*, philosophical productions] to the Royal Society, except a paper about a new way of suddenly producing flame by the bare contact of three bodies, each of them actually cold."*

His declining health obliged him to forego active participation in the affairs of the Royal Society, and to resign from the governorship of the New England Company. It must have been about this time that he issued another advertisement (given below), for a copy at the Royal Society is endorsed " Written since y[e] Revolution."

" *An Advertisement*.

Mr. Boyle finds himself oblig'd to intimate, to those of his Friends & Acquaintance that are wont to do him the honour and favour of visiting him. 1. That he has by some unlucky accidents (whereof he has given notice to the Public)† had many of his writings corroded here & there, or otherwise so maim'd, that w[th]out he himself fill up the *Lacunae* out of his Memory or Invention, they will not be intelligible. 2. That his age & sicklines haue for a good while admonish'd him to put his scatter'd and partly Defac'd writings into some kind of order, that they may not remain quite useles. And 3. That his skilfull & friendly Physician, seconded by Mr. Boyles best Friends, has pressingly aduis'd him against speaking daily wth so many Persons as are wont to visit him, representing it, as That w[ch] cannot but much Waste his Spirits, and by obliging him to sit a great deal too much for a Person subject to the stone of the Kidnys, and on several other accounts, Jmpair His Health, and Disable him for holding out long. And He is also oblig'd further to intimate, That by these & other Jnducements, He dos at length thô unwillingly find himself reduc'd to deny himself part of the satisfaction frequently brought him by the Conversation of his Friends & other Jngenious Persons, and to desire to be excus'd from receiving visits (unless upon occasions very extraordinary) two Dayes in the week, namely on the Forenoon of Tuesdays & Frydays (both foraigne Post-dayes) and on Wednesdays and Saturdayes in the Afternoon, that He may haue some time both to recruit his Spirits, to range his Papers & fill up the *Lacunae* of them, and to take some care of his Affaires in Jreland, w[ch] are very much disorder'd & haue their face often chang'd by the public Calamities there."‡

---

* *Works Fol.*, V, p. 246; *Works*, **VI, p. 61.**

† May 1688.

‡ B.P., **XXXVI.**

One of the copies of this advertisement in the Royal Society's archives has the following note by Henry Miles added at the bottom of the sheet:

> " NB. There was a board put over the door with an Inscription thereon to signifie when he did & did not receive Visits, which was lately* in being & I hope to obtain it."

In spite of being handicapped by bodily indisposition, Boyle was able to complete some further works for the press. In 1690 two works of his were published; one, with the rather misleading title o *Medicina Hydrostatica*, was the first treatise in English devoted to the determination of specific gravity; the other was Part I of *Th Christian Virtuoso*, which was devoted to the theme that " there is no Inconsistence between a Man's being an Industrious Virtuoso, and a Good Christian," that is to say, there is no conflict between science and religion. Part II of the latter work (as far as it was possible to recon struct it from the confused remains of Boyle's surviving papers in th next century) first appeared in 1744 in the collected edition of hi works edited by Thomas Birch. The last work to be published b Boyle in his lifetime was *Experimenta et Observationes Physicae*, which is a miscellaneous collection of observations relating to lodestones diamonds, colours, medicines and medical case histories.

---

* This would be about 1742.

V

'The Lamp Fails ...'

(1691)

About the middle of the year 1691 Robert Boyle became aware of his continued failing health, and resolved to draw up his last will and testament, which he signed, 18 July 1691. In connection therewith, he made the following declaration on the same day:

> " Whereas I am this day about to perfect my last will and testament, I do hereby, to prevent and secure myself from all scruples, solemnly protest and declare, that I do not intend by signing and sealing the said will, or any other will or codicil that I may hereafter sign, to abridge myself of any power that law or equity, or the nature of a will, do or can give me, to dispose freely of all or any of my temporal concerns, even those assigned to pious or charitable uses therein mentioned; and that I reserve to myself a full liberty, when this or any other testament shall be perfected, to annul, revoke, or alter the whole will, or any part of it, and dispose otherwise of my concerns, as freely as if the said will, or any part or codicil of it, had never been signed and published, or so much as intended by me. Witness my hand this 18th day of July 1691.
>
> Robert Boyle.
>
> Witness John Warr."*

Various codicils were added during the months preceding his death. Some memoranda concerning his testamentary dispositions are to be found on small, odd sheets of paper,† and are in the nature of reminders. One such reads:

> " Whatever Papers be found of mine relating to Theodora I desire may be burnt without fail."

---

* *Works Fol.*, I, *Life*, p. 84; *Works*, I, p. cxxxiv.

† B.P., **XXXVI**: R.S. MSS. 194.

As Henry Miles remarked, " orders of that kind committed to writing were not always executed;"* and these particular papers relating to the unpublished part of *The Martyrdom of Theodora and Didymus* did in fact escape destruction, for they are listed among the titles of manuscripts not inserted in Birch's edition of Boyle's works,† though since that time they have disappeared. Another paper on atomical philosophy, similarly endorsed, has survived.‡

Robert Boyle ordered his papers. Lists were prepared of his philosophical writings not yet printed as at 3 July 1691, and of his theological writings.§ An inventory dated 17 September 1691 gives the contents of boxes in his bedroom.|| Once more we find instructions for the destruction of certain papers:

> " Copys of Letters of his [R. Boyle's] to Mr. Oldenburg to be rifled and the rest burnt.
> A bundle of other Letters to be rifled and the rest burnt."

Notwithstanding the deterioration in his health, Boyle still maintained an interest in experimental work, for occasional entries thereon were recorded during July and August of that year. The last known letter written by Boyle bearing date 8 October 1691, and addressed to Dr. Daubeny Turbeville, the famous oculist, reveals that Robert and his sister Katherine were then extremely ill.

> " Worthy Doctor,
> If you knew how very ill my sister and I have been, since you left this place, you would not wonder, that you have not had from me an enquiry, how you bore your journey to Salisbury; and whether you now enjoy the benefit of it, by the recovery of your health, wherein very many, and I most particularly, are concerned. Excuse then, I beseech you, my past silence, and give me, if you can, the contentment of an assurance, under your own hand, that you are well, which will be a great cordial to me. If I had known of your going so soon, I should have furnished you with spirit of hartshorn, and some other trifles, that you are pleased to let me provide you here. And therefore, if you want any thing now of that kind, be pleased to command it of me. Your constant kindness to me, and the benefits I have received from your skill and experience in your profession, urge me to acquaint you with the distemper in my eyes, that did much surprise me, and does still

* Add. MSS. 4229, fol. 112. Letter of Henry Miles to Andrew Millar dated 29 September 1741.

† *Works Fol.*, I, *Life*, p. 151: *Works*, I, p. ccxxxvii.

‡ B.P. **XXXV**: R.S. MSS. 186.

§ B.P. **XXXVI.**

|| B.P. **XXVI**, fol. 162.

much afflict me, having continued with me for about a month. The case in short is this; in the day time, I see, thanks be to God, as I use to do, and so till five o'clock in the afternoon; but then, as soon as candles are brought in, I find a very sensible decay in my sight; so that, though I can see all the same gross objects as I did before, and could, if I durst, read printed books, as I have often tried, yet the reflection from those objects is not vivid, as it was wont to be; and if I look upon somewhat distant objects, methinks I see them through a thin mist, or a little smoke; but when the candles are newly snuffed, and so the light increased, I see far better, for a little while, till it begin to have more snuff: this distemper continues, as long as I make use of candle-light, but the next morning, by God's goodness, I find myself as before, only now and then there seems to fall slowly down, sometimes in one eye, and sometimes in another, a faintly shining vapour, which immediately disappears. I have such apparitions of late, for these two or three years, without any bad consequence. What this distemper may proceed from, I know not, though I remember I have heard you more than once take notice of the narrowness of my pupil. Sight is a thing so dear to all men, and especially to studious persons, that I earnestly beg, you would be pleased to consider my case deliberately, and acquaint me with your thoughts of the cause; and more particularly, to send prescriptions of the receipts you would have me employ, and your directions what else you would have me do towards the cure of it. I have had too much trial of your friendship, to doubt, that, upon so important an occasion, you will refuse me the effects of it which will exceedingly add to your former favours, and oblige me to study the ways of shewing myself,

Sir,

your most affectionate

and most humble servant.

I have, of yours, the water for clearing the sight, which I much esteem, and have more than once used; but sometimes use, upon your authority, a little honey diluted with succory; but I much want the constant eye-water, to be used every night, or oftner, especially if it may be fitted for the present state of my eyes; and therefore beg, that you would be pleased to prescribe me one, that I may make here. I forgot to tell you, that, for some months last past, I have been much troubled with what they call vapours, or fumes of the spleen, and with some scorbutic disaffections."*

* *Works Fol.*, V, p. 246: *Works*, VI, p. 61.

On 20 October 1691 Boyle transferred £100 Hudson Bay stock to Edmond Doughty, leaving himself £50 stock which he still held at his death.*

A letter dated 21 October 1691 from John Locke reveals that he was then assisting Boyle to prepare the latter's *History of the Air* for the press. He says:

> " By this bearer, my servant, who comes to town on purpose, you will receive your papers concerning the titles of the air. I have read them all over very carefully, numbered them according to the titles they belong to, and laid them in that order, the best I could, according to the state they are in. I have besides corrected many of the mistakes of your amanuenses; but yet, for all this, they are not in a condition to be sent to the printer, or put into a bookseller's hand, in order to publishing. For besides that they are not all laid in the order of the titles they belong to, it often happening, that those of different titles are writ in the same paper, which I would not venture to cut in pieces, without your direct order; and some of them being writ on both sides the leaf, cannot be cut; and therefore I have contented myself only to mark in the margin the title each belongs to: besides this, I say, there are some faults in the writing, which, without consulting you, I durst not correct; and in other places defects and omissions, which I could not supply; and many things, which suggest queries, which I think fit to advise with you about . . ."†

The work was finally seen through the press by Locke after Boyle's death.‡

Robert Boyle's sister, Katherine, died 23 December 1691, and was buried three days later in the chancel of the church of St. Martin-in-the-Fields, Westminster.§ The loss of a sister|| with whom he had

---

* Hudson's Bay Record Society, **VIII**, p. 332.

† *Works Fol.*, **V**, p. 571: *Works*, **VI**, p. 543.

‡ Part of the MS is in the Lovelace Collection. Bodleian MSS. Locke. c. 37. The agreement between the printers and J. Locke regarding the copyright of this work is in the same collection. Bodleian MSS. Locke. b.1. fol. 163.

§ City of Westminster Public Library. Account of Mr. Christopher Cock, Senior Churchwarden of the Parish of St. Martin for the year 1691. Entry under 26 December 1691.

|| Katherine, Viscountess Ranelagh, was highly esteemed by her contemporories. Sir John Leeke said of her, " . . . a more brayve wench or a Braver spiritt you have not often mett wth all. She hath a memory that will hear a sermon and goe home and penn itt after dinner verbatim . . . " (Verney, **I**, p. 203). John Milton in a letter to her son wrote, " . . . nam et mihi omnium necessitudinum loco fuit." (Milton's *Prose Works*. Bohn edition. **III**, p. 511). The Duke of Ormonde, in reference to the great influence she exerted on her

lived for so many years, and for whom he had so much affection was certainly a great sorrow for Robert, and undoubtedly hastened his own death, which occurred at " three quarters of an hour after twelve at night " during the night of Wednesday, 30 December 1691, *i.e.*, he died in the early morning of 31 December. " Mr. Thomas Smith, apothecary in the Strand, [this at the time of publication of *Works Fol.* (1744)] who lived seventeen years with Mr. Boyle, was with him at his death "*

Robert Boyle's physician, Sir Edmund King, in a letter to Christopher, Viscount Hatton, has left the following account of the last days.

> " My being out of my bed 2 nights together wth Mr Boyle, who's death I foretold the first day he complained of an alteration, wch was Tuesday; and that day I tolde my Ld Rochester† and others

* *Works Fol.*, **I**, Preface, p. [iii]: *Works*, **I**, p. ii.

† Laurence Hyde, 1st Earl of Rochester, (1641–1711). (*D.N.B.:* G.E.C.).

family, said that " as for the other branches, she governs them very absolutely." (R.C.H.M. Ormonde. N.S. VI, p. 138). Gilbert Burnet in a sermon said, " She lived the longest on the publickest Scene, she made the greatest Figure in all the Revolutions of these Kingdoms for above fifty Years, of any Woman of our Age. She imployed it all for doing good to others, in which she laid out her Time, her Interest, and her Estate, with the greatest Zeal and the most Success that I have every known. She was indefatigable as well as dextrous in it: and as her great Understanding, and the vast Esteem she was in, made all Persons in their several turns of Greatness, desire and value her Friendship: so she gave her self a clear Title to imploy her Interest with them for the Service of others, by this that she never made any use of it to any End or Design of her own. She was contented with what she had: and though she was twice stript of it, she never moved on her own account, but was the general Intercessor for all Persons of Merit, or in want: This had in her the better Grace, and was both more Christian and more effectual, because it was not limited within any narrow Compass of Parties or Relations. When any Party was down, she had Credit and Zeal enough to serve them, and she imployed that so effectually, that in the next Turn she had a new stock of Credit, which she laid out wholly in that Labour of Love, in which she spent her Life: and though some particular Opinions might shut her up in a divided Communion, yet her Soul was never of a Party: She divided her Charities and Friendships both, her Esteem as well as her Bounty, with the truest Regard to Merit, and her own Obligations, without any Difference, made upon the Account of Opinion.

She had with a vast Reach both of Knowledg and Apprehensions, an univeral Affability and Easiness of Access, a Humility that descended to the meanest Persons and Concerns, an obliging Kindness and Readiness to advise those who had no occasion for any further Assistance from her; and with all these and many more excellent Qualities, she had the deepest Sense of Religion, and the most constant turning of her Thoughts and Discourses that way, that has been perhaps in our Age. Such a *Sister* became such a *Brother;* and it was but suitable to both their Characters, that they should have improved the Relation under which they were born, to the more exalted and endearing one of *Friend.*" (*A Sermon Preached at the Funeral of . . . Robert Boyle.* London, 1692, p. 33).

of his friends and desir'd an other physician to be sent for, wch was y^{t} night; and at 10, as I told yor L^{p}, I was sent for to be in the house all night, but did not let him kno it, for fear of surprizing him (he was up all Tuesday). Wednesday morn was much better, but exceeding low and faint; would rise in the afternoon, and at 5 we met and desir'd him to goe to bed at 7. Dr Stockholm* was to sit up; but at 10 or 11 he grew worss, so that my Lord Ranalough† and Lady Thanet Dowager‡, etc., was sent for, and they sent for me. I was just hot in my bed, after something I had taken for my great colde; and, tho' I had not a graine of hope he would live till I came, yet, considering it was the last attempt I could make to serve won who for many years past had great affection for me and rely'd under God, as he often told me, upon my care, I was resolv'd to goe to him. It was one a clock in the night; but he was dead before I cam. His lamp went out for want of oyle; soe did his sister's too . . . He has been kept alive, by God's blessing, upon the dint of care severall years, to a woonder."§

Another letter, from Sir Charles Lyttelton,|| to the Viscount Hatton says,

" . . . He was not sick above 3 houres, but s^{d} his heart was broke when she [his sister] died . . . There is an odd report goes that, when Lady Ranelagh lay dying, there was a flame broke out of one of ye chimneys, wch being observed by ye neighbours gave notice of it, and, the chimney being looked [into], there was no cause found for it in y^{e} inside, yet appeared to flame for som time to those wthout; and y^{e} same thing happened when Mr Boyl died; . . . "¶

He was buried close to his sister in the chancel of the church of St. Martin-in-the-Fields, Westminster, 7 January 169$\frac{1}{2}$, when Gilbert Burnet, Bishop of Salisbury, preached the funeral sermon, taking as his text, *For God giveth to a man that is good in his fight, wisdom, knowledge, and joy.* Eccles. ii. 26.

---

* William Stokeham.

† Richard Jones, 3rd Viscount Ranelagh; RB's nephew.

‡ Elizabeth, widow of the 3rd Earl; RB's niece.

§ Camden Society. [N.S.], Vol. **XXII**, 1878. *Correspondence of the Family of Hatton.* Edited by E. M. Thompson. Vol. **II**, p. 166. The letter can be assigned to 8 January 169$\frac{1}{2}$, and not 2 January as printed.

|| (1629–1716). M.P., for Bewdley. Succeeded as second baronet.

¶ Camden Society. [N.S.], Vol. **XXII**, 1878. *Correspondence of the Family of Hatton.* Vol. **II**, p. 168.

Sir Edmund King tells us that " There was a vast crow'd of persons of Qualitie in the church y^{t} stay'd out the whole time."* According to John Evelyn " His funeral (at which I was present) was decent, and, though without the least pomp, yet accompanied with a great appearance of persons of the best and noblest quality, besides his own relations."† Boyle's request that there should be no unnecessary pomp at his funeral seems to have been respected, if we may judge by the churchwarden's account which records the payment of fees of 10*s*. 5*d*., including 2*s*. for the pall, which was not the best one.‡

Numerous eulogies and panegyrics appeared after Robert Boyle's death. In addition to what he said in the funeral sermon, Gilbert Burnet in his " Rough Draught of my own Life " wrote of Boyle in the following terms:

> " . . . he had the purity of an angell in him, he was modest and humble rather to a fault. He despised all earthly things, he was perhaps too eager in the pursute of knowledge, but his aim in it all was to raise in him a higher sense of the wisdome and glory of the Creator and to do good to mankind, he studied the Scripture with great application and practised universall love and goodnes in the greatest extent possible, and was a great promoter of love and charity among men and a declared enemy to all bitternes and most particularly to all persecution on the account of religion."§

In his *History of my own Time* Burnet says Boyle

> " was looked on by all who knew him as a very perfect pattern; he was a very devout Christian, humble and modest, almost to a fault, of a most spotless and exemplary life in all respects. He was highly charitable; and was a mortified and self-denied man, that delighted in nothing so much as in the doing good. He neglected his person, despised the world, and lived abstracted from all pleasures, designs, or interests . . . "||

Various necrological sheets and elegies were printed.¶ John Evelyn confided to his diary,

---

* B.M. Add. MSS. 29,585. Letter dated 23 January $169\frac{1}{2}$ to Christopher Hatton.

† Evelyn (Bray), Vol. **III**, p. 352. Letter dated 29 March 1696 to William Wotton.

‡ City of Westminster Public Library. Account of Mr. Christopher Cock, Senior Churchwarden of the Parish of St. Martin for the year 1691. Entry under 7 January $169\frac{1}{2}$.

§ H. C. Foxcroft. *A Supplement to Burnet's History of my own Time*. Oxford, 1902. p. 464.

|| G. Burnet. *History of my own Time*. Edited by O. Airy. 2 vols., Oxford, 1879–1900. Vol. **II**, p. 343.

¶ Fulton, Nos. 304–311.

> " This last week died that pious admirable Christian, excellent Philosopher, & my worthy Friend Mr. Boyle, a greate losse to the publique, & to all who knew that vertuous person."*

At a later date, in a letter to William Wotton, who had dallied so long with the task of writing an account of the life and work of Robert Boyle, John Evelyn wrote,

> " But by no man more have the territories of the most useful philosophy been enlarged, than by our *hero* [R. Boyle], to whom there are many trophies due. And accordingly his fame was quickly spread, not only among us here in England, but through all the learned world besides. It must be confessed that he had a marvellous sagacity in finding out many useful and noble experiments. Never did stubborn matter come under his inquisition but he extorted a confession of all that lay in her most intimate recesses; and what he discovered he as faithfully registered, and frankly communicated; in this exceeding my Lord Verulam, who (though never to be mentioned without honour and admiration) was used to tell all that came to hand without much examination. His was probability; Mr. Boyle's suspicion of success . . . those many other treatises relating to religion, which indeed runs through all his writings upon occasion . . . show how unjustly that aspersion has been cast on philosophy, that it disposes men to atheism . . . . He was the furthest from it in the world, and I question whether ever any man has produced more experiments to establish his opinions without dogmatising. He was a *Corpuscularian* without Epicurus; a great and happy analyzer, addicted to no particular sect, but as became a generous and free philosopher, preferring truth above all; in a word, a person of that singular candour and worth, that to draw a just character of him one must run through all the virtues, as well as through all the sciences. And though he took the greatest care imaginable to conceal the most illustrious of them, his charities and the many good works he continually did, could not be hid. It is well known how large his bounty was upon all occasions . . . he was the most facetious and agreeable conversation in the world among the ladies, whenever he happened to be so engaged; and yet so very serious, composed, and contemplative at all other times; though far from moroseness, for indeed he was affable and civil rather to excess, yet without formality.

* Evelyn (de Beer), Vol. V, p. 81. 1 January $169\frac{1}{2}$.

As to his opinion in religious matters and discipline, I could not but discover in him the same free thoughts which he had of philosophy; not in notion only, but strictly as to practice, an excellent Christian; and the great duties of that profession, without noise, dispute, or determining; owning no master but the Divine Author of it; no religion but primitive, no rule but Scripture, no law but right reason. For the rest, always conformable to the present settlement, without any sort of singularity The mornings, after his private devotions, he usually spent in philosophic studies and in his laboratory, sometimes extending them to night: but he told me he had quite given over reading by candle-light, as injurious to his eyes. This was supplied by his amanuensis, who sometimes read to him, and wrote out such passages as he noted, and that so often in loose papers, packed up without method, as made him sometimes to seek upon occasions, as himself confesses in divers of his works. Glasses, pots, chemical and mathematical instruments, books and bundles of papers, did so fill and crowd his bed-chamber, that there was but just room for a few chairs; so as his whole equipage was very philosophical without formality. There were yet other rooms, and a small library (and so you know had Descartes), as learning more from men, real experiments, and in his laboratory (which was ample and well furnished), than from books . . . .

In his diet (as in habit) he was extremely temperate and plain;* nor could I ever discern in him the least passion, transport, or censoriousness, whatever discourse or the times suggested. All was tranquil, easy, serious, discreet and profitable; so as, besides Mr. Hobbes, whose hand was against everybody and admired nothing but his own, Francis Linus excepted (who yet with much civility wrote against him), I do not remember he had the least antagonist.

In the afternoons he was seldom without company, which was sometimes so incommodious, that he now and then repaired to a private lodging in another quarter of the town, and at other times (as the season invited) diverted himself in the country among his noble relations . . . .

* "Mr. Boyl never drinks any strong drink: hee every morning eats bread and butter, with powder of eyebright spread on the butter. His supper is water-gruel and a couple of eggs: his dinner is mutton, or veal, or a pullet, or walking henne, (as hee calls them,) which goe to the barne door when they will." [This about 1673]. (Diary *of the Rev. John Ward, A.M.*, arranged by C. Severn. London, 1839, p. 135.)

In his first addresses, being to speak or answer, he did sometimes a little hesitate, rather than stammer, or repeat the same word; imputable to an infirmity, which, since my remembrance, he had exceedingly overcome. This, as it made him somewhat slow and deliberate, so, after the first effort, he proceeded without the least interruption, in his discourse. And I impute this impediment much to the frequent attacks of palsies, contracted, I fear, not a little by his often attendance on chemical operations. It has plainly astonished me to have seen him so often recover, when he has not been able to move, or bring his hand to his mouth: and indeed the contexture of his body, during the best of his health, appeared to me so delicate, that I have frequently compared him to a chrystal, or Venice glass; which, though wrought never so thin and fine, being carefully set up, would outlast the hardier metals of daily use: and he was withal as clear and candid; not a blemish or spot to tarnish his reputation: and he lasted accordingly, though not to a great, yet to a competent age; . . . and to many more he might, I am persuaded, have arrived, had not his beloved sister . . . with whom he lived, a person of extraordinary talent and suitable to his religious and philosophical temper, died before him. But it was then that he began evidently to droop apace . . . "*

Abraham de la Pryme† wrote as follows in his diary,

"Alas! who can refrain from tears, what learned man can but lament at the sad newse that came the other night, viz., the death of the famous and honourable Mr. Boyle, a man born to learning, born to the good of his country, born to every pious act, whose death can be never enough lamented and mourned for. England has lost her wisest man, wisdom her wisest son, and all Europe the man whose writeings they most desired, who well deserved the character that the ingenious Redi‡ gives him, who calls him *Semper veridicus, et quavis sublimi laude dignus!* I have heard a great deal in his praise and commendation. He was not only exceeding wise and knowing, but also one of the most religiousest and piusest men of his days, never neglecting the public prayers of the church or absenting himself therefrom upon any occasion. He was exceeding charitable to the poor and needy, and thought whatever he gave to them too little! He was a mighty promoter

* B.M., Add. MSS. 4229, fol. 58. Evelyn (Bray), Vol. **III**, p. 348–352. Letter dated 29 March 1696.

† (1672–1704), antiquary; F.R.S. (*D.N.B.*).

‡ Francesco Redi, (1626–1698), Italian philospher. (Bompiani, *Autori.*)

of all pious and good works, and spent vast sums, as I have heard, in getting the Bible and several more religious books to be translated and printed in Irish and spred about that country, that his poor countrymen might see the light of the Gospel. He was a mighty chemist, etc."

" . . . [Mr. Boyle] having been the [most] learned, wisest, and godliest man that England every brought forth. He was a mighty strict, pious man . . . Some condemns him for being too credulous and giving too much heed to the relations of his informers in philos[ophical] matters, but this springs from nothing but ignorance and envy."*

It is regrettable that Robert Hooke's diary covering this particular period is missing; the forthright comments of this rather crabbed individual might be revealing in view of the very long association he had with Boyle.

Thomas Platt, the unofficial Resident for Tuscany in London, reported Boyle's death in a letter dated 8 January 169½:

" Mori in questa città . . . il famoso sig[r] Ruberto Boile, grandemente lamentato da' letterati, aveva campato per arte più di 20 anni essendo stato sempre macilentissimo, e per gl'ultimi due anni non era mai uscito di camera, essendo spento, consumato per mancanza d'oglio, ma con vivacità di spirito fin all'ultimo: ha lasciato i suoi beni in terreni alla sua famiglia ed i danari contanti ch'erano 8000 lire a poveri ed altre opere pie, la sua pietà essendo tanto più singolare che in questo paese i letterati hanno ordinariamente sentimenti molto differenti, e viene stimato per santo da protestanti, essendo sempre stato di vita esemplare."†

Samuel Pepys wrote to John Evelyn saying,

" . . . forasmuch as (Mr Boyle being gone) wee shall want your helpe in thinking of a man in England fitt to bee sett up after him for our Peireskius‡ . . . "§

Constantyn Huygens,|| when referring to the receipt by the Royal Society of another treatise by Anthony van Leeuwenhoeck, wrote from Whitehall to his brother Christiaan to say,

---

* *The Diary of Abraham de la Pryme.* Publications of the Surtees Society. (1869), Vol. **LIV**, p. 21 and 24. Entries during January 169½.

† Archivio di Stato di Firenze, *Mediceo*, 4247. Quoted from A. M. Crinò. " Robert Boyle visto da due contemporanei." *Physis*, (1960), **II**, 319.

‡ Nicolas Claude Fabri de Peiresc, (1580–1637), a scholar of varied and profound learning.

§ *Private Correspondence and Miscellaneous Papers of Samuel Pepys 1679–1703.* Edited by J. R. Tanner. 2 vols. London, 1926. Vol. **I**, p. 51. Letter dated 9 January 169½.

|| The son, (1628–1697); secretary to William II and William III of Holland.

" Son portrait [i.e. de Leeuwenhoeck] estoit (ce disent ils aussi) tousjours dans la ruelle du lict de feu Monsr Boyle, decedé depuis quelques jours et mort de Phtisie."*

Christiaan Huygens in a letter to Leibniz mentioned Boyle's death and commented on his work.

" Mr. Boyle est mort, comme vous scaurez desia sans doute. Il paroit assez etrange qu'il n'ait rien basti sur tant d'experiences dont ses livres sont pleins; mais la chose est difficile, et je ne l'ay jamais cru capable d'une aussi grande application qu'il fait pour establir des principes vraisemblables. Il a bien fait cependant en contredisant à ceux des Chymistes."†

In reply Leibniz said,

" Ainsi je voudrois que Mons. Boyle nous eut laissé ses conjectures. Mais c'est encor plus dommage que ses plus curieuses experiences le plus souuent ne sont rapportées qu'a demy. Tantost il s'excuse parce qu'un amy ne luy donne pas le pouuoir de les publier; tantost sur quelqu'autre raison . . .

Il est vray que le Chancelier Bacon scavoit quelque chose de l'art de faire les experiences et de s'en servir; mais ce que vous dites de feu Mr. Boyle, est encor veritable à son egard, qu'il n'estoit pas capable d'une assez grande application pour pousser les consequences autant qu'il faut."‡

At an earlier date, when writing to Oldenburg, Leibniz had said, in regard to the activities of detractors:

" Illustri Boylio rogo me data occasione commendes; nolim Virum eximium scriptis eorum quos nuperrimè ejus Pneumatica experimenta ac ratiocinationes aggressos audio, diverti ab illis, quibus multo melius mereri de publico potest, Chymicis experimentis quæ utinam ne diutius publicis precibus neget. Intactum est hoc doctrinæ genus, saltem philosophis. Primus est Boylius qui non dicam nugari desiit, sed demonstrare coepit. A quo si corpus quoddam Chymicum impetrare poteris; obligabis profectò genus humanum: dici enim non potest quanti ad omnem vitam momenti sit Chymia. Ego certè sæpe pro valetudine Viri vota facio, nam vereor ne aliquando jacturam irreparabilem faciamus culpa quorundam obtrectatorum, qui sæpe Viros publico bono natos a suis publicandis absterrent."§

---

* Huygens. Vol. **X**, p. 232. Letter dated 18 January 1692 [N.S.].

† *Ibid.*, p. 239. Letter dated 4 February 1692 [N.S.].

‡ *Ibid.*, p. 263. Letter dated 19 February 1692 [N.S.].

§ Newton. *Correspondence*, **I**, 314. Letter dated 15 July 1674 (N.S.).

Notices of Boyle's decease were not wanting in the news journals of the day. *The Gentleman's Journal** reported,

> " The Honourable Mr. *Boyle*, that *Columbus* of the Philosophical World, hath ended his Life with the Year, to the unspeakable loss of the Learned. A learned and eloquent Pen will give us his Life."

This projected life by the " eloquent Pen " of Gilbert Burnet did not materialize. The *Europische Mercurius*† carried the following note.

> " den 9den overleeden, en den 17den met een treffelijke Lykreden, opgesteld en uitgesproken door den Bisschop van Salisbury, betreurd de geleerden Heer Robert Boyle, een Philosooph van een byzondere karakter, die zich byna niet en had bemoeid als met Physica, en de Natuur zodanig doorkoopen, dat nooit iemand de zelve beter heest gekend."

The *Mercure Historique et Politique*‡ reported,

> " Le fameux M. Boyle est mort, regreté de tous les savans; Quoi qu'on dût s'attendre à ce coup, car il étoit déjà assez vieux, cela n'a pas laissé de fraper ceux qui connoissoient son mérite. C'étoit un Philosophe d'un caractére assez particulier. Il ne s'étoit attaché qu'à la Physique, bien qu'il ne méprisât pas les autres parties de la Philosophie. On peut dire que jamais homme n'a mieux connu la nature qu'il la connoissoit. Il l'avoit étudiée avec un attachement prodigieux, & avoit fait tant d'experiences, pour expliquer les Phenoménes que nous y voyons arriver, qu'il y en a peu aujoud'hui dont on ignore les veritables causes. M. l'Evêque de Salisburi fit son Oraison funebre, le 17 du mois passé."

With regard to Robert Boyle's scientific work there are several contemporary, or near contemporary, comments. In addition to the one quoted above, Leibniz had written to Christiaan Huygens very shortly before Boyle's death as follows.

> " Je m'etonnerois si M. Boyle qui a tant de belles experiences, ne seroit arrivé à quelque theorie sur la Chymie, apres y avoir tant medité. Cependant dans ses livres et pour toutes consequences qu'il tire de ses observations, il ne conclut que ce que nous sçavons tous sçavoir, que tout se fait mecaniquement. Il est peut-estre trop reservé. Les hommes excellens nous doivent

---

* January 169½., p. 16.

† January 1692, [N.S.], p. 73.

‡ February 1692, [N.S.], Vol. **XII, p.** 208.

laisser jusqu'à leur conjectures, et ils ont tort, s'ils ne veuillent donner que des verités certaines."*

It is of interest to compare this opinion with that given many years earlier by Margaret Cavendish.† She wrote,

" . . . concerning his experiments, I must needs say this, that, in my judgment, he hath expressed himself to be a very industrious and ingenious person; for he doth neither puzle Nature, nor darken truth with hard words and compounded languages, or nice distinctions; besides, his experiments are proved by his own action. But give me leave to tell you, that I observe, he studies the different parts and alterations, more then the motions, which cause the alterations in those parts; whereas, did he study and observe the several and different motions in those parts, how they change in one and the same part, and how the different alterations in bodies are caused by the different motions of their parts, he might arrive to a vast knowledg by the means of his experiments; for certainly experiments are very beneficial to man."‡

Francesco Redi § described Boyle as " Litterato di alta fama, dotto, diligente, e sempre veridico, e meritevole d'ogni lode più sublime." || but took him to task for his credulity.

" Chi avrebbe mai pensato, che il Boile, che oggi negli scoprimenti delle Cose Naturali è il più grand' uomo, che sia nell' Europa, e che mai vi sia stato, e che forse anco vi sia per essere, chi dico avrebbe mai pensato, che anch'egli fosse infetto della peste della credulità? Io per me non lo avrei mai sognato. N'ebbi però qualche leggier sospetto nello scorrere agli anni passati il suo Libro delle gemme;¶ ma ora essendomi capitato il nuovo, ed ultimo suo Libro** . . . Non solamente mi è cresciuto il sospetto, ma mi sono totalmente avveduto, che ancor egli è credulo, ma di questa così fatta credulità non ne do la colpa a lui, ma al Paese, nel quale egli è nato . . . in questo Libro la sua credulità è troppo manifesta . . . Egli è un Libro, che chiaramente si vede, che è lavoro, a fattura di un grand' uomo, ed io lo

---

* Huygens, Vol. **X**, p. 228. Letter dated 29 December 1691.

† Duchess of Newcastle (1624?–1674), authoress.

‡ *Philosophical Letters*, London, 1664. Sect. IV. Chapt. 22, p. 495.

§ (1626–1698). Natural philosopher, physician, poet. (Bompiani. *Autori.*)

|| F. Redi. *Opere*. 7 vols Venice, 1742–1760. Vol. **II**, p. 17. The comment is in his *Esperienze intorno a diverse cose naturali*. 1st edition Florence, 1671.

¶ R. Boyle. *An Essay about the Origine & Virtues of Gems*. London, 1672.

** R. Boyle. *Of the Reconcileableness of Specifick Medicines to the Corpuscular Philosophy*. London, 1685. Latin edition, London, 1686.

rassomiglierei ad un Quadro di Tiziano, in cui questo grande Artefice avesse voluto dipingere le sua Innamorata, e traportato dallo affetto l'avesse caricata di tante, e così belle fattezze, che avesse fatta sì con tutte le eccellenze del disegno, e del colorito una bellissima figura, ma però in alcune parti non simile alla vera . . . "*

David Abercromby in discussing the authors of the New Philosophy said, [as for] " *Descartes*, *Verulam*, the Honourable *Robert* Boyle, who in not a few things, has out-done them both, and is desrvedly styl'd abroad, *The* English *Philosopher;* he being indeed, the Honour of his Nation, as well as of his Family."†

In amplification of the opinion of Leibniz quoted above, that of John Freind‡ can be added:

" No Body has brought more Light into this Art [Chemistry] than Mr. *Boyl*, that famous Restorer of Experimental Philosophy: Who nevertheless has not so much laid a new Foundation of Chymistry, as he has thrown down the old; he has left us plentiful Matter, from whence we may draw out a true Explication of things, but the Explication it self he has but very sparingly touch'd upon."§

In this same year, 1712, John Hughes||, in an essay contributed to *The Spectator*¶, wrote:

" The Excellent Mr. *Boyle* was the Person, who seems to have been designed by Nature to succeed to the Labours and Enquiries of the extraordinary Genius, [Lord Bacon]. By innumerable Experiments He, in a great Measure, filled up those Plans and Out-Lines of Science, which his predecessor had stretched out. His Life was spent in the Pursuit of Nature, through a great Variety of Forms and Changes, and in the most rational, as well as devout, Adoration of its Divine Author."

Herman Boerhaave**, the famous Dutch physician and chemist, unknowingly, was credited with giving high praise to Boyle:

" that prodigy of knowledge, and application, [who] excels not less as a chemist, than as a natural philosopher. His character is one

* F. Redi. *Opere*. Venice, 1742–1760. Vol. IV, p. 189. Fragment of a letter. As Redi refers to the Latin edition of 1686, the date of the letter cannot be earlier.

† *Academia Scientiarum* . . . London, 1687, p, 156.

‡ (1675–1728), M.D., physician.

§ J. Freind. *Chymical Lectures*. London, 1712, p. 4.

|| (1677–1720). Poet, author and translator.

¶ No. 554. 5 December 1712.

** (1668–1738). Professor at Leyden.

of the most amiable, and desirable in the world: No body ever made more experiments, or with more care and caution, than he; nor does any body relate the events thereof with more candour, and fidelity. He is ever exceedingly wary and reserved in drawing conclusions from his experiments; and rarely concludes too much. He held a very amicable correspondence with all the chemists, and other *Virtuosi* of his age; and thus effected a kind of commerce, or communication of secrets. Add, that he was always perfectly frank in imparting what he had learnt, or discover'd, to the public.

"He gives abundance of good things relating to *fossils*, their origin, separation, preparation, purification, and the ways of fitting them for human uses: as also to *vegetables:* but he is chiefly admired as to what relates to *animals*, and the analysis of their parts; in which he follow'd the path trod by *Helmont*, in his book *de Lithiafi*. A noble instance hereof we have in his history of *human blood:* which contains abundance of the most valuable, and exquisite things; and on every occasion he endeavours to found his reasonings on experiments. His treatises of the *unforeseen failure of experiments*, and *sceptical chemist*, shew his great modesty, and moderation; and how far he was from that common vice of the chemists, boasting, and promising more than he could perform.

"He is author of a vast number of pieces, all composed in the same spirit. They are so many, and printed so separately, that it is exceedingly difficult to procure a compleat Collection. His style is a little loose, and diffusive; and the course of his writings sometimes interrupted with digressions, and particularitites not very essential.

"One objection is made to him in his chemical character, *viz*. that taking the virtues of his preparations on the report of other people, he commends some of them too much, and attributes virtues to them which experience does not avow. For, as he did not practise physic himself, his way was, upon making any new preparation, to give it some physicians to make trial thereof; and they, it seems, out of complaisance, would speak more largely of it than it deserved. Hence, those profuse commendations he bestows on the spirit of human blood and hart's-horn in phthisical, and on the *Eus Veneris* in rachitical cases.

"As to the point of credulity commonly charged on Mr. *Boyle;* there does not seem to be any great foundation for it: Nobody inquires into things more severely, or speaks of them more dubiously than he. True, a long, and intimate acquaintance with the writings of *Paracelsus*, and *Helmont*, might possibly have some effect on the bent of his mind; and induce him to give too much belief to many of the *arcana* they pretend to: and the more so, considering the great number of strange incidents, and events which had fallen within his own observation."*

Eustace Budgell† expressed the opinion

> " That no Philosopher, either before, or after him, ever made so great a *Number* of curious and profitable Experiments. He very rightly judged, that this was the only proper Method to become a Master of the *Secrets of Nature;* and there is one Particular, for which he can never be too much admired or commended; it is evident, that he made all his Experiments without any Design to confirm or establish any particular System. He is so much in earnest in his Search after Truth that he is wholly indifferent where he finds it. We may truly say, That he has *animated* Philosophy; and put in *Action* what before was little better than a *speculative* Science."‡

William Brande§, writing for the *Encyclopaedia Britannica*, expressed his opinion as follows :

> " Boyle has left voluminous proofs of his attachment to scientific pursuits, but his experiments are too miscellaneous and desultory to have afforded either brilliant or useful results; his reasoning is seldom satisfactory; and a broad vein of prolixity traverses his philosophical works. He was too fond of mechanical philosophy to shine in Chemistry, and gave too much time and attention to theological and metaphysical controversy to attain any excellence in either of the former studies. He who would do justice to Boyle's scientific character, must found it rather upon the indirect benefits which he conferred, than upon any immediate aid which he lent to science. He exhibited a variety of experiments in public, which kindled the zeal of others more capable than himself. He was always open to conviction; and courted opposition and controversy, upon the principle that truth is often elicited by the conflict of opinions . . . "||

---

* H. Boerhaave. *A New Method of Chemistry*. Translated by P. Shaw and E. Chambers. London, 1727. p. 46–47.

† (1686–1737).

‡ E. Budgell. *Memoirs of the Lives and Characters Of the Illustrious Family of the Boyles*. Third edition. London, 1737, pp. 119–120.

§ William Thomas Brande (1788–1866). Chemist, F.R.S. (*D.N.B.*).

|| " The Progress of Chemical Philosophy ". Third Dissertation, *Encyclopaedia Britannica*, Supplement (1824), p. 15.

To this may be added the conclusion of George Wilson:*

> "We know no natural philosopher with whom, in quality of intellect, and habits of working, Boyle can exactly be compared. We could compound him, however, pretty well out of Dr. Joseph Black and Dr. Priestley. He had the versatility, energy, and unsystematic mode of carrying on research of the latter. Priestley and Boyle were constantly experimenting on all kinds of things, and made many trials without a definite object, or precise expectation as to the result. Both stood before the oracle, putting endless unconnected and isolated questions to the priestess, anxious for an answer, but without preconception what the answer would be. Boyle, however, paid much more attention to the reply than Priestley did, and understood its meaning a great deal better. Both were equally ingenious in devising experiments, and successful in performing them, but Priestley often totally misunderstood the phenomena he brought to light, and was led completely astray by his own experiments. Boyle resembled Black in the accuracy with which he observed results, in the caution with which he drew conclusions, and the skill with which he interpreted the phenomena he witnessed. He had the energy and versatility of Priestley, and the caution and logic of Black, but he was less versatile than Priestley, and more incautious and less logical than Black."†

It is rather surprising to discover that no lapidary memorial appears ever to have been erected to Robert Boyle or to his sister in the church where they were buried. There is no mention of such in John Le Neve's *Monumenta Anglicana;*‡ and we may take it as certain that Edward Hatton in *A New View of London,* § which was published while the old church of St. Martin-in-the-Fields was still standing, would have recorded any monumental inscriptions to them had they existed. It is of interest to note that the mural tablet to Robert's niece, Frances Jones, which Hatton records as being placed on the south side of the chancel, is still to be seen in the crypt of the present church. ||

---

* (1818–1859). M.D., Chemist; professor of technology, University of Edinburgh.

† G. Wilson. *Religio Chemici.* London and Cambridge, 1862, pp. 247–248.

‡ 5 vols. 8vo. London, 1717–1719.

§ 2 vols. 8vo. London, 1708. The reference numbers used by him in Vol. **I**, pp. 341 *et seq.*, provide the key to the numbers used by George Vertue on his engraved Plan of the Church of St. Martin-in-the-Fields. I am not aware that this fact has been previously noticed.

|| The inscription is given in Hatton's *New View*, Vol. **I**, p. 342, as well as in the London County Council Survey of London, Vol. **XX**, *The Parish of St James, Westminster, Part I, South of Piccadilly*, London, 1960, p. 32.

When the old church was demolished shortly after 1720,* no systematic record was made of the disposal of the remains of bodies interred there. The matter was investigated by Tilden,† to whose account nothing can be added. The present edifice was built between 1722 and 1724; it contains no memorial to Robert Boyle, or to his sister, Katherine; nor does it contain the remains of Robert Boyle, whose final resting place is unknown.

This is an appropriate point to mention a quotation which describes Robert Boyle as the " Father of Chemistry ", and which is sometimes alleged to be on his tombstone. In view of what has been said above, it is quite clear that this cannot be so. In fact, the earliest appearance of the quotation known to me occurs in *The British Quarterly Review* for 1849,‡ where, in an essay review, George Wilson, writes,

> " A late distinguished professor, indeed, guiltless of any purpose of jesting or playing upon words, once gravely summed up the memorabilia of Boyle's history in the singular epitome, that he was ' the son of the Earl of Cork and the father of modern chemistry '."

The quotation frequently appears in garbled form.

At an even earlier date James Price (1752–1783), the chemist who committed suicide after failing to establish his claim to convert mercury into silver and gold, had called Robert Boyle " the Venerable Father of English Philosophic Chemistry." § We have here, perhaps, an ancestral version of the quotation cited by Wilson.

---

* L.C.C. Survey of London, Vol. **XX**, p. 24.

† William A. Tilden. " The Resting-place of Robert Boyle." *Nature*, (1921), **108**, 176.

‡ Vol. **IX**, p. 216. The essay is reprinted in George Wilson's *Religio Chemici*, London, 1862; and the quotation will be found on p. 189.

§ J. Price. *An Account of some Experiments on Mercury, Silver and Gold.* Oxford, 1782, p. 2. The passage is not in the second edition of 1783.

VI

Last Will and Testament

Robert Boyle's last will and testament is printed in Appendix I *in extenso* from the original. Certain matters concerning his testamentary dispositions are dealt with here; other minor notes will be found in the appendix.

The disposal of Robert Boyle's personal estate was not long delayed after the grant of probate, 26 January $169\frac{1}{2}$, and this part of his total estate ultimately realized about £10,000*.

Library

The diary of a non-conformist divine, possibly Zachary Merrill, contains the following entry under date 1 April 1692,

> "Yesterday I took M^{r} Ch. and Mr Holworthy to visit Mr Boyles Library. we saw a very strong digestive that would Supple the hardest bone. a good Collection of Mineralls and oars & earths. I had before seen his Library containing 330 fol. 801. q^{tos} 2440 Oct and 12. most of y^{m} well bound they may be had for 3 or 400^{l} (tho' worth 1000) because they must not be sold by Auction. He was a most unaffected Gentlem. His lodging Room and the other were very plaine etc: "†

Evidently, the intention was to sell the collection *en bloc;* but the following advertisements in *The London Gazette* at the beginning of July indicate that the hope was not realized.

> "The Library of the Honorable Robert Boyle Esq; lately deceased, will be sold by Retail, on Monday the 11th of this instant July. Any Gentlemen or others may have the Opportunity to view, and buy what they please at reasonable Rates. Attendance will be given at the late Lady Ranelaughs House in

---

* Public Record Office, Chancery Proceedings. Reynardson Division. C 9/457/104. 26 June 1693.

† Bodleian MSS, Rawlinson D. 1120 fol. 62 *b*.

the Pall-Mall, near St. James's, on Monday, Tuesday, and Thursday, from 9 in the morning till 7 at night."*

"The Library of the Honorable Robert Boyle Esq; lately deceased, will continue to be sold (for the Satisfaction of Gentlemen and others) this present Thursday the 14th, and on Friday and Saturday the 15th and 16th of July, and no longer, at the late Lady Ranelaughs House in the Pall-Mall, near St. James."†

The remaining, unsold books probably languished at Pall Mall until the disposal of Lady Ranelagh's house had taken place. Robert Hooke was concerned in this matter, and under date 5 January 169^{2}_{3} he records in his diary, "Mr. Boyles house lett."‡ Any property remaining in the house had then to be removed. As regards the books, Hooke has recorded their fate. Some had obviously been bought by a dealer and were exposed for sale on secondhand book-stalls in Moorfields; and others were sold at auction.§ These are the entries in Hooke's diary,

21 *March* 169^{2}_{3}. "In MF [=Moorfields] I saw neer 100 of Mr. Boyles high Dutch Chymicall books ly exposed in Moorfields on the railes."

23 *March* 169^{2}_{3}. "... in MF saw many of Mr. Boyles German chemicall books."

28 *March* 1693. "Read a *Catal.* of Morgans and Boyles books at Toms."||

The clue to this catalogue is found in an advertisement that appeared in *The London Gazette.*

"Bibliotheca Morganiana, or a Catalogue of the Library of Mr. Silvanus Morgan, containing most of the English Histories, a good Collect. of Books of Heraldry, Genealogies, Numismat, Inscriptions, &c. To which is added the Library of an Honourable Gentleman deceased, consisting of Divinity, History, &c. in Gr. Lat. and English, many of which are large Paper Gilt and Adorned: To be sold by Auction at Tom's Coffee-house near Ludgate on Wednesday the 5th of April next, at 3 Afternoon. Catalogues may be had tomorrow at Mr. Patridges at Charing-Cross, Mr. Hargraves in Holborn, and at the place of sale."¶

---

* No. 2782, 7–11 July 1692.

† No. 2783, 11–14 July 1692.

‡ Gunther, Vol. **X**, p. 196, 12 December 1692: and p. 203.

§ D. McKie. "Boyle's Library." *Nature*, (1949), **163**, 627.
R. E. W. Maddison. "Robert Boyle's Library." *Nature*, (1950), **165**, 981.

|| Gunther, Vol. **X**, p. 223, 224, 225.

¶ No. 2856. 23–27 March 1693. Repeated with slightly different wording in No. 2858. 30 March–3 April 1693.

When this catalogue* is examined, it is a disappointment to find that the two collections of books have not been listed separately, but have been gathered together under their different sizes and subjects. Notwithstanding the totally different interests of Morgan† and Boyle, no clear separation can be made with a view to establishing original ownership. So far, only five books have been identified with certainty as having belonged to Robert Boyle.‡ He used no book-plate, and rarely put his name in a book. From his correspondence and other papers it is possible to compile a short list of books that were certainly in Boyle's possession, but it by no means reaches the total number of volumes mentioned in the quotation given above (p. 198).

As far as Boyle's other household possessions are concerned there is one unexpected reference in the will of Sarah Fowler, widow of the Rev. Jacob Fowler, who bequeathed to " Mr. Joseph Barnardiston late of Chancery Lane, Stationer my half pint Silver Mugg formerly the Honourable Mr. Boyles."§

---

* " Bibliotheca Morganiana:/or a/ CATALOGUE/ Of the Library of/ Mr. Silvanus Morgan./ Containing all the Valuable and Scarce English Historians,/ with a Curious Collection of Books of Heraldry, Ge-/nealogies, Antiquities, Medals, Coins, Inscriptions &c. in/GREEK, LATIN and ENGLISH./ To which is added the Latin Part of the Library of an/Honourable Gentleman, lately deceased, consisting of /Divinity, History, Geography, &c./ The whole are generally gilt on the back many of them of/the Large Paper curiously adorned./ To be sold by Auction at Tom's Coffee-house adjoyning/to Ludgate, on Wednesday the 5th of April, 1693./ Rule/ Conditions of SALE./ 1 The highest Bidder is the Buyer./ 2 The Books for ought we know are perfect, if any appear otherwise, the Buyer may take or leave them./ 3 That every Person be obliged to give in his Name and Place of Abode, paying also 5s. in the Pound for/ what he buys; and be obliged to take his Books away within 3 days after the Sale is ended./ Catalogues are distributed at Mr. Partridges at Charing-/ Cross; the Coffee-houses about the Inns of Court; Bat-/sons Coffee-house against the Exchange, and at the Place/ of Sale."

† Sylvanus Morgan (1620–1693), arms painter and author. (*D.N.B.*)

‡ R. Birley. " Robert Boyle's Head Master at Eton." *N. & R.*, (1958), **13**, 104: R. Birley. " Robert Boyle at Eton." *N. & R.*, (1960), **14**, 191: Fulton, p. iv.

§ Somerset House, P.P.R., P.C.C., 55 Tyndall (1766) Will proved 4 February 1766. The testatrix also gave and presented " to the Gentlemen of the Royal Society who are in the Government thereof the Original Picture of the Hon[ble] Mr. Boyle once President of the said Society to be by them placed in their Usual place or room of assembling." Notice of this gift was communicated to the Royal Society by Dr. Ducarel, F.R.S., but there is no evidence that the picture was ever delivered to the society. (Royal Society MSS. 419. *Donations of Natural History &c.*, 6 February 1766.) I am indebted to Mr. L. P. Townsend, Archivist of the Royal Society, for directing my attention to this reference. The picture has not been identified or traced.

Mineral Collection

Henry Hunt, who was Robert Hooke's assistant, as well as operator and general factotum to the Royal Society, " acquainted the Society, that he had received the Minerals of Mr Boyle which his Executors had Sent in Deal boxes for the Repository." This report was made on 29 June 1692.* On 14 July following,

> " Mr Hunt acquainted the Society with what progress had been made in the Examination of Mr Boyles Minerals, to wit that there were found some of them written upon, to Signify what they were, and that Some other were Marked only with a Letter Seeming to refer to Some Catalogue of them written in paper, but these Catalogues being not found with the said Minerals Dr Hook was desired to Enquire of the Executors, whether Such Catalogues were not to be found amongst Mr Boyle's papers, and either to get them, or some Copies of them for the Repository."†

On 20 July

> " Dr Hooke Reported that he had according to the Societys' Desire Spoken with Mr Warr about Notes or Catalogues of the Minerals of Mr Boyle, that he would make Inquiry concerning them, and if any such could be found, That he would procure them for the Society. There being some discourse concerning the having some of those minerals examined by Chymicall Tryals; Sr John Hoskyns‡ related that Mr Moult§ would very freely make them for the use of the Society." ||

The Royal Society MSS 200 comprises two quarto paper-covered note books, one of which on the first page is headed " Mr. Boyls Minerales. R. H. Hunt." I think these two items are the catalogues of the Royal Society's former collection of minerals.

When the Royal Society moved from the house it occupied in Crane Court into rooms provided by the Government at Somerset House the available space was found to be insufficient for the Society's needs. So, the decision was made to offer the repository of rareties to the Trustees of the British Museum, who accepted the offer, and acknowledged receipt of the collection in 1781.¶ Robert Boyle's mineral specimens, together with other items presented by him to the Royal Society

---

* Journal Book of the Royal Society, **VIII**, p. 123.

† Journal Book of the Royal Society, **VIII**, p. 126.

‡ (1634–1705) 2nd baronet; barrister; P.R.S. (*D.N.B.*).

§ George Moult, elected F.R.S., 1689.

|| Journal Book of the Royal Society, **VIII**, p. 127.

¶ C. R. Weld. *A History of the Royal Society*. 2 vols, London, 1848, Vol. **II**, p. 125.

during his lifetime,* are now presumably dispersed amongst the various collections of the British Museum (Natural History), in so far as they have survived.

Sealed Papers

Various sealed papers that Robert Boyle had deposited with the Royal Society during his lifetime were opened at a meeting of the Council on 9 February 169$\frac{1}{2}$. They were three in number.

1. Dated 30 September 1680. Deposited with Robert Hooke as Secretary to the Royal Society, 14 October 1680.† The title is " A Method of Preparing the Phosphorus of Humane Urine." The paper was published in the *Philosophical Transactions* (1691–3), **17.** No. 196, p. 583.‡

2. 30 November 1683. " Mr. Boyle sent a depositum of an *arcanum* to remain sealed in the custody of the secretary."§ The paper deals with a chemical method for testing the sweetness of distilled sea-water (see p. 153), and was read before the Royal Society on 17 February 169$\frac{1}{2}$, when Hans Sloane gave a successful demonstration of the method. The paper was published in the *Philosophical Transactions* (1691–3), **17.** No. 197, p. 627.||

3. Deposited 6 September 1684. This paper contained " Titles for the Natural History of Mineral Waters."¶

There was a fourth sealed paper which Boyle lodged with the Royal Society on 13 February 166$\frac{7}{8}$,** concerning which there seems to be no subsequent mention.

Portrait

Robert Boyle's portrait by Johann Kerseboom†† was presented by his executors to the Royal Society, and on 16 November 1692, " The President was desired to Return the Societie's Thanks to the Executors

---

* Some of these will be found listed in N. Grew's *Musaeum Regalis Societatis*, London, 1681.

† Royal Society. Classified Papers, 1606–1741, **XI** (1), 21.

‡ *Works Fol.*, **V**, p. 198: *Works*, **V**, p. 743.

§ Birch. *History*, Vol. **IV**, p. 231. Francis Aston was the secretary.

|| *Works Fol.*, **V**, p. 199: *Works*, **V**, p. 744.

¶ Royal Society. Council Minutes, 9 February 169$\frac{1}{2}$.

** Birch. *History*, **II**, 247: *Works Fol.*, **V**, p. 384; *Works*, **VI**, p. 268. Letter of H. Oldenburg to RB dated 18 February 166$\frac{7}{8}$: *Works Fol.*, **V**, p. 254; *Works*, **VI**, p. 74. Letter of RB to H. Oldenburg dated 21 February 166$\frac{7}{8}$.

†† R. E. W. Maddison. "The Portraiture of the Honourable Robert Boyle, F.R.S." *Annals of Science*. (1959, published 1962), **15**, p. 163 and Plate 13.

of Mr. Boyle for their Present of Mr. Boyle's picture this Day presented."* It is still in the possession of the Society.

Chemical Papers

By his will Robert Boyle entrusted his sister Katherine with the proper disposal of his manuscripts and other writings. As she predeceased her brother, he appointed Sir Henry Ashurst in her stead, and enjoined all his executors to have especial regard to "that part of my will, that I had particularly recommended to the care of my dearest sister Ranelagh deceased." Apart from these instructions in his will, he also left a memorandum concerning his chemical papers, to the effect that they should be examined by three physicians nominated by him. A letter written by John Warr, Senr, to his son John Warr, Junr, one of the executors, reveals that

> "The Doctors met according to appointm^{t} and at first they desired to see Mr Boyle's note†, impowering and directing them what to do. I told them it was not in my custody: but at last they took out one Box of Chemical processes, and spent some hours in writing, but went not through above half of it, but appointed to meet ag^{n} Tuesday next. Mr Sinclair‡ and myself did attend them all the while. They desired me to send to you for two things, one was the note of Directions mentioning their three names & the 2^{d} was for the key or keys of ye Chemical terms, without which they could order nothing."§

The identity of the three physicians is disclosed in John Locke's letter of 26 July 1692 to Isaac Newton. He says:

> "Mr Boyle has left to Dr Dickison || Dr Cox¶ & me the inspection of his papers. I have here inclosed sent y^{u} the transcript of two of them that came to my hand because I knew y^{u} desired it. Of one of them I have sent y^{u} all there was: of the other only the 1st period, because y^{t} was all y^{u} seemed to have a minde to. If y^{u} desire the other periods I will send y^{u} them too. If I meet with any thing more concerning the process he communicated to y^{u} y^{u} shall have it. And if there be any thing more in relation

---

* Journal Book of the Royal Society, **VIII**, 1690–1696.

† This note or memorandum has not been found.

‡ Robert St. Clair, one of RB's servants.

§ Letter dated 16 July 1692. Original not found. Quoted from an extract made by Henry Miles. (B.M., Add. MSS. 4313, fol. 90.) The reply of John Warr, Junr, is missing, too. Miles gives the following extract from it: "The note that the D^{rs} desire to see, cannot, I doubt be produc'd till I come up, but they may rest assur'd of the matter of Fact."

|| Edmund Dicki[n]son. (1624–1707). M.D. (*D.N.B.*).

¶ Presumably Daniel Coxe. (?–1730). M.D., F.R.S.

to any of Mr Boyles papers or any thing else wherein I can serve yu be pleased to Command."*

What the result was of the examination of these papers by the three physicians we do not know. Boyle's papers became shared between his elder brother Richard, Earl of Burlington, and John Warr, Junr, who between them seem to have kept a fairly close hold over the papers.† After the death of John Warr in 1715, his executor, Thomas Smith, came into possession of those papers held by him. Smith offered them back to the then Earl of Burlington, who declined them.‡ The chemical papers that Smith made available to Miles were sent by the latter for perusal to Peter Shaw, who had published an abridgment of Boyle's philosophical works.§ William Wotton, for his projected but unrealised life of Boyle received some papers for perusal from Warr. They were not returned, and Miles had extreme difficulty in recovering them from Wotton's son-in-law, William Clarke.|| After they had been surrendered Miles found an . . . "abundance of Chemical papers besides those which were examined by Dr. Shaw . . . "¶

How many of Boyle's scientific papers were lost before coming into the possession of Miles, it is impossible to say. Those which he did receive, together with letters and other manuscript material, were presented by his widow in 1770 to the Royal Society. The correspondence is contained in seven volumes, and the other papers in forty-six volumes. Some other papers remained in the possession of

* Facsimile in *Catalogue of the Newton Papers sold by Order of The Viscount Lymington*. Sotheby & Co., Sale 13–14 July 1936, Lot No. 143. Printed in D. Brewster. *Memoirs of the Life, Writings, and Discoveries of Sir Isaac Newton*. Second edition. 2 vols, Edinburgh, 1860. Vol. **II**, p. 366: also in *The Correspondence of Isaac Newton*, edited by H. W. Turnbull. Cambridge, published for the Royal Society, Vol. **III**, (1961), No. 390, p. 216. Many years previously RB had communicated to both Locke and Newton certain information about a process for multiplying gold. Newton had lost part of it, and had asked Locke if he could supply the missing part from RB's papers. *Ibid.*, No. 389, p. 215. Newton gave his opinion of the process to Locke in a letter dated 2 August 1692. *Ibid.*, No. 391, p. 217. See Brewster, *loc. cit.*, p. 76.

† B.M., Add. MSS. 4229, fol. 44, p. 18.

‡ B.M., Add. MSS. 4229. fol. 110. Letter of H. Miles to [? A. Millar] dated 24 September 1741. "Mr. Smith has a legal demand as executor to Mr. Warr to the papers: for he kept those I have, by the consent of Mr. Boyle's next relations, to whom he made the offer of 'em, many years since, but they refused to accept it—after having examined a good number of them."

§ Peter Shaw. *The Philosophical Works of the Honourable Robert Boyle Esq.* 3 vols. 4to., London, 1725. Second edition, London, 1738.

|| (1696–1771) Antiquary; divine. (*D.N.B.*)

¶ B.M., Add. MSS. 4314 fol. 90. Letter of H. Miles to [? T. Birch] dated 10 August 1743.

Thomas Birch, who bequeathed them with his own papers to the British Museum in 1756.*

The Boyle Lecture

Robert Boyle's will clearly sets out his intentions in the establishment of this Lecture. That the trustees appointed by him very quickly gave consideration to their obligations, is shown by John Evelyn who, in his diary under date 13 February 169½, tells us that he went that morning to a meeting with his co-trustees, when they made choice of Richard Bentley,† a chaplain to the Bishop of Worcester, as the first Boyle Lecturer. On Monday, 7 March 169½, at St. Martin-in-the-Fields he preached his first sermon of the series on *The Folly and Unreasonableness of Atheism.*‡

The majority of the trustees wished to appoint Bentley as the lecturer for 1693 as he "had so worthily acquitted himselfe", but with "much reluctancy", in order to "gratifie Sir Jo: Rotheram", admitted Richard Kidder,§ Bishop of Bath and Wells in his stead.‖ However, Bentley was appointed for 1694.¶

A list of the Boyle Lecturers with the titles of those sermons that have been printed will be found in J. F. Fulton's *Bibliography of the Honourable Robert Boyle***. One of the most popular of the series was William Derham's *Physico-Theology* (1713), which went through numerous editions, both in English and in translation.

It would seem that there were difficulties in remunerating the lecturers in accordance with the disposition of Boyle's will, which provided a rent charge on a dwelling house in the City of London. Because "the House sometimes stood empty, and Tenants brake, or failed in due Payment of their Rent, therefore the Salary sometimes remained long unpaid, or could not be gotten without some Difficulty."†† Some trouble had already been experienced in 1694, for Evelyn

* The vicissitudes of RB's papers are dealt with more fully by R. E. W. Maddison in "A Summary of Former Accounts of the Life and Work of Robert Boyle." *Annals of Science.* (1957, published 1958), **13**, Section X, p. 99.

† (1662–1742). D.D., F.R.S. *D.N.B:* Evelyn (de Beer), Vol. **V**, p. 88.

‡ This was the general title used in 1693 to cover the collection of eight sermons in any edition. For Bentley's discussion with Isaac Newton on matters discoursed of in some of the sermons see *The Correspondence of Isaac Newton,* edited by H. W. Turnbull. Cambridge, for the Royal Society, Vol. **III**, (1961), Nos. 405, 406.

§ (1633–1703) *D.N.B.*

‖ Evelyn (de Beer), Vol. **V**, p. 123, 14 December 1692.

¶ *Ibid.*, p. 160.

** 2nd Edition, p. 197 ff.

†† Note by W. Derham at the end of the dedication to his lectures. *Collected Boyle Lectures,* 3 vols. London, 1739, Vol. **II**.

records that he spoke with Bishop Tenison with a view of procuring an Act of Parliament " to rectifie some things deficient in Mr. Boyles settlement."* Nothing seems to have developed on that occasion, for on 2 May 1696 Evelyn dined at Lambeth, and met his co-trustees " to consult about settling Mr. Boyles Lecture for a Perpetuity, which we concluded upon, by buying a Rent Charge of 50 pounds per Annum."† In May 1697 the Attorney General filed a Bill of Complaint against the executors of Boyle's will in which he complained amongst other things that negotiations with Philip Neeve of the Inner Temple for the purchase of a rent charge in fee simple of £50 for the Boyle Lecture had not been completed. He described the executors' reasons for not having done so as " frivolous and vaine ". He prayed the Lord Chancellor to compel the executors to complete the terms for establishing the Boyle Lecture. In their answers the defendants expressed their willingness to go on with the purchase of an estate for providing the £50 remuneration for the Boyle Lecturer.‡ Through the efforts of Bishop Tenison the yearly stipend of £50 per annum for ever, to be paid quarterly, was secured by a charge on a farm in the parish of Brill, co. Bucks.§ The lecture is still given.

Bequest for the Advance or Propagation of the Christian Religion amongst Infidels

It is evident that some neglect on the part of Robert Boyle's executors was responsible for one of them, namely, Richard, Earl of Burlington, taking action in the Chancery Court against his co-executors Sir Henry Ashurst, Bt., and John Warr, the latter of whom seems to have failed the trust reposed in him by his master, Robert Boyle.

The Bill of Complaint, dated 26 June 1693, after stating that Robert Boyle died possessed of " a very great personal estate consisting in ready money, jewels, plate, rings, household stuff, and other things " amounting to about £10,000, apart from real estate, recites the main intentions of the last will and testament, and refers to Boyle's charge to John Warr to take upon himself the more onerous part of the settlement of the estate.|| It continues by saying,

> " And the said John Warr at first did take upon him more immediately the troublesome part of the Execution of the said Will according to the engagement of his said Master; and for

---

* Evelyn (de Beer), Vol. V, p. 197, 29 November 1694.

† *Ibid*, p. 238.

‡ P.R.O. Chancery Proceedings; Mitford Division. C8/564/41. 17 May 1697. P.R.O. Chancery Decrees and Orders. C33/428.

§ W. Derham. *loc cit.* E. Carpenter. *Thomas Tenison.* London, 1948, p. 34, note.

|| P.R.O. Chancery Proceedings. Reynardson Division. C 9/457/104.

sometime did follow the direction and appointment of your said Orator, and of the said Sr. Henry Ashurst in the Execution of the said Will according to his duety and the trust in him reposed; But the said John Warr haveing in his owne custody, and possession, or power not onely the greatest part of the ready money yet undisposed off that did belong to the said Robert Boyle amounting to a very great sume, but also most of the mortgages, Bonds, and other Securities before mencōned for the money due to the said Mr. Boyle in his life time; and to his Executors since his death, and severall other deeds and Writings wch doe belong to the Estates that were of the said Mr. Boyles; He the said John Warr doth now by agreement with the said Sr. Henry Ashurst refuse to give a particular Account of the said money, Securitys, and Writings to your Orator, and will not deliver them upp to him, or inform him where they are, that hee may have recourse thereunto, and absolutely refuses to follow the directions of your Orator concerning the said Mr. Boyles will, whereby your Orator is altogether disabled from performing the same, and the trust thereby in him reposed by his dear and honoured Brother the said Mr. Robert Boyle to his great trouble and disquiet; For which Reasons, and for that the said John Warr* is a Person of noe considerable Estate or Ability in the World, and in the Judgment of your Orator, as well as of the Testator not to bee solely trusted with the said money, Securities and Writings, your said Orator prays that the same may bee brought before some Master of this Court, To the intent they may bee deposited in a third able & secure hand, soe that all the Executors aforesaid may joyntly report thereunto, and mutually execute and performe the last Will and Testament of the said Mr. Boyle; And your Orator likewise prayes that the said John Warr may upon his corporall Oath set forth what money hee hath received since the death of the said Mr. Boyle that did belong unto him, & where and in whose hands the said moneys now are, and likewise what deeds, Writings, Securities for moneys, and also what other Things hee the said John Warr ever had, or now hath that did belong to the said Mr. Boyle at the time of his death, and where or in whose custody or possession the same now are; and that hee and the said Sr. Henry Ashurst may answere the premisses, and your Orator bee relieved according to equity and good conscience being remediless at Law, May it please your

* However, Sir Peter Pett (1630–1699; original F.R.S.) who had dealings with John Warr said: " I believe him to be a discreet & pious man, & such an one as the French call tout à fait honet homme." (B.M. Add. MSS. 4229. fol. 44v.)

> Lsp to grant unto your Orator their Maties most gratious Writ of Subpena to bee directed to the said Sr. Henry Ashurst and John Warr thereby commanding them, and either of them at a certaine day and under a certaine paine therein to be limitted p[er]sonally to bee and appeare before your Lsp in this honobie Court, then and thereupon their corporall oaths true full, direct and p[er]fect answere to make to all and singular the p[re]misses; and further to stand to & abide such Order, decree, and Judgement touching the p[re]misses as shall seeme meete to yor Lsp."

The answers, if any, made by the defendants are not attached to this bill. It would seem, that, for the purpose of devoting the greater part of Robert Boyle's residuary estate to the propagation of the gospel amongst infidels, the executors agreed to purchase the Manor of Brafferton, co. York, for the sum of £5,400. There was evidently some unjustifiable delay in completing the settlement, for the Attorney General lodged a Bill of Complaint against the executors with a demand that they be "compelled" to proceed with the purchase of the Manor.* The executors answered this Bill of Complaint, 15 July 1695, and the cause was heard before the Lord Keeper the following day and on other days in August. The defendants agreed to lay out the sum of £5,400 in lands by the purchase of the Manor of Brafferton, and to apply the yearly rent in promoting the Christian religion amongst infidels.† Settlement was announced by the court 21 November 1695. The annual rent charge of £90 for ever was granted to the Company for Propagating the Gospel in New England; and the income from the estate in excess of £90 after deducting administrative charges was allocated to propagating the gospel in Virginia.‡

After the purchase of the Manor and lands of the Brafferton estate had been completed, the executors by indenture dated 13 August 1695 granted the above stated rent charge of £90 *per annum* to the Company for Propagating the Gospel in New England§. One half of this sum was directed to be applied by the Company for the salary of two ministers to instruct the natives in or near New England in the Christian religion; the other half was directed to be placed at the disposal of the President and Fellows of Harvard College, Cambridge, New England, to be employed as salary for two more ministers to teach the natives in or near the College the Christian religion. By indenture of lease and release, dated 30 and 31 August 1695 respectively,

* P.R.O. Chancery Proceedings. Reynardson Division. C 9/273/7.
† P.R.O. Chancery Decrees and Orders. C 33/283. Ff 714; 737; 750v; 859.
‡ P.R.O. Chancery Decrees and Orders. C 33/285. Ff 223v; 344; 432v; 792.
§ City of London. Guildhall MSS. 8002. Indenture tripartite.

the Manor and estate of Brafferton so charged were conveyed to the Mayor, Commonalty and Citizens of London upon trust for the performance of the intentions of the charity that had been established. The net surplus of the income from the estate was directed to be paid to the agent in London for the President and Masters of the College of William and Mary in Virginia for the advancement and propagation of the Christian religion amongst infidels in Virginia. The legal aspects of the case lingered on,* and it was not till 9 June 1698 that the final *Rules and Orders* respecting the charity were established by an order of the Lord Chancellor.†

It is to be noted that Robert Boyle himself did not specify that his bequest should go to Harvard College,‡ or to the College of William and Mary. The choice of these institutions was made by the executors themselves. It is not surprising that the New England Company appears in the administration of the bequest seeing that Boyle was once Governor of the Company, and Sir Henry Ashurst, one of the executors, was a member. The only specific bequest made by Boyle to the Company was one of £100 to be employed as a stock for the relief of poor Indian converts.

The subsequent history of this charity is rather checquered. The Brafferton rent charge of £90 was on the whole regularly received by the Company, though the same cannot be said in regard to Harvard College, whose governing body in 1710 had to ask for its share of the charity, which had been used by the Company for the salaries of preaching ministers. The deliberations of the governing body of Harvard College on the disposal of its share received from the Company's Commissioners in Boston are available. Sometimes the money was used to educate persons at the College to prepare them for preaching to the Indians, when no ministers were actually engaged in preaching to them.§

---

* P.R.O. Chancery Decrees and Orders. C 33/289/648.

† The *Rules and Orders* are given in Appendix II.

‡ The earliest Indian graduate of Harvard College has provided an epistolary curiosity by his letter, written in Latin, to RB. (B.L., **II**, 12. Letter undated, but 1663, of Caleb Cheeshahteaumauk (A.B., of Harvard, 1665; died 1666) to RB. A facsimile is in S. E. Morison. *Harvard College in the Seventeenth Century*. Cambridge, Mass. (1936), p. 345.

§ *Publications of the Colonial Society of Massachusetts. Collections*. (1925), **15** and **16**. Harvard College Records. Corporation Records 1636–1750.

W. Kellaway. *The New England Company, 1649–1776*. London, 1961, p. 173–5.

S. E. Morison. *Harvard College in the Seventeenth Century*. Cambridge, Mass. (1936), p. 485–6.

F. L. Weis. "The New England Company of 1649 and its Missionary Enterprises." *Publications of the Colonial Society of Massachusetts*. Transactions *1947–1951*. (1959), **18**, 134.

James Blair*, who had been elected agent for the College of William and Mary, was sent to London, where he arrived early September 1691, for the purpose of securing a charter and endowment. It is claimed that through Bishop Burnet he obtained an introduction to the Earl of Burlington, whom he entreated to direct Robert Boyle's bequest " for pious and charitable uses " to the support of an Indian school at the College.† It has been noted above that the residue of the income from the Brafferton estate was indeed directed to be paid to the agent in London for the President and Masters of the College of William and Mary in Virginia. In 1723 a brick house was built close to the college out of the proceeds received from the charity, and it has since been known as the " Brafferton building ". It was originally used for the Indian children; but the Indian school has long ceased to exist, and the building has been used as a dormitory.‡

In a letter dated 11 August 1732 from William Dawson§ to the Bishop of London|| we read: " . . . now my lordship if our humble Proposal to lay part of the Brafferton money wch is in Mr. Perrys¶ hands for the Purpose of furniture and books meets with approbation and encouragement from your Lordship, we have a very convenient room for a library over the Indian School. My Lord Burlington I am informed has promised to present us with the Hon[ble] Mr. Boyles Picture, which we intend to hang up in the aforesaid Library . . . "** The portrait, which is one of the replicas of the Johann Kerseboom painting, was presented by the 3rd Earl in 1732.†† This picture survived the fire at the college in 1859, and now hangs in the library.‡‡

---

* (1656–1743) Scottish divine; sent to Virginia as a missionary, 1685; secured charter for a college in Virginia. *D.N.B.*

† L. G. Tyler. *The College of William and Mary in Virginia.* Richmond (Virginia), 1907, p. 11.

‡ Tyler, *loc cit.*, p. 23, 62.

An account of RB's charity and the Brafferton purchase is also given in [Anon] *The History of the College of William & Mary from its Foundation, 1660 to 1874.* Richmond (Virginia), 1874.

§ President of the College of William and Mary, 1743–1752.

|| Edmund Gibson (1699–1748). *D.N.B.*

¶ Micajah Perry (?–1753) Haberdasher; agent in London for the President and Masters of the College of William and Mary. M.P., for London, 1727–1741; Lord Mayor of London, 1738–1739.

** "Unpublished Letters at Fullham, In the Library of the Bishop of London." *William and Mary College Quarterly Historical Magazine*, (1901), **9**, 218.

†† J. M. Jennings. "Notes on the Original Library of the College of William and Mary in Virginia." *The Papers of the Bibliographical Society of America*, (1947), **41**, 239.

‡‡ R. E. W. Maddison. " The Portraiture of the Honourable Robert Boyle, F.R.S." *Annals of Science.* (1959, published 1962), **15**, p. 165 and Plate 14.

The Chancery Decree of 1698 remained effective until the revolt of the American colonies. Relationships between England and America deteriorated, and finally in May 1779 at a meeting of the Court of the New England Company it was resolved, " that this Court do not think themselves warranted by their Charter from the Crown in remitting Money to New England so long as that Country continues in Arms against his Majesty & their fellow Subjects etc."*

When peaceful relations had been re-established between the two countries, the College of William and Mary made a claim for the payment of accumulated arrears, which with the proceeds of timber sales amounted in 1790 to nearly £14,000. The claim was resisted by the then Bishop of London; and his successor, Beilby Porteus,† instituted a suit in the Chancery Court on an information to the Attorney General. The consequent proceedings against the Mayor, &c of the City of London, the College of William and Mary, and others have been described in full detail in a very informative article elsewhere.‡ The bishop was successful in his suit, the Lord Chancellor directing that a plan for appropriating the charity to some other purpose should be submitted. As the title of Earl of Burlington was extinct,§ Bishop Porteus claimed to be the only person entitled to form, alter, and vary the regulations of the charity. He had a great interest in missionary activities, so he proposed that the funds in question be devoted to such work in the West-India Islands. His scheme was approved by the court, and " The Society for the Conversion and Religious Instruction and Education of the Negro Slaves in the British West-India Islands " was formed.||

As a result of the prohibition of further importation of negro slaves from Africa, and their subsequent emancipation in 1834, the scope of Beilby's society was extended, and its name altered to " The Society for Advancing the Christian Faith in the British West India Islands

* W. Kellaway. *The New England Company, 1649–1776*. London, 1961, p. 277.

† (1731–1808) D. D., Cambridge. Translated from Chester to London, November 1787.

‡ H. L. Ganter. " Some Notes on " The Charity of the Honourable Robert Boyle, Esq., of the City of London, Deceased.' *William and Mary College Quarterly Historical Magazine*, (1935), Series 2, **15**, 1; 207; 346.

§ Richard Boyle (1695–1753), 3rd Earl of Burlington and 4th Earl of Cork.

|| Robert Hodgson. *The Life of the Right Reverend Beilby Porteus, D.D.* 2nd edition London, 1811, p. 111 *et seq.*

Beilby Porteus. *A Letter to the Governors, Legislatures, and Proprietors of Plantations, in the British West-India Islands*, London, 1808, p. 2.

A. Caldecott. *The Church in the West Indies*. London, 1898.

F. W. B. Bullock. *Voluntary Religious Societies, 1520–1779*. St. Leonards-on-Sea, 1963, p. 151.

and elsewhere within the Dioceses of Jamaica and of Barbadoes and the Leeward Islands and in the Mauritius."*

With regard to that part of the income from the Brafferton estate remaining under the control of the New England Company a chancery decree directed the funds to be applied " For the advancement of the Christian religion among infidels in divers parts of America under the Crown of the United Kingdom."† The missionary operations were transferred to Canada.‡

The Executors' Charitable Schemes

Mention has been made above (p. 206) of the Attorney General's Bill of Complaint filed with the Lord Chancellor in May 1697 in which he complained of the executors' tardiness in making provision for the £50 fee to the Boyle Lecturer. In their answers the defendants, after dealing with various other matters, gave their proposal for disposal of the ultimate residue of Robert Boyle's estate. They had agreed between themselves that if the ultimate residue amounted to £3500, then the Earl of Burlington should receive £1800, Sir Henry Ashurst £1000, and John Warr £700 for charitable uses. John Warr's share was to be fixed at £700, but the other two executors were to increase or decrease their shares in proportion as the residue exceeded or fell short of £3500. Schemes for the charitable application of these sums were put forward. The proposed schemes were duly considered by the Court of Chancery, and approved by a decree pronounced by Lord Chancellor Somers on 1 July 1697.§

These three schemes, which have been considered in detail elsewhere|| by the present writer, will be described briefly here.

1. *The Earl of Burlington's Scheme*

The Chancery decree of 1697 directed the Earl to " build an Almshouse for men and a schoole house or one of them att Bolton Abby in ye County of Yorke and purchase lands of Inheritance in that or some adjacent County for ye perpetuall support and maintenance of the poore therein and ye schoolmaster of ye s^d schoole or either of them." Richard, Earl of Burlington, died shortly afterwards in January 1698, and was succeeded by his grandson Charles, who built a schoolhouse

* Patent Rolls. 6 Gul IV. Part 10. No. 3. 11 January 1836.

† [H. W. Busk]. *A Sketch of the Origin and the Recent History of the New England Company*. London, 1884, p. 19. Decree of 23 April 1792.

‡ Special Correspondent. " The New England Company. Three Centuries of Missionary Endeavour." *The Times*, 27 July 1949, p. 5.

§ P.R.O. Chancery Decrees and Orders. C 33/287. Ff. 653; 654.

|| R. E. W. Maddison. " The Charitable Disposal of Robert Boyle's Residuary Estate." *N. & R.*, (1952), **10**, 15. Full references are given in this article.

and school on his estate at Bolton Abbey. The land near the ruins of the Abbey together with the buildings thereon were, in 1700, conveyed to trustees to keep a grammar school there to be known as " The Free School of the honourable Robert Boyle, deceased ". Certain other properties were purchased in order to provide an endowment, from which the usher of the school was to receive £12 p.a., and the schoolmaster the balance of the income. If no usher were appointed, the whole income was payable to the schoolmaster for his own maintenance, and for the repairs to the schoolhouse. In 1701 statutes and regulations for the government of the school were made by Charles, Earl of Burlington, with the consent and advice of the Archbishop of York. These provided that " the school should be open for the children of all noblemen, gentlemen and others, able to give their children a liberal education in the Latin and Greek tongues, upon such terms as the parents and the master of the school might agree upon, with preference to the children of the tenants of the Earl and his heirs, and other inhabitants within" certain specified localities. Provision was made for certain other children to be instructed in Latin and Greek free, or on payment of one shilling per quarter per child according to circumstances. Further provision was made for other poor children to be instructed in English, writing, and arithmetic on payment of the like sum.* The master and usher were required to attend church with their scholars every Sunday; to instruct them in the catechism; and to read prayers in school each morning and evening. The position of usher in the Boyle School seems to have been held jointly with that of parish clerk. The names of the ushers have not been preserved; a certain Robert Keith was probably the first master of the school from 1702–8.*

Richard, 3rd Earl of Burlington, in 1715 gave £5 for ever " to be paid to the writing master of Bolton to teach the boys writing for six weeks in June and July."† It is interesting to note that the number of scholars doubled itself during those months when writing and arithmetic were taught; in the remaining part of the year the number did not exceed 15 to 20. From its foundation down to the time of the Brougham Charity Commission, in the early part of the nineteenth century, the curriculum of the school was almost entirely devoted to the classics, although the statutes did make provision for instruction in general elementary subjects.

---

* Reports of the Commissioners for inquiring concerning Charities in England and Wales. (Known as Lord Brougham's Charity Commission.) Folio. London, 1920. **III**, 509, and Appendix 253.

† A. P. Howes. *The Parish Register of Bolton Abbey 1689 to 1812.* Skipton, 1895.

In 1863 it was proposed to sell the school and site, and with the proceeds to provide a new school in a more convenient situation. Nothing was done until 1872 when the Charity Commissioners empowered the trustees to make the sale, and to apply the proceeds in the establishment of a new school to be regulated by a scheme to be made under the Endowed Schools Act of 1869. In 1874 the erection of the new school was started at Beamsley. After rather protracted negotiations a scheme was established for the regulation of the Boyle School. It received approval in 1887, and united in one foundation, to be called the Boyle and Petyt Foundation, the endowments of the Boyle Free School with part of Sylvester Petyt's Charity, namely, the school at Hazelwood. The scheme of 1887 was modified in respect of the governing body and application of income by the Board of Education's scheme of 1923, which still regulates the Boyle and Petyt Foundation for the benefit of Bolton Abbey and neighbouring places in the West Riding of Yorkshire.

The site of the original Boyle School is part of the base court of Bolton Abbey. Some of the monastic buildings were probably modified to form the school premises; whilst the schoolmaster's house was established on the site of the priory kitchens, and the Boyle Room on the site of the infirmary.* It was customary for the Duke of Devonshire† to present the headmastership of the school to the chaplain of the Abbey, so that the schoolhouse served as the residence for the one individual who held both appointments. This arrangement continued until the new school at Beamsley was erected. The schoolmaster's house was subsequently assigned as a rectory. Over the porch of this building still appears the coat of arms of the Boyle family differenced by a martlet. The following inscription in stone is over the porch:

SCHOLA BOYLIANA
ROBERTUS BOYLE ARM^R SUMPTUS ET STIPENDIA
HUIC SCHOLAE FUNDANDE PERPETUANDE
QUAE LEGAVIT
CAROLUS COMES
BURLINGTON ET CORKE
FUNDUM LIGNUM LAPIDES ET ALIA
DESIDERATA AD AEDES ERIGENDAS
MUNIFICE DONAVIT
A.D. MDCC.

* For a full description of the Priory of St. Mary see A. H. Thompson, "History and Architectural Description of the Priory of St. Mary, Bolton-in-Wharfedale." *Publications of the Thoresby Society* (1928), **30**.

† The title of Earl of Burlington became extinct at the death of the 3rd Earl. His youngest daughter and heir, Lady Charlotte Boyle, married the Marquis of Hartingdon, who became the 4th Duke of Devonshire.

On the south-east side at right angles to the Rectory is a building which was used as schoolroom till 1878; it is now used as a Parish Room, and is known as "The Boyle Room".

2. *Sir Henry Ashurst's Scheme*

The Chancery decree directed Sir Henry Ashurst

> "To purchase ground rents in ye City of London and Westm[inster] or either of them or in the suburbs thereof or Lands Tenemts or Hereditamts elsewhere in ye Kingdome of England in fee simple out of ye rents issues and p[ro]fitts whereof ye severall Annuall sumes of 40s & 8£ are to bee pd to ye Company of Merchant Taylors in Lond and to ye Mayor Bayliffes and Comonalty of ye City of Oxford or to either of ye sd Corporacons ye sd yearly sumes of 40/- to bee layd out Annually forever in the buying of English Bibles to bee distributed or given to & amongst ye Children of poore freemen of either or both ye said Corporacons & ye sd yearly sume of eight pounds to bee distributed & given Annually forever to & amongst four poore freemen aged 60 yeares or upwards and 4 poore widdowes of freemen aged 50 yeares or upward of either or both ye sd Corporacons to each of ye sd poore freemen and widdowes the yearly sume of 20s apeece & ye residue ande remaind[er] of ye rents issues and p[ro]fitts thereof to bee yearly laid out for ever in ye teaching and instructing poore Children to read & write to buy them bibles or towards ye releife and mayntenance of poore widdowes of Viccars in the County of Oxon & in Bridport in Dorsettshire or in either of ye sd places".

Sir Henry's share of the residuary estate amounted to £348 6*s*. 8*d*,* which sum was used to purchase fee-farm rents in the county of Lancaster. The charity was established in 1701, when it was transferred to trustees. At the time of the Brougham Commission the net income available to this charity amounted to £12 8*s*. 0*d*. The descendants of Sir Henry Ashurst still appointed the Corporation of Oxford to receive the charity; £10 being distributed equally amongst four poor freemen and four poor freemen's widows, and the balance of 48*s*. to some poor man nominated by the Ashurst family.

Between 1822 and 1871 this Oxford Boyle Charity acquired in some unknown manner the income from £55 10*s*. 6*d*. Consolidated £3% annuities. In 1884 the Charity Commissioners made a scheme for regulating the Oxford Municipal Charities, as a result of which four branches of charitable activity were created for the Robert Boyle

* On this basis the Earl of Burlington's share was £627 instead of the hoped-for £1800. With Warr's share fixed at £700, the approximate total of the ultimate residue of RB's estate was £1675.

Charity, namely, the Loan Branch; the Apprenticing Branch; the Almshouse and Annuity Branch; and the General Branch. The income allocated to the Apprenticing Branch was used for placing out as apprentices to any suitable trade, occupation, or service, the children of poor persons resident in the City of Oxford; or for the instruction of children with the view to their entering upon trade or handicraft. The income of the General Branch was used for the benefit of deserving and necessitous persons resident in the City of Oxford.

In 1907 the scheme regulating these charities was amended. The Apprenticing Branch was abolished, and in its stead two branches were established: (1) The Apprenticing and Advancement in Life Branch; (2) The Education Branch. The Education Branch of the Oxford Municipal Charities is regulated by a scheme of the Board of Education made in 1930, whilst the charities, apart from the Education Branch, are regulated by a scheme of the Charity Commission made in 1932, varied by a scheme made in 1949.

3. *John Warr's Scheme*

At the time when the Chancery decree was made in 1697 a large part of Robert Boyle's personal estate had not been got in by the executors. A short while before it had been agreed that " the Earle of Burlington and John Warr shall and may out of the first moneys that they shall severally have and receive as part of their respective proporc̄ons goe with the Building of the Almshouse and Schoole house or Schoole houses before there be any purchase made of Lands for the endowing thereof; hopeing by that means that the Almshouse and School house or Schoolehouses may be neare ffinished by the time the purchases are made ".* The Earl of Burlington's successor promptly discharged his duty; he built the school, purchased lands for an endowment, and conveyed the whole to trustees in 1700. Neither Sir Henry Ashurst nor John Warr was equal to this efficiency. Both in May 1698 applied to the Chancery Court for an extension of time in which to settle the regulations of their proposed charities.† Sir Henry Ashurst's scheme was finally established on a proper legal basis in 1701.

John Warr was directed by the Chancery decree of 1697 " to build a schoole house for boyse att Yetminster near Sherbourne in Dorsettshire & to p[ur]chase lands of Inheritance in ye s^{d} County or near it for ye mayntenance and encouragem^{t} of ye m[aste]r thereof for ye tyme being ". In 1699 John Warr acquired about 40 acres of land at Knighton, and in 1712 another acre. He had also purchased a small estate of his own, on which he had erected a school with a dwelling

* P.R.O. Chancery Proceedings; Mitford Division. C 8/564/41.
† P.R.O. Chancery Decrees and Orders. C 332/89/531. 31 May 1698.

house for the master.* The earliest mention of this school that has come to my notice occurs in the annual accounts of charity schools for 1711, where the following entry appears:

> "Yetminster, *Dorsetshire*. A Sch[ool] part of the Charity of the honourable *Robert Boyle*, Esq; for 20 poor Boys. An House built, and £20 a Year settled."†

The entry is starred, denoting either that the school had been set up since the previous year's account, or had not been mentioned before for lack of information. It is interesting to note that the Yetminster Foundation was a Charity School, whereas the Earl of Burlington's Foundation at Bolton Abbey was a Grammar School.

It is regrettable that John Warr, in whom Robert Boyle had shown so much confidence, completely neglected to establish trusts for his charitable scheme. Why, or how, he failed to do so can only be a matter of speculation. After his death in 1715‡ steps were taken to establish a trust upon a proper legal basis; and in 1717 the school premises with about 41 acres of land were conveyed to trustees.* The school then entered upon a long period of apparently uneventful existence, until, with the passing of the Endowed Schools Act of 1869, a revised scheme for the management of the school was made in 1873. Other orders modifying the scheme were made in later years. In 1947 the school was amalgamated with the Yetminster County School, where all the teaching takes place. The premises of the old Boyle School were used as a canteen and dining-room for the children attending the amalgamated school. A further scheme was made by the Minister of Education in 1955 in accordance with which the Boyle School is discontinued; the foundation and its endowment are now administered under the name of Boyle's Educational Foundation, and the income is devoted to other educational purposes, with preference for boys resident in the Parish of Yetminster.§

A replica of Johann Kerseboom's portrait of Robert Boyle hangs in the school-house.||

* Lord Brougham's Charity Commission. (1837), **XXX**, 142.

† *An Account of Charity-Schools in Great Britain and Ireland: With the Benefactions thereto; and of The Methods whereby they were set up and are governed.* The Tenth Edition, with large Additions. London, 1711.

Idem. The Eleventh Edition. London, 1712. An account of the government, orders to be observed by the master and scholars, etc., is given on pp. 3–10.

‡ He died intestate. Letters of administration were granted to Thomas Smith in February 1716. This is undoubtedly the same Thomas Smith who was with RB at his death.

§ Scheme of the Board of Education. No. 3758P. 30 March 1955.

|| R. E. W. Maddison. "The Portraiture of the Honourable Robert Boyle, F.R.S." *Annals of Science.* (1959, published 1962), **15**, p. 170 and Plate 22.

In conclusion, it is perhaps desirable to emphasize that Robert Boyle did not give any specific instructions for the establishment of the three charities described above. They were proposed by the three executors, whose proposals were approved by the Court of Chancery. As with the passage of time so many charities have disappeared, either through the negligence or dishonesty of trustees, or the operations of the law, it is gratifying to find that Robert Boyle's benefactions are still active, and perform works which would certainly give satisfaction to their pious founder. The value of certain of these benefactions has been whittled down by the sale of the original real estates under various official schemes of reorganisation, and investment of the proceeds in trustee stocks. Administrative convenience does not necessarily operate to the advantage of a testator's intentions.

VII

Miscellanea

Robert Boyle's Personal Appearance and Health

In his youth Boyle's personal appearance was that of a normal, healthy growing lad. Isaac Marcombes, who was his guardian and tutor during his residence abroad during the years 1639–1644, refers several times to Robert's rapid growth, saying that he is very fat, has rosy cheeks, and " is an Eale, a quarter and halfe high ". According as we take the French or the Genevan ell as a basis, Robert was then between 5′ 2″ and 5′ 4½″ tall.* Subsequent to his return to England from the continent in 1644, Robert Boyle developed " a sickly constitution ", as he himself describes it. At an early date (1647) he was " disquieted " with the stone, for which condition he received treatment at various times over a number of years from G. Stirke (1649), J. Unmussig (1654), and probably also from G. Boate. An attack of the ague, which lasted three to four weeks, was treated by Dr. H. Davies (1649); and he had recurrent attacks in later years (1657, 1659). Dr. Coxe gave him an electuary for *sputo sanguinis* (1652 or 1654).

Throughout his adult life Boyle complained of " the great weaknesse of my sight ", or the " unhappy distempers of my eyes ", or " the flame of a candle is offensive to my weak eyes." For his eyes he consulted William Harvey and the well-known oculist, Daubeney Turberville. His last letter to the latter has been quoted above (p. 180). His serious " paralytic distemper " (1670) has already been described (p. 145).

Boyle has provided an interesting account of his own case in the preface to the posthumously published *Medicinal Experiments*.† He says:

* For references see, R. E. W. Maddison. " The Portraiture of the Honourable Robert Boyle, F.R.S." *Annals of Science*. (1959, published 1962), **15**, 178.

† London, 1692. This preface is not present in the first edition of the work, which was undoubtedly printed for private distribution. See R. E. W. Maddison. " The First Edition of Robert Boyle's *Medicinal Experiments* ". *Annals of Science*. (1962, published 1964), **18**, 43.

" . . . many may think it strange, as I my self have been prone to do, that I should presume to recommend Medicines to others, who for divers Years have been so infirm and sickly myself. And some, 'tis like, will upbraid me with, *Medice, cura teipsum*. But on this Occasion, I may represent, that being the thirteenth or fourteenth Child of a Mother, that was not above 42 or 43 Years old when she died of a Consumption, 'tis no wonder I have not inherited a robust or healthy Constitution. Many also have said, in my Excuse, as they think, that I brought my self to so much Sickliness by over-much Study. But I must add, that tho' both the foremention'd Causes concurred, yet I impute my infirm Condition more to a third, than to both together: For the grand Original of the Mischiefs that have for many Years afflicted me, was a Fall from an unruly Horse into a deep Place, by which I was so bruised, that I feel the bad Effects of it to this Day. For this Mischance happening in Ireland, and I being forc'd to take a long Journey before I was well recover'd, the bad Weather I met with, and the Mistake of an unskilful or drunken Guide, who made me wander almost all Night upon some wild Mountains, put me into a Fever and a Dropsy (*viz.* an Anasarca:) For a compleat Cure of which I pass'd into England, and came to London; but in so unlucky a Time, that an ill-condition'd Fever rag'd there, and seiz'd on me among many others; and though thro' God's Goodness I at length recover'd, yet left me exceeding weak for a great while after; and then for a Farewel, it cast me into a violent Quotidian or double Tertian Ague, with a Sense or Decay in my Eyes, which during my long Sickness I had exercis'd too much upon critical Books stuff'd with Hebrew, and other Eastern Characters. I will not urge that divers have wonder'd that a Person in such bad Circumstances, by the Help of Care and Medicines, (for they forget what ought to be ascrib'd to God) should be able to hold out so long against them. But this after the foregoing Relation may well be said, that it need be no great Wonder, if after such a Train of Mischiefs, which was succeeded by a scorbutick Cholick that struck into my Limbs, and deprived me of the Use of my Hands and Feet for many Months, I have not enjoyed much Health, notwithstanding my being acquainted with several choice Medicines; especially since divers of these I dare not use, because by long sitting, when I had the Palsy, I got the Stone, voiding some large ones, (as well as making bloody Water) and by that Disease so great a Tenderness in my Kidneys, that I can bear no Diureticks, tho' of the milder Sort, and that I am forc'd to forebear several Remedies for my

> other Distempers, that I know to be good ones . . . This short Narrative may, I hope, suffice to shew that my personal Maladies and Sickliness cannot rightly infer the Inefficacy of the Medicines I impart or recommend . . . "

Apart from any intrinsic interest in the subject itself, it is hardly surprising that Boyle should have engaged in much self-medication, a practice which was probably encouraged by some of his own unfortunate experiences from physic prescribed by physicians. His interest in medico-chemistry was further stimulated by the desire to alleviate pain and suffering of others, with the result that he was an ardent collector of recipes and seeker after new or improved remedies. In this he was certainly encouraged by his sister, Katherine.* Boyle acquired considerable renown as a prescriber, and was ever ready to proffer a remedy to known or unknown enquirers.†

Besides Faithorne's drawing (Plate I), which gives evidence of a person not in the best of health, we have the confirming statements of contemporaries. Abraham Hill‡ records that "Mr. Boyle is a person somewhat sickly, and taken up with many affairs, so that he much declines foreign correspondence."§ Caspar Lindenberg's account of his meeting with Boyle is recorded as follows:

> "Er [Lindenberg] rühmte unter andern die Gelehrsamkeit und die grosse Höflichkeit des Roberti Boyle, und versicherte, dass er von allen Engelländern, so er gesprochen, das beste und recht gut Latein geredet. Als er sich bey ihm habe melden lassen, habe er ihm zur Antwort wissen lassen, es solte ihm sein Zuspruch gar angenehm seyn, dafern er nur nicht kürzlich an einem Orte, da die Pocks, oder Kinder-Pocken grassiret, gewesen: denn er habe sich

---

* Her medical receipt book is in the B.M., Slo. MSS. 1367.

† A certain Jo. Cooke, employed in the government service, invoked RB's help for his poor child afflicted with the King's Evil. (*Works Fol.*, **V**, p. 635; *Works*, **VI**, p. 642. Letter dated 13 August 1664.). RB recommended Huygens to take *Aqua simplex paralyseos* as a cure for sleeplessness. (Huygens. *Oeuvres*, **VIII**, No. 2086. Letter of H. Oldenburg to C. Huygens dated 7 February $167\frac{5}{6}$). He prescribed for George, 1st Earl of Berkeley, (B.L., **I**, 62. Letter dated 29 September 1680); and for Hooke's impaired sight during 1689 he recommended an infusion of *Crocus metallorum*, and on another occasion, clarified honey with rosemary flowers (Gunther, **X**, pp. 124, 130.). On 13 April 1689 he examined the eyes of a young man who had cataract in his left eye, and saw colours with his right eye (B.P. **XXI**, item 84).

‡ R. E. W. Maddison. "Abraham Hill, F.R.S. (1635–1722)," in *The Royal Society, Its Origins and Founders*, edited by Sir Harold Hartley, London, 1960, pp. 173–182.

§ Abraham Hill, *Familiar Letters*, London, 1767. p. 229. Letter of A. Hill to Dr. Nicolaus Witte at Riga in reply to a letter from the latter dated 17 July 1663.

vor dieser Krankheit sehr gefürchtet. Herr Boyle seye auch sehr schwächlicher Constitution gewesen, und habe so dürr und elend, wie ein Sceleton, ausgesehen."*

James Yonge† was taken by Robert Hooke to visit Boyle in the autumn of 1687. He described him as "a gentleman of vast knowledge and learning, and great courtesy. He entertained me with free discourse above an hour in his chamber. He is a thin man, weak in his hands and feet, almost to a paresis; soft voice, pleasant though pale countenance, somewhat long faced, and a long straight nose somewhat sharp."‡

John Evelyn described Boyle as "rather tall and slender of stature, for most part valetudinary, pale and much emaciated; nor unlike his picture in Gresham College."§ (Plate XXX).

Notwithstanding his "sickly constitution", Robert Boyle lived a full life. Apart from his philosophical pursuits, his attention to the affairs of the Royal Society, the New England Company, and other public bodies, he took part in hawking, setting and angling.|| He claimed to be "very Musically given", though it is not known if he played any musical instrument.¶ He was not above using his chemical knowledge "for recreation sake, and to affright and amaze Ladies" by burning paper impregnated with weak spirit of wine;** or to play a joke with the staining property of silver nitrate.††

Boylaen Eponymy

The Sign of "Boyle's Head"; and Boyle's Head Court

It is hardly surprising that a person of Robert Boyle's fame should, after his death, have featured on tradesmen's signs. The following users of the sign of "Boyle's Head" have come to my notice, and it is possible that there are others.

1. Thomas SMITH. (?–1742) Apothecary. He lived from 1700, or perhaps late 1699, till his death at his dwelling house

* Z. C. von Uffenbach. *Merkwürdige Reisen durch Niedersachsen, Holland und Engelland.* 3 vols. Ulm and Memmingen, 1753–1754. Vol. **II**, p. 68.

† (1647–1721), Surgeon.

‡ *The Journal of James Yonge.* Edited by F. N. L. Poynter. London, 1963, p. 200.

§ Evelyn. *Diary* (Bray), Vol. **III**, p. 351. Letter of J. Evelyn to William Wotton dated 29 March 1696.

|| *Works Fol.*, **I**, p. 203; *Works*, **I**, p. 316; *Works Fol.*, **II**, p. 182 *et seq*, Section IV; *Works*, **II**, p. 391 *et seq*, Section IV.

¶ *Works Fol.*, **I**, p. 541; *Works*, **II**, p. 181.

** *Works Fol.*, **I**, p. 212; *Works*, **I**, p. 331.

†† *Works Fol.*, **II**, p. 34; *Works*, **I**, p. 714.

at the sign of " Boyle's Head " opposite Salisbury Street in the Strand, London. He it was, who " lived seventeen years with Mr. Boyle, and was with him at his death."

2. John WHISTON. (?–1780) Bookseller. His shop was in Fleet Street, London. William Whiston's *Six Dissertations*, 8vo London, Printed for John Whiston at Mr. *Boyle's* Head in *Fleet-Street*, 1734 has on the title-page a vignette obviously based on the Smith mezzotint by John Smith,* but reversed. Another book by the same publisher bearing this vignette is *Remarks on Spenser's Poems*, 8vo London, 1734. It appears also on the *Harveian Orations* of 1746 and 1747, but not on subsequent ones published by Whiston.
3. Lancelot SHADWELL. Chymist. In Leadenhall Street, London, about 1760†.
4. Nathaniel CONANT. Bookseller.

This sign gave the name of Boyle's Head Court to a cul-de-sac between Bull Inn Court and Oliver's Alley on the north side of the Strand opposite Salisbury Street.‡ The court is shown with this name on Rocque's map of 1746. On Morden and Lea's map of 1682 it is called Shearman's Entry, and appears as such in [E. Hatton's] *A New View of London.*§

In the rate books of the parish of St. Martin-in-the-Fields, the name of Boyle's Head Court does not occur; in 1705 the name is Glove Alley; in 1742 and 1749 it is Stationer's Alley; and at other times it is simply noted as Alley. The court disappeared during subsequent rebuilding on this site.||

Boyle's Hell

This is the name that has been acquired by a certain piece of chemical glassware employed for the preparation of *mercurius praecipitatus per se*, *i.e.*, mercuric oxide prepared by heating mercury in air. The term *enfer* was in use in France by the middle of the seventeenth century for a particular shape of matrass having a flat broad base.¶ Robert

---

* R. E. W. Maddison. " The Portraiture of the Honourable Robert Boyle, F.R.S." *Annals of Science* (1959, published 1962), **15**, 161.

† Private communication dated 8 February 1952 from the late Sir Ambrose Heal.

‡ *London and its Environs Described.* 6 vols. 8vo., London, 1761. Vol. **I**, p. 343.

§ London, 1708, Vol. **I**, p. 75.

|| L.C.C. Survey of London. *The Strand.* Vol. **XVIII**, (1937), p. 127.

¶ Nicaise Le Febure (Nicolas Le Fèvre). *Traicté de la Chymie.* Paris, 1660. pp. 833–4, 835.

Boyle was not responsible for the term " hell ". Furthermore, when he refers to the preparation of precipitate *per se*, he uses the expressions " convenient glass vessel ", or " conveniently shaped glass "; and in connection with his experiments on the effluxions of fire he speaks " of those infernal glasses (as they call them)."* It is only in certain chemical writings by French authors of the seventeenth and eighteenth centuries that we find the term *enfer*. The most explicit description is given by Lemery.† The earliest use of *enfer de Boyle*, that I have noticed, occurs in Diderot's *Encyclopaedia*‡, where in Volume V issued in 1755 we find the following:

> " *Enfer de Boyle*, vaisseau circulatoire d'un verre fort, composé de plusieurs pièces, qui toutes ensemble font une espèce de matras, ayant le col long & étroit & le globe très-applati, imaginé par le célébre Anglois dont il porte le nom, pour faire ce qu'on appelle *le mercure fixé per se*."

Subsequently, the term is used quite generally by French writers, for example, Lavoisier and Fourcroy. It will even be found in the standard French dictionary of Littré.§

The term " Boyle's Hell " does not appear in any contemporary, or near contemporary, English, German, or Italian, chemical literature that I have consulted. Its appearance in early nineteenth century English works derives from quotation from French writers.

This subject has been more fully described by the present writer elsewhere.||

Boyle's Law

At a meeting at Gresham College, 18 September 1661, " Mr. Boyle brought in his account in writing of the experiment made by him of the compression of air with the quicksilver tube; which was ordered to be registered."¶ It was duly entered 2 October 1661.** The results were published in *A Defence Of the Doctrine touching the Spring and*

---

* *Works Fol.*, **I**, p. 271; **III**, pp. 620, 346; *Works*, **I**, p. 427; **IV**, p. 308; **III**, p. 716.

† Nicolas Lemery. *Cours de Chymie*. Nouvelle édition par M. Baron. Paris, 1757, pp. 199–200.

‡ Denis Diderot et J. le Rond d'Alembert. *Encyclopédie, ou Dictionnaire raisonné des sciences, des arts et des métiers*. 35 vols. Paris, 1751–1780.

§ E. Littré. *Dictionnaire de la Langue Française*. 7 vols. Paris, 1956–8.

|| R. E. W. Maddison. " Boyle's Hell ". *Annals of Science*. (1964, published 1965), **20**, 101.

¶ Birch. *History*, **I**, p. 45.

** D. McKie. " Boyle's Law." *Endeavour* (1948), 7, 148-151. An illustration of the relevant entry in the MS Register Book of the Royal Society is given there on p. 150.

Weight Of the Air (appended to the second edition of *New Experiments Physico-Mechanical, Touching the Air* (1662)) where we find the experimental data confirming the " *Hypothesis*, that supposes the pressures and expansions to be in reciprocal proportion."* This hypothesis had been suggested to Boyle (as he acknowledged) by Richard Towneley (1628–1707). The latter together with Henry Power (1623–1668) had previously made experiments with the Torricellian tube in respect of rarefaction, and had obtained results agreeing with constancy of the product of pressure and volume.†

At the time when Boyle made his experiments, he was fortunate in having Robert Hooke as his valuable assistant; and there is evidence that the latter had as much to do with verifying the generalization now known as Boyle's Law, as did Boyle himself. The position is appropriately summed up by I. B. Cohen, who says, " this is a law discovered by Power and Towneley, accurately verified by Hooke, accurately verified again by Boyle (aided in some degree by Hooke), first published by Boyle, but chiefly publicized by Mariotte, in short, the ' law of Power and Towneley, and of Hooke and Boyle, and—to some degree—of Mariotte '. This example certainly shows the complexities in making precise the individual contributions in the common multiply-discovered scientific law and thus argues the folly of eponymy in science."‡

Although Boyle's Law is frequently called Mariotte's Law, it is to be noted that the latter in his essay *De la Nature de l'Air* (1679) did not claim its discovery.§ Priority of publication rests with Boyle.

Boyle Point (*or Temperature*)

Boyle's Law ($pv = \text{constant}$), which was established from experiments on *air*, is strictly valid only for an *ideal* gas. For a real gas

$$pv = A + Bp + Cp^2 \ldots .$$

where A, B, C etc. are virial coefficients.|| For a molar weight of an

---

* *Works Fol.*, **I**, p. 100, Chap. V; *Works*, **I**, pp. 156 ff.

† C. Webster. " Richard Towneley and Boyle's Law ". *Nature* (1963), **197**, 226.

E. Gerland. " Die Entdeckung der Gasgesetze und des absoluten Nullpunktes der Temperatur durch Boyle (nicht Townley) und Amontons." Contained in P. Diergart, *Beiträge aus der Geschichte der Chemie dem Gedächtnis von Georg W. A. Kahlbaum*. Leipzig, 1909, p. 350.

‡ I. B. Cohen. " Newton, Hooke, and ' Boyle's Law ' ". *Nature*, (1964), **204**, 618.

§ McKie, *loc cit:* J. R. Partington. *A History of Chemistry*, Vol. **II**, (1961), p. 523, n. 2: J. R. Partington. *An Advanced Treatise on Physical Chemistry*. 5 Vols. London, 1949–1954, Vol. **I**, p. 553.

|| J. R. Partington. *An Advanced Treatise on Physical Chemistry*. 5 vols. London, 1949–1954, Vol. **I**, p. 577.

ideal gas $pv = RT$, where R is the general gas constant and T is the absolute temperature on the Centigrade gas scale; hence $A = RT$. The coefficients B and C are of influence at high pressures. Deviations from the simple Boyle's Law are determined by B, which depends on temperature. That temperature at which B becomes zero is called the Boyle Point (or Temperature), and the gas will obey Boyle's Law up to fairly high pressures (though it is still not an ideal gas) so long as other virial coefficients do not have effect.*

The Robert Boyle Lecture

There are two lectures of this name. The first was founded by Robert Boyle himself in accordance with the terms of his last will and testament, and has been considered above (p. 205).

The other was instituted by the Oxford University Junior Scientific Club in 1892,† and the inaugural lecture was given by Sir H. W. Acland. A list of the lecturers with the titles of their addresses will be found elsewhere.‡

The Robert Boyle Medal

The Royal Dublin Society instituted in 1899 a medal " to be awarded from time to time with a view of encouraging worth in the different branches of science." Robert Boyle was chosen to be commemorated by this medal for he " did more for science than any other of the great Irishmen who have passed away."§ The medal was executed by Allan Wyon.‖ The obverse shows a profile of Robert Boyle facing to the right taken either from the marble bust at Trinity College, Dublin, or the plaster bust at the Royal Dublin Society, and bears the inscription —IN HONOREM ROBERTI BOYLE ET AUGMENTUM SCIENTIARUM. FELIX QUI POTUIT RERUM COGNOSCERE CAUSAS. At the left is the name ROB/ERT/VS/BOY/LE. The reverse of the medal shows a figure of Minerva modified from the seal of the Society, and bears the inscription—REGALIS : SOCIETAS : DUBLINENSIS : CONDITA : A : A : MDCCXXXI.

The first recipient of this medal was Dr. G. J. Stoney, F.R.S., in 1899 for his contributions to science, and his personal influence on

---

* *Ibid.*, 672.

† *J. Oxford Jun. Sci. Club.* (1892), 1, 68.

‡ Fulton. pp. 202–204.

§ Henry F. Berry. *A History of the Royal Dublin Society*. London, 1915, p. 373 *et seq.*

‖ (1843–1907). Chief engraver of H. M. Seals.

scientific advancement in Ireland. A list of subsequent recipients and illustrations of the medal have been given elsewhere.*

The Air-Pump

(i) *Machina Boyleana*

Reference has already been made to the development of the air-pump by the joint efforts of Boyle and Hooke (chapter III). The machine they evolved was the first air-pump to be made in England, so it is not surprising that the machine possessed various imperfections and inconveniences. The history of the development of the air-pump in England may be divided into four stages under the following dates.†

> " 1659. The construction of a pneumatical engine, consisting of a single-barrelled pump, with a solid piston moved by a rack and pinion, and a globular glass receiver directly communicating with the cylinder, which had an aperture in it, closed and opened by a plug moved by the hand, and playing the part of a valve.
>
> " 1667. The separation of the glass receiver from the cylinder, and introduction of the air-pump plate, on which bell-jars could be placed and used as receivers. The pump was still single-barrelled, and wrought by a rack and pinion, but with an aperture in the piston, instead of in the cylinder, furnished with a moveable stopper.
>
> " 1676. The introduction of a double-barrelled pump, with self-acting valves in the cylinders and pistons, and with piston-rods suspended at opposite ends of a cord passing over a pulley.
>
> " 1704. The combination of the rack and pinion of the first and second air-pumps, with the two barrels, twin-pistons, and self-acting valves of the third."

From what we know of Hooke's inventive genius, there can be no doubt but that he must be accorded the merit for carrying through the

* R. E. W. Maddison. " The Portraiture of the Honourable Robert Boyle, F.R.S." *Annals of Science.* (1959, published 1962), **15,** 208 and Plate XLVI.

† G. Wilson. " On the Early History of the Air-Pump in England." *Edinburgh New Philosophical Journal.* (1849). This lucid account seems to have been quite forgotten. It is not even mentioned by Fulton. Some of the material in the account was used by Wilson in his article on RB in *Religio Chemici*, London, 1862.

Other references to this subject are:
Gunther **VI,** p. 70 *et seq:* E. N. da C. Andrade. " The Early History of the Vacuum Pump." *Endeavour* (1957), **16,** 29: *idem.* " The History of the Vacuum Pump." *Advances in Vacuum Science and Technology*, edited by E. Thomas. 2 vols Oxford, 1960. Vol **I,** p. 14–20: H. D. Turner. " Robert Hooke and Boyle's Air Pump." *Nature* (1959), **184,** 395.

mechanical realization of Boyle's first air-pump, which is shown in the illustration Plate XVIII. The distinctive features of this pump were the provision of one barrel only; and the direct connection of the cylinder to the receiver. The cylinder had a capacity of about 126 in^3, and the receiver about 1732 in^3. The brass barrel of the pump was about 14 inches long, and 3 inches internal diameter. The solid piston or sucker was operated by a rack and pinion turned by a handle. The upper end of the cylinder had a hole, fitted with a plug, which could be put in, or taken out, by hand, and which constituted the only valve in the engine. The top of the receiver was provided with a large opening for the insertion of objects into it. The opening could be made smaller by fitting a brass ring containing a brass stopper. The bottom end of the receiver had a narrow neck cemented into a brass stopcock, which in turn fitted into an opening in the upper end of the cylinder near the stopper-valve. The whole was placed upon a wooden tripod support with the glass receiver uppermost.

Evacuation of the receiver was a tedious task. In the starting position the stopcock of the receiver was open, the valve of the cylinder closed, and the sucker at the bottom of the cylinder. The stopcock was then closed, the valve opened, and the sucker pushed to the top of the cylinder. The valve was then closed, the stopcock opened, and the sucker restored to the bottom of the cylinder. This sequence of operations was repeated as often as required. Condensation of air into the receiver was achieved by reversing the order of closing and opening the stopcock and valve.

Boyle's preferred name for this apparatus was "Pneumatical Engine", though he sometimes called it "pneumatic pump" or "air-pump". It was also known as a "Rarefying Engine" and "Boyle's Cylinder". On the continent it was known as *Machina Boyleana*. It is a matter of regret that this important specimen of the early vacuum pump has not survived.

Boyle's first air-pump was, at the suggestion of Hooke, incorporated into the background of the engraving prepared by Faithorne from his drawing of Boyle.* (Plate XXI).

A Latin poem, *Experimenta Machinae Pneumaticae Ab Honoratissimo D. D. Boylaeo Inventae*, written by Henry Stephens† was published in the second volume (1699) of the *Musarum Anglicanarum Analecta.*

---

* *Works Fol.*, V, p. 535; *Works*, VI, p. 499. Letter of R. Hooke to RB assigned to 8 September 1664.

† (1672?–1739). M.A., Fellow of Merton College, Oxford. Eventually canon of Winchester. (*D.N.B.*)

" INCLUSAM quâ vi Terram circùm ambiat Aër,
Urgens mole gravi; errantes quo pondere cogat
Particulas; quanto oppressus sese explicet ictu
Carceris impatiens, primus, BOYLÆE, recludis;
Nam te Naturæ deserta per ardua dulcis
Raptat amer; tibi res intactæ Laudis & Artis
Debemus; tibi fas soli cognoscere caufas
Abstrusas, Verique imos penetrare recessus.
O Terræ Cœlique potens! qui Limen utrumque
Exploras, geminique aperis commercia Mundi:
Tu reseras quicquid secum meditatur in Aulâ
Aëriâ Juno; Imperii tu detegis oras
Cærulei, liquidamque plagam, & pellucida regna.
QUIN Hyalum profer, latum quod crescit in orbem,
Quâque Tubo aptatur, cerâ viscoque tenaci
Unge fovens, ne se insinuet penetrabilis Aër.
Atque ubi jam spatio expressa eluctatur inani
Aura levis, tenuesque in ventos hausta recedit:
Continuò fragor auditur; ruit omnis ab alto
Incumbitque Aër Phialæ; & densissima torto
Verbere tempestas circùm furit; illa fatiscens
Solvitur, & sparso fcintillant fragmina vitro.
Sin sit convexa, & curvum lunetur in arcum,
Stat contrà obluctans, partesque gravata coërcet
Firmiùs, æthereum pondus confisa figurae
Contemmit, laterique illisam decutit auram.
SIC sese attollit nativo fornice septa
Testudo in concham crescens, sic grandia saxa
Fert tergo impunè, & serratos sustinet orbes
Vulneris haud metuens; nequicquam concava pulsu.
Testa sonat, solidumque terit rota stridula dorsum.
ACCIPE nunc, quantas vires tenuissima quævis.
Particula aëria emittat, quali impete sese
Expediat vinc'la indignans: quin obice rupto
Non jam se capit expatiata; ita turbine magno
Evolat, & veterem poscit revoluta figuram.
Nam si Vesicam Phialæ alvo includis, & auram
Chrystallo errantem educis, ventosa tumore
Protinus assurgit Pellis, majorque videri
Vitrum explet, raucoque crepat displosa fragore.
Scilicet urgentis cœli gravitate remotâ
Funiculi aërii exiliunt, & murmure circum
Claustra fremunt, tentantque aditum & latera undique pulsant.

NON vi majori tormentis emicat ardens
Sulphur, non ruptâ fulmen de nube trisulcum
Irruit, evertitque imis radicibus ornos.

HAUD aliter vento, quos presso implevit Ulysses,
Turgescunt utres gravidi, dum flatibus intus
Hinc fævit Boreas, furit illinc turbidus Auster,
Explorantque viam: donec, mox carcere fracto,
Erumpunt omnes turmatim, atque æquora miscent.

AT fragilem saliens cùm Rana intraverit Urnam,
Machinaque ingenti gyro circùm acta rotetur:
Ecce! inflatum animal turgenti pectore anhelat,
Miraturque novam molem, atque immania membra;
Tum furiis accensum Hyali cava mœnia tundit,
Singultusque ciet, mediâque in morte recumbens
Distinctam ostendit maculis squalentibus alvum.
Illi Natura effinxit mirabile textum
Pulmonum, alternam tumidis qui sollibus auram
Accipiant reddantque: illi altas ire sub undas
Indulgens dedit, ambiguâque senescere vitâ:
Nequicquam; tenuis nescit puri ætheris haustus
Erigere inflatu, & vacuas distendere cellas.

SIC ubi captivus limosâ educitur undâ,
Et patitur cœlum Piscis; mox frigore siccum
Horrere aspicias, durisque rigescere squamis.

ID metuens serâ Teneriffam nocte Viator
Ascensu cautus superat, quando aëra vesper
Humentesque umbræ densant, fumique volucres:
At sub mane ruit, montisque è vertice præceps
In valles fertur, dubiâ cùm luce rubescit
Nondum pura dies; nempe orto Sole vapores
Diffugiunt, simplex terrenâ fæce relictâ
Rarescit, non jam ulterius spirabilis, Æther.

QUID referam, in Vitro Sonus ut vanescit, & aures
Debilis arrectas fallit, longeque recedens
Paulatim moritur? Radios ut vibrat inertes,
Et tremula incerto languescit lumine Tæda;
Ut Lympha ebullit magno commota tumultu,
Exultantque intus latices, dum roscida fontem
Mentitur, crebrâque aspergine subsilit unda
Irriguas simulans scatebras. Sic Cynthia Nerei
Turbatas attollit aquas; furit æstus arenis,
Atque agitata imo fervescunt æquora fundo."

The following by ' Junius ' appeared in *The Gentleman's Magazine*, (1740), **10**, 194.

The AIR-PUMP.

" *Domitian*, as old story rings,
(That most ridiculous of kings)
Was wont, whole days, to divertise
In slaught'ring hosts of puny flies,
Preferring to all courtly joys
Sports only fit for butchers boys.
But had the monarch learn'd the knowledge
Since practis'd by our modern college,
Of using their *pneumatic engine*,
'Twould have afforded pleasure swinging;
The sight of ev'ry rare experiment
Had given his heart unusual merriment.
For instance—To have seen a mouse,
Shut fast within its crystal house,
And thence the air exhausted all,
To view the creature gasp and sprawl;
At ev'ry suction of the pump
Observe him pant from head to rump,
Spew, kek, and turn him on his back—
T'had been, ye powers, a mighty knack!
What arts of choking, tort'ring, killing,
Adepts to teach him had been willing:
All nature he'd have known, no doubt,
He would have *pump'd* her secrets out.
Dogs, kittens, ev'ry four-legg'd thing,
Had been game royal for the king;
He'd been with *lice*, and scrubby vermin,
Familiar as a cousin-german,
Diverted with each day a new-whim,
No *toad* had come amiss unto him.
Perhaps, by novelty excited,
Fresh objects had this prince delighted.
Known had but been th'invention then,
He would have try'd his pump on men,
Have found receivers apt and fit,
T'have made the operation hit.
Mercy! what sights! well worth his prying;
A quite new way of courtier trying!
The beau when fast included there,
More light than wind, that child of air,

Soon grown convuls'd, would drop and tire,
And with a pump or two expire.
The belle, within a little second
Wou'd die, it safely may be reckon'd;
Creature, that least confinement bears,
She cannot live without her *airs*,
Meer butterfly, all gay and light,
For ever flutt'ring in your sight.
Dull politicians, tools who seem
Made solely up of earth and phlegm,
Like moles that deeply shrouded dwell,
Perhaps might stand the tryal well.
Flatt'rers, those ear-wigs put by th'lump in,
Would yield an endless fund for pumping.
The empty coxcombe, in that cloyster
With skull more thick than senseless oyster,
Could there no alteration know,
He always lives *in vacuo*.
Death! cry the angry virtuosi,
Shall a young stripling here but nose high,
Presume to sport and hurl defiance
At grave adepts, and piss on science?
The saucy youngster for his chattering
Deserves another pump—a watering.
Loath to offend, yet prone to jest,
I let th'advent'rous subject rest;
Lest haply while thus matters stand ill,
To my own fate I lend a handle."

(ii) *Antlia Pneumatica*

A constellation in the southern hemisphere bears this name. The *Machina Pneumatica* of Boyle, " though lost to earth, has been translated to the heavens, where in the southern hemisphere

micat inter omnes
Boylium *sidus*, *velut inter ignes*
Luna minores."*

* R. T. Gunter. *The Old Ashmolean*. Oxford, 1933, p. 101. A modification of the lines in Horace, *Carmina*, Lib. I, Ode XII, 46–48, where the reference is to *Julium sidus*. Cf., Virgil, *Georgica*, I, 32–35, on making a new constellation for Caesar which will doubtless be called after him, and saying that the *Scorpion* will draw in his claws to make room for him.

After his visit to the Cape of Hood Hope in 1751–52 for the purpose of surveying the southern sky, N. L. De La Caille (1713–1762) drew up a star catalogue* in which he introduced several new constellations, including *La Machine Pneumatique*. The original large scale map of the southern celestial hemisphere on which the new constellations appeared is preserved at the Académie des Sciences, Paris. The planisphere was subsequently published in his *Coelum Australe Stelliferum*, 4to Paris, 1763; (in which the names of the new constellations were turned into Latin); it will be found also in J. Fortin's *Atlas Céleste de Flamsteed*, 2nd edition, Paris, 1776, Plate 29. The latter version is reproduced in Basil Brown's *Astronomical Atlases, Maps & Charts*, London, 1932, Plate X. *Antlia* remains the accepted name for the constellation in question.†

The available evidence does not make it clear if De La Caille intended to commemorate Boyle's version of the air-pump, or Otto von Guericke's very inefficient pistons with which he nevertheless produced a vacuum and performed the spectacular demonstration known as the experiment with the Magdeburg Hemispheres. The former intent seems more probable seeing that the term *Machine Pneumatique* (*Machina Pneumatica*), by which Boyle's machine was known on the continent, was adopted. The constellation is always represented on eighteenth and nineteenth century celestial globes by an air-pump of eighteenth century pattern.‡ Specimens may be seen in the Museum of the History of Science, Oxford, and elsewhere.

Chemical and Pharmaceutical Products

There are several products in this category to which Boyle's name has been attached. In view of the fact that he was not the first to describe some of them, one cannot escape the conclusion, as far as pharmaceutical products are concerned, that his name was used to enhance their reputation as well as their vendibility.

(i) *Electuarium Boyleanum*

℞. *Sem. Papav. albi, Hyoscyami albi pul. ā ℥ss. Syr. Papav. errat. Cons. Ros. rub. ā ℥i ss. m.*

---

* *Memoires de l'Academie Royale des Sciences à Paris*, 1752 (Published 1756), p. 539. The planisphere is on Plate 20 at p. 592. See also the edition issued at Amsterdam, 1761.

† R. H. Allen. *Star-Names and their Meanings*. New York, 1899, p. 42. E. Delporte. *Atlas Céleste*. Cambridge, 1930.

‡ There are significant variations in the star groups of the constellations *Antlia*, *Microscopium*, and *Telescopium* on celestial globes and charts, which reflect the changes in the design of the instruments in question. See D. J. Warner, ' Celestial Technology ', *Smithsonian Journal of History* (1967) **2**, 35.

Sanguinem refrigerat; serum tenue incrassat, acre contemperat; Vasculorum ora occludit; Fibrillarum irritationes, & spasmos consopit. Nob. Boyle mire laudat ad Haemoptoen. Dos. qu. Jugland. bis in die, post Phlebotomiam, & lenem Catharsin ritè celebratas.

Thomas Fuller (1654–1734) gave the above receipt in his *Pharmacopoeia Extemporanea**. It was still given in the fifth English edition (1740).

The electuary is listed in *Catalogus Medicinarum . . . Preparatarum per Jacobum Goodwin & Robertum Carter*, London, *ca* 1710.

(ii) *Liquor fumans Boylei*

In his *Experiments and Considerations Touching Colours* (1664)† Boyle describes the preparation of " a Volatile Tincture of Sulphur, which may probably prove an excellent Medicine." His method is: " take of common Brimstone finely powdred five Ounces, of Sal-Armoniack likewise pulveriz'd an equal weight, of beaten Quick-lime six Ounces, mix these Powders exquisitely, and Distill them through a Retort plac'd in Sand by degrees of Fire, giving at length as intense a Heat as you well can in Sand." Boyle referred to the product, (which is a mixture of ammonium hydrosulphide and ammonium polysulphides, and smells of ammonia and hydrogen sulphide), because " it is very pertinent to our present Subject, The change of Colours. For though none of the Ingredients be Red, the Distill'd Liquor will be so: and this Liquor . . . will upon a little Agitation of the Vial first unstop'd (especially if it be held in a Warmer hand) send forth a copious Fume, not Red, like that of Nitre, but White; . . . yet the Liquor itself being touch'd by our Fingers, did immediately Dye them Black."

Although Boyle's name has become associated with this product, he was not the discoverer, for similar products had been prepared previously by Basil Valentine; Jean Beguin (whose product was later known as *spiritus fumans sulphuratus Beguini* or *spiritus sulphuris volatilis Beguini*); van Helmont; F. Hoffman (whose product for medicinal use was known as *tinctura sulphuris volatilis Fr. Hoffmanni*).‡ In later times the product has been known as volatile liver of sulphur, sulphure of ammoniac, sulphuret or hydrosulphuret of ammonium.

* Editio altera. London, 1702. I have not seen the first edition.

† *Works Fol.*, **II**, 59. *Works*, **I**, 754.

‡ L. Gmelin. *Hand-book of Chemistry*. 15 vols. London, 1848–1861. Vol. **II**, p. 454: J. W. Mellor. *A Comprehensive Treatise on Theoretical and Inorganic Chemistry*. 16 vols. London, 1922–1937. Vol. **II**, p. 645: T. S. Patterson, " Jean Beguin and his *Tyrocinium Chymicum*," *Annals of Science* (1937), **2**, 261–2: J. R. Partington, " Joan Baptista van Helmont." *ibid.* (1936) **1**, 383.

Fourcroy has recorded the accidental discovery of the violent reaction between this product and concentrated sulphuric acid.* Brande describes the preparation of the product, which he says is used as a test for metals, and employed in medicine, principally in diabetes and in gout.† Hoefer makes the interesting remark in connection with Boyle that " malgré les immenses services qu'il a rendus à la science et au progrès, sa mémoire est aujourd'hui à peu près oubliée. Les chimistes et les physiciens eux-mêmes, qui lui doivent tant, ne connaissaient guère son nom que par celui de *liqueur fumante deBoyle*."‡

It is of interest to note that Boyle's Fuming Liquor was used, no doubt on account of its pungency, by Mongez and de Lamanon§ on the Pic de Teneriffe, as well as by de Saussure (*fils*) on the Roche Michel|| in connection with experiments on the volatilization of liquids at high altitudes.

(iii) *Pilulae Lunares Boylei*

Concerning these pills Boyle himself has declared, " I do not pretend to be the Inventor " of them. He gives the method of their preparation in *Some Considerations Touching the Usefulness of Experimental Philosophy* (1663)¶, but his very detailed instructions may, for our purpose, be shortened by quoting from *The Edinburgh New Dispensatory*.**

> " Dissolve pure silver in aquafortis, . . and after due evaporation, set the liquor to crystallise. Let the crystals be again dissolved in common water, and mingled with a solution of equal their weight of nitre. Evaporate this mixture to dryness, and continue the exsiccation with a gentle heat, keeping the matter constantly stirring till no more fumes arise."

He then directs the product to be made into pills with moistened bread crumbs. They were considered by Boyle to be a specific for dropsy, and a remedy for those " troubled with serious Humours, and also the palsy."

These pills are listed in *Catalogus Medicinarum . . . Preparatarum per Jacobum Goodwin & Robertum Carter*, London, *ca* 1710.

---

* A. F. Fourcroy. *Elements of Chemistry and Natural History.* 5th edition, 3 vols. Edinburgh, 1800. Vol. **II**, p. 156.

† W. T. Brande. *A Manual of Chemistry.* 6th edition, 2 vols, London, 1848., Vol. **I**, p. 409.

‡ F. Hoefer. *Histoire de la Chimie.* 2nd edition, 2 vols. Paris, 1869., Vol. **II**, p. 177.

§ *Journal de Physique* (1768), **29**, 151.

|| H. B. de Saussure. *Voyages dans les Alpes.* 8 vols. Neuchâtel, 1803–1796., Vol. **V**, p. 134–6.

¶ *Works Fol.*, **I**, p. 556; *Works*, **II**, p. 204.

** Edinburgh, 1786, p. 512.

(iv) *Syrupus Boyleanus*

℞. *Rad. Symph. maj.* ℥*vj. Herb. Plantag. M xij. contusis exprimatur succus, cui adde Sacch. pondus aequale, & f. Syr. s. a.*

Hospitem in Medicinâ eum esse oportet, cui non innotescit hic Syrupus (instar Authoris sui) nobilis; & quàm fortitèr Haemoptoicis subvenire solet.

Thomas Fuller (1654–1734) gave the above receipt in his *Pharmacopoeia Extemporanea**. It was still given in the fifth English edition (1740), which was probably the source for its quotation in *The Family Receipt-Book* in the early part of the nineteenth century (London, N.Y. [*ca* 1814]).

Literary Style

On the subject of literary style Robert Boyle has made some interesting remarks. The Introductory Preface to *Occasional Reflections* (1665) gives his views on the writing of this kind of short meditative essay, of which he was not the originator. Apart from the novelty of writing some of the reflections in the form of dialogues, he initiated the style of writing which spiritualizes common things and events. His first imitator in this direction seems to have been William Spurstow with *The Spiritual Chymist* (1666), which was followed by similar works by John Flavel and a host of others. The popularity of this kind of composition, coupled, perhaps, with the quaintness of Boyle's style gave rise to the well-known satire by Jonathan Swift entitled *A Meditation upon a Broom-Stick*, and Samuel Butler's *An Occasional Reflection on Dr. Charlton's Feeling a Dog's Pulse at Gresham-College.* Swift's satire was roundly condemned in highly censorious terms by John Weyland in the preface to his edition of Boyle's *Occasional Reflections* (1808), which have been quoted above (p. 117).

Contemporary, or near contemporary, opinions on Boyle's style are, as might be expected, rather varied. Margaret Cavendish, Duchess of Newcastle, praised it, saying: " he is a very civil, eloquent and rational writer; the truth is, his style is a gentleman's style."† At a much later date John Evelyn wrote: " . . . of his style . . . which those who are Judges, think he was not altogether so happy in, as in his Experiments: I do not call it Affected; but doubtless not answerable to the rest of his great parts; and yet, to do him rights, it was much Improv'd in his *Theodora* & later writings."‡ With regard to this work Samuel

* Editio altera. London, 1702. I have not seen the first edition.

† *Philosophical Letters.* London, 1664, p. 495.

‡ Evelyn (Bray), Vol. **III**, p. 350: B.M. Add. MSS. 4229, fol. 59. Letter to William Wotton dated 29 March 1696.

Johnson said: " The ornaments of romance in the decoration of religion, was, I think, first made by Mr. Boyle's ' Martyrdom of Theodora ', but Boyle's philosophical studies did not allow him time for the cultivation of style; and the completion of the great design was reserved for Mrs. Rowe."*

The Proemial Essay to *Certain Physiological Essays and other Tracts* (1661) is even more revealing; for, besides giving his views on the writing of what are now called scientific communications, it shows that Boyle was conscious of his own occasional verbosity, and use of over-long complex periods.

> " . . . as for the style of our experimental essays, . . . I have endeavoured to write rather in a philosophical than a rhetorical strain, as desiring, that my expressions should be rather clear and significant, than curiously adorned . . . perspicuity ought to be esteemed at least one of the best qualifications of a style; and to affect needless rhetorical ornaments in setting down an experiment, or explicating something abstruse in nature, were little less than improper . . . But I must not suffer my self to slip unawares into the common place of the unfitness of too spruce a style for serious and weighty matters; and yet I approve not that dull and insipid way of writing, which is practised by many chymists, even when they digress from physiological subjects. For though a philosopher need not be sollicitous, that his style should delight its reader with his floridness, yet I think he may very well be allowed to take a care, that it disgust not his reader by its flatness, especially when he does not so much deliver experiments or explicate them, as make reflections or discourses on them: for on such occasions he may be allowed the liberty of recreating his reader and himself, and manifesting, that he declined the ornaments of language, not out of necessity, but discretion . . .
>
> " . . . as for exotick words and terms . . . I expect, that persons not conversant in the philosophical composures written . . . in our language will be apt to suspect me for the willing author of divers new words and expressions . . . I have not at all affected them, but have rather studiously declined the use of those, which custom has not rendered familiar, unless it be to avoid the

---

* Review of *Miscellanies on Moral and Religious Subjects, in prose and Verse* by Elizabeth Harrison. S. Johnson. *Works*. Edited by R. Lynam. London, 1825., Vol. V, p. 702.

Mary Deverell's *Theodora and Didymus. An Heroic Poem*, London, 1784, was inspired by Boyle's work. The libretto of G. F. Handel's oratorio *Theodora* (1749) is also based on Boyle's work, as well as, possibly, P. Corneille's tragedy *Théodore* (1645).

frequent and unwelcome repetition of the same word . . . or for some peculiar significancy of some such word, whose energy cannot be well expressed in our language, at least without a tedious circumlocution . . . But besides the unintentional deficiencies of my style, I have knowingly and purposely transgressed the laws of oratory in one particular, namely, in making sometimes my periods or parentheses over-long: for when I could not within the compass of a regular period comprise what I thought requisite to be delivered at once, I chose rather to neglect the precepts of rhetoricians, than the mention of those things, which I thought pertinent to my subject . . ."*

John Ray when writing to John Aubrey said: " You are not ignorant how Mr. Boyl hath been κωμωδούμενος for some new-coyn'd words such as ignore & opine."† Eustace Budgell said " that his Stile is far from being correct; that it is too *wordy* and *prolix;* and that though it is for the most Part plain and easy, yet, that he has sometimes made use of harsh and antiquated Expressions: Yet, under all these Disadvantages, so curious is his Matter, and so solid are his Observations, that the hardest Thing we can say of his most careless Piece, is, That it appears like a beautiful Woman in an Undress."‡ Martin Wall wrote: " His style may perhaps be deemed by some incorrect or too diffuse; but that was less his own imperfection than that of the age in which he lived, when a florid and luxurious mode of writing was considered as the most elegant and expressive."§ In reference to the quaintness of Boyle's style, James Price said, " It much resembles the massive furniture of ' *other days* ' made cumbrous by its own ornaments." ||

Robert Boyle in Poetry

With the possible exception of Isaac Newton there appear to be more allusions in poetry to Robert Boyle and his works than to any other English scientist. They are to be found even in works written during his lifetime. In most cases the quality of these poetical effusions is not of the highest, yet they are of interest in providing evidence of the

* *Works Fol.*, **I**, p. 195–6: *Works*, **I**, p. 304–5.

† *Further Correspondence of John Ray*, edited by R. W. T. Gunther. The Ray Society, Vol. **114**. London, 1928, p. 170. Letter dated 22 September 1691.

‡ E. Budgell. *Memoirs of the Lives and Characters Of the Illustrious Family of the Boyles*. 3rd edition. London, 1737, p. 125.

§ M. Wall. *An Inaugural Dissertation on the Study of Chemistry*, 7 May 1781, Oxford, p. 74.

|| J. Price. *An Account of some Experiments on Mercury, Silver and Gold*, Oxford, 1782, p. 2.

impact of science on literature.* Seventeenth century literature was responsible to no small extent for spreading the concepts of the new philosophy. John Dryden in his *Essay of Dramatic Poesy* (1668) wrote, " . . . nothing spreads more fast than Science, when rightly and generally cultivated ". The frequent association of Boyle with Newton is a noticeable feature in the quotations given below.

It is very noticeable that Boyle has almost entirely escaped the attention of satirists and humourists. There is not even a clerihew devoted to him. However, writers of light verse (quasi-poetry) on modern astronomical discoveries have not failed to seize the opportunity of riming Boyle with Hoyle.†

The specific references to Robert Boyle in poems written from 1661 onwards, which have come to the notice of the present writer, are set out below.

[1661] *In praise of the choice company of Philosophers and Witts who meet on Wednesdays weekly, at Gresham College.*‡

[Stanza 8] " To the Danish agent late was showne
That where noe aire is ther's noe breath.
A glass this secrett did make knowne,
Wherein a catt was putt to death
Out of the glass the air being screw'd,
Pus dyed and never so much as mew'd.

[Stanza 9] The selfe same glass did likewise cleare
Another secrett more profound,
That nought but aire unto the eare
Can be the medium of the sound;
For in this glass emptied of aire,
A striking watch you cannot heare."

---

* D. Bush. *Science and English Poetry*. New York, 1950.

D. Bush. " Science and Literature." In *Seventeenth Century Science and the Arts*, edited by H. H. Rhys, Princeton, 1961.

W. Eastwood. *A Book of Science Verse*, London, 1961.

† I. Robinson, A. Schild & E. L. Schucking. *Quasi-Stellar Sources and Gravitational Collapse*. Chicago and London, 1965, p. 471.

Fred Hoyle (1915–), F.R.S., Plumian Professor of Astronomy in the University of Cambridge.

‡ F. Sherwood Taylor. " An Early Satirical Poem on the Royal Society." *N. & R.* (1947), 5, 37. See also D. Stimson. *Scientists and Amateurs*, London, 1949, p. 56.

There are several manuscript copies of this poem in existence. Sherwood Taylor suggests that the author was William Godolphin (1634?–1696; F.R.S., 1664.) Annotations on this poem are given by Sherwood Taylor, *loc. cit.*

1662. John DRYDEN.* *To My Honoured Friend Dr. Charleton.* Printed in Walter Charleton, *Chorea Gigantum, or the most famous antiquity of Great Britain, Stonehenge, standing on Salisbury Plain, restored to the Danes.* London, 1663.

"Gilbert shall live, till loadstones cease to draw
Or British fleets the boundless ocean awe,
And noble Boyle, not less in nature seen,
Than his great brother,† read in states and men."

1671. Daniel Georg MORHOF.‡

Robert Boyle's *Tracts Written About The Cosmicall Qualities of Things* (1671) was translated into Latin by Morhof, who contributed the following complimentary poem to the edition published at Amsterdam and Hamburg, 1671.

"Aërii dudum tibi, Vir celeberrime, tractus
Finibus immensis quà patuere, patent.
Nunc imos penetrat terraeq[ue], marisq[ue], recussus
Gloria radices hîc habitura suas.
Nam quae naturae, quae sunt primordia rerum
Nominis, ô BOILI, sunt elementa tui.
Alta fuit, fuit ampla; modos implerat ut omnes,
Nunc etiam, BOILI, fama profunda Tibi est."

[1671] Martin KEMPE.§ *Damons Schreiben an seinen Floridan und die Pegnitz-Hirten.* Printed in J. H. Amarantes. *Historische Nachricht von des löblichen Hirten- und Blumen-Ordens an der Pegnitz Anfang und Fortgang.* Nürnberg, 1744. p. 314.

"Ein Stern der ersten Gröss Herr *Boyl* war mir lieber,
Als aller Nymphen Gunst, die als ein Strom vorüber
Offt mit Betrübnis rauscht: Kunst-Werke nutzen mehr,
Und bringen dem, der sie begreifet, Lehr und Ehr."

* John Dryden (1631–1700) Poet; original F.R.S.

† Roger Boyle, Earl of Orrery.

‡ D. G. Morhof (1639–1691) Professor of poetry, Rostock, 1660; of rhetoric, Kiel, 1665; of history, 1673.

He travelled in England in 1670, when he met RB by whom he was much impressed.

§ Martin Kempe. (1637–1682); member of the Fruchtbringende Gesellschaft and the Pegnesischer Hirten- und Blumenorden, language and literary societies founded in the seventeenth century. He travelled in England in 1671, and met RB.

[1677] Martin KEMPE. *Appendix de Societate Regia Londinensi.* Printed in M. Kempe. *Charismatum Sacrorum Trias, Sive Bibliotheca Anglorum Theologica.* Königsberg, 1677. p. 667.

[Stanza 12] " Wer das Wunder dieser Zeit,
Mein Herr Boyle darin wirket,
Wird von keiner Sterblichkeit
Unterdrükket noch bezirket;
Seine Bücher die er schreibt,
Hat Erfahrung fest bewähret:
Wer ohn sie ein Ding erkläret,
Irret oft, weil er leicht gläubt.
Wenn ein Prinz ist Salomon,
Kröhnet Weisheit seinen Thron."

1680. Henry Lavor. *In Noctilucam viri amplissimi summique philosophi, R.B.*

EPIGRAMMA

" Aemulus en! nostri surgit Britannicus heros;
Nostraque partitur munera, Phoebus ait.
Non jam de coelo poscent sibi lumina noctes;
I nunc per nemora, & fige, Diana, feras.
O vivat, vigeat longaevus, plurima ut ultra
Proferat in lucem splendida scripta, precor.
Hunc modo parca favens multos conservet in annos,
Altior humanâ sorte feretur homo."

The above was sent to Boyle by John Beale with the accompanying remarks. " Sir, this is from our quaker, a man exquisitely learned in all the arts and sciences, that are yet known amongst us; a profound admirer of all your volumes; very reserved, and of few words; so that this is merely enforced from his great respects, and strong affections. He gave me epigrams upon some of your other volumes, which I sent to Mr. Oldenburg, as I received them, in his own hand; so that I have no copy."*

* *Works Fol.*, V, p. 504; *Works*, VI, p. 443. Letter from J. Beale to RB dated 2 March 168$^{0}_{1}$.

1686. Edmund ARWAKER.* *Fons Perennis. A Poem On the Excellent and Useful Invention of Making Sea-Water Fresh.* London, 1686. p. 11.

" But precious Ointments of Eternal Fame,
Embalm great *Boyle's* most celebrated Name.
Boyle the bless'd *Moses* of our happy Land,
Who from the Ocean does *fresh Springs* command;
By whose safe Conduct we new Worlds may know,
Worlds which with more than *Canaan's* Plenty flow:
And *England* now may vie with *Israel's* Bliss,
Our's scarce inferior to Their *Moses* is;
The Skill their's had to *Ægypt* was confin'd,
Our's leaves that *Ægypt*, and the World behind;
And is with Nature so familiar grown,
She has no Secret left to him unknown.
For the strict Searches of his piercing Eye,
Earth has no Place too low, nor Heav'n too high;
His Knowledge sinks into the deep-hid Mine.
And soars to Heighths of Mysteries Divine,
Of which he does such near Idea's draw
As if unveil'd he the bright Objects saw.
Nor does he yet approach too rudely near,
Kept at just Distance by an awful Fear;
But into all does, like the Angels, pry,
With trembling Dread, and blushing Modesty;
And when he treats of their *Seraphick Love*,
None but such Transports his Affections move;
Whose stronger Heat, from his refin'd Desires,
Repels the infl'ence of Inferior Fires;
But when exalted, and employ'd on High,
His ravish'd Soul dissolves in Ecstasie,
Nor wou'd from that Sublimer Bliss descend,
Nor one short Thought on this mean World mispend,
Did not the Pow'r, he does Above revere,
Display his Splendor in his Creatures here;
Through which he does, with strict Enquiry, pass,
And, in their Beings, their Creator trace.
He knows each Creature's Vertue, and its Use;

* Edmund Arwaker, the younger, (?–1730) M.A., Dublin. Held various ecclesiastical positions: (H. Cotton. *Fasti Ecclesiae Hibernicae.* Dublin, 1847–1860).

An account of the invention and RB's connection with it is given on p. 147 *et seq.*

And from the Worst can Excellence produce.
By him the Waters, *Acid* and *Marine*,
Are purg'd and freed from their Destructive Brine:
The Sailor now to farthest Shores may go,
Since in his Road these *lasting Fountains* flow;
The Sea, corrected by this wondrous Pow'r,
Preserves those now, whom it destroyed before:
No more with Thirst the Feav'rish Sea-man dyes,
The Briny Waves afford him fresh Supplies.
The mighty *Boyle* does by his pow'rful Art,
The Ocean to a Well of Life convert;
Whose Fame had *Israel's* thirsty Monarch heard,
He had these Springs to *Bethel's* Well preferr'd;
And their Diviner Vertue had (if known)
Excus'd the Risque he made *three Worthies* run:*
Had these in *Naaman's* Days been understood,
Jordan's fam'd Stream had scarce been thought so Good;†
Nor wou'd their Influence, more truely Great,
Require he shou'd the Healing Bath repeat.
Boyle, our good Angel, stirs the Sov'reign Pool,
That makes the Hydropic-Leprous Seaman whole;
And now, who first shall put to Sea, they strive,
Since safer there, than on the Shore they live:
And, when to Coasts remote they boldly steer,
Proclaim the Worth of their Preserver there.

We shall to *India* be in Debt no more
For the rich Fraights we carry from its Shore:
This far more precious Secret left behind
Will amply pay for all the Wealth we find;
The Purchace of that Treasure with this Art,
Our former Course of Traffick will invert:
The *Indians* now, for Gold, shall buy our Store,
As we their Gold, for Trifles, heretofore.

The tender Mother now who, for her Son,
Storm'd Heav'n with Pray'rs and Repetition,
Does her remaining Breath to Praise convert,
To Celebrate this Life-preserving Art:
And the glad Wife, wrapt in her Husband's Arms,
For his Return, applauds its wondrous Charms."

---

* 1 Chronicles 11, 17-19.
† 2 Kings 5, 10–14.

1692. After Robert Boyle's death several eulogies in verse were published in memory of the deceased.

(i) [Anon]. *Lachrymae Philosophiae.* An Elegy.
(ii) [Anon]. *Memento Mori.* A necrological broadside.
(iii) [Anon]. *Natura Lugens.* A necrological broadside.
(iv) [Anon]. *Epigram.* In *The Works of the Learned*, [edited by J. de la Crose], January 169½.
(v) [Anon]. *An Elegy.* In *The Athenian Mercury* (1692) **6,** No. 2.
(vi) [Matthew Morgan]. *An Elegy.* Oxford, 1692.

Bibliographical details of these items will be found in Fulton, 2nd edition, Nos 304, 305, 306, 307, 308 and 310 respectively. No. 305 is reproduced on Fig. 24 of that work; and No. 306 is reproduced on Fig. 2 of II Fulton.

[1699]. Henry STEPHENS. *Experimenta Machinae Pneumaticae Ab Honoratissimo D. D. Boylaeo Inventae.* Printed in *Musarum Anglicanarum Analecta*, 2 vols, Oxford, 1692–9. Vol. **II** (1699), p. 16 *et seq.*

This poem has already been quoted; see pp. 229–30.

[1700]. Sarah FYGE. *Poems on Several Occasions, Together with a Pastoral.* London.

" *On the Honourable* Robert Boyl's, *Notion of Nature.*"

" Tis bravely done, great *Boyle* has disenthron'd
The Goddess Nature, so unjustly Crown'd,
And by the Learn'd so many Ages own'd.
Refuge of Atheists, whose supine desire,
Pleas'd with that Stage, no farther will aspire:
It damps the Theists too, while they assign,
To Nature, what's done by a Power divine.
We know not how, nor where, to ascribe events,
While she's thus Rival to Omnipotence;
Sure that alone, the mighty Work can do,
The Power that did create, can Govern too:
It is not like our sublunary Kings,
That must be circumscrib'd to place, and things,
Whose straighten'd Power, doth Ministers Elect,
That must for them remoter business act,
The Omnipotence, of the Power Divine,
Argues it need no Deputies assign;

Nor is't beneath the Glory of his State,
To Rule, Protect the Beings he create:
But stop my Pen, blush at thy weak pretence,
Tis *Boyle*, not thee, that must the World convince;
Boyle the great Champion of Providence.
Whose conquering Truths in an Inquiry drest,
Have celebrated Nature dispossest;
Not the Vice gerent of Heavens settled Rules,
But nice Idea of the erring Schools.
Fate, Fortune, Chance, all notional and vain,
The floating Fictions of the Poet's brain;
The World rejects, yet stupidly prefers,
This wild Chimera of Philosophers:
This more insinuating Notion lay,
Unquestion'd till you made your brave Assay,
Which doth the daring Sceptick more confute,
Than a suspected Orthodox dispute.
They can't pretend Int'rest, thy Lines doth Bribe.
With which they censure, the Canonick Tribe:
'Twas Love of Truth alone, thy Pen did move,
Nor none but thee, could so successful prove.
Methinks I all the School-mens Shades espy,
Tending thy Tryumphs of Philosophy,
And all the pregnant Naturist of Yore,
From the Great Stagarite, to descartes and more;
Resigning their Gigantick Notions now,
And only what you write for Truth allow.
See they have all their renounc'd Volumes brought,
(Bidding Mankind believe, what you have Taught;)
Asham'd they've been, renown'd so many Years,
Each from his blushing Brow his Laurel tares:
With their own Hands, in one just Wreath they twine,
Adorning that victorious Head of thine.
And shall my Female Pen, thy Praise pretend,
When Angels only, can enough commend,
In songs, which like themselves, can know no End."

1704. *The Tryal of Skill: or, A New Session of the Poets. Calculated for the Meridian of Parnassus, in the Year MDCCIV.* London.

This poem contains a reference to Boyle, but I have not seen it.

1709. [John REYNOLDS.] *Death's Vision Represented in a Philosophical Sacred Poem.* London, 1709.

This poem contains several lines which the accompanying notes show to have been inspired by Boyle's works.

1724 October 10. Samuel BOWDEN.* *To Dr. Morgan on his Philosophical Principles of Medicine.*†

"Then shone the learned, the industrious BOYLE
And sought out Truth with an unweary'd toil;
BOYLE on Experiment alone rely'd,
And Nature, which he lov'd, was still his guide."

1727. James THOMSON.‡ *Summer. A Poem.* London, 1727.

In this, the first edition of the poem, we find:

"Nor be thy *Boyle* forgot; who, while He liv'd,
Seraphic, sought TH'ETERNAL thro' his Works,
By sure Experience led; and, when He dy'd,
Still bid his Bounty argue for his God,
Worthy of Riches He!"

1730 and 1735. *Idem.*

In the third and fourth editions of the poem the above puotation becomes:

"What need I name thy *Boyle*, whose pious search
Still sought the great *Creator* in his works,
By sure experience led?"

I have not seen the second edition.

1746. *Idem.*

In the author's last edition of this poem the quotation is again altered:

"Why need I name thy Boyle, whose pious Search
Amid the dark Recesses of his Works,
The great Creator sought?"

---

* This poem is contained in Thomas Morgan's *Philosophical Principles of Medicine*, London, 1725.

† Samuel Bowden. (fl. 1733–1761). Physician. (*D.N.B.*).

‡ James Thomson (1700–1748). Poet.

1733. Queen Caroline, consort of George II, erected two typical eighteenth century conceits in the Royal Gardens at Richmond. One was known as Merlin's Cave, and the other as the Grotto or Hermitage. The central round room of the latter contained busts of Isaac Newton, John Locke, Samuel Clarke, and William Wollaston; whilst in a recess, on a pedestal and higher than the rest, was the bust of Robert Boyle with a rayed sun behind. Numerous poems were written on the Hermitage and the busts. A full account of the marble bust of Boyle, which was placed in the Hermitage, with quotations from the poems alluding to it has been given by the present writer elsewhere,* so the numerous items will not be repeated here.

1736. [Philip BENNET]. *The Beau Philosopher. A Poem. London,* 1736.

" The Beau's Design in the Poem [is] to unite Politeness and Philosophy by teaching Philosophers to write so as to be understood by Beaus and Ladies."

" To trace each *Philosophic* Maze with Care,
And through perplexing Windings lead the *Fair:*
To form, without a Cent'ry's studied Toil,
A Coquet NEWTON, or a powder'd BOYLE."

1739. [ANON]. *The Star Gazer: Or, Latitude for Longitude, A Poem.* Printed by J. Jones.

" . . . Could call the Twinklers by their Names,
Count *Lunar's* spots and *Phoebus*' beams;
Locke, Bacon, Boyle he had by heart
Perfect in *Euclid* ev'ry part;
On *Ticho Brahe* did often pore,
Gassendi and *DesCartes* read o'er,
Besides hard names a hundred more."

1740. 'Junius'. *The Air-Pump. Gentleman's Magazine,* (1740), **10**, 194.

This poem has already been quoted; see pp. 231-2.

---

* R. E. W. Maddison. " The Portraiture of the Honourable Robert Boyle, F.R.S." *Annals of Science* (1959 published 1962), **15**, 196 *et seq.*

1743. [Ashley COWPER]. *The Progress of Physic, by* Timothy Scribble. Second edition London, 1743.

The author after dealing with the search for the " Grand Elixer " and " Gold " continues, on page 18,

" Not so a *Boyle* his useful Hours employ'd;
The *True Philosopher's* unerring [(1)] *Guide*—
Boyle, for no *Systems* anxious to contend,
Experiment, the *Means*—and *Truth*, his *End*—
Who wak'd to *Life*, and into *Action* brought,
The *Child of Fancy*, slumb'ring but in *Thought:*
Who *Chymic Aids* to *Physic* best apply'd,
Nor boasted *Pow'rs* that Angels are deny'd:
Thro' all his *Works*, the *One Supreme* confess'd,
And in the *Creature* the *Creator* bless'd—
Who wisely knew to *shun* the wild *Extremes*
Of *Sceptic-Doubtings*, and *Scholastic Dreams;*
To surer *Science* pointed out the *Road*
Which *Boerhaave*, late—and ev'n a *Newton*[(2)] trod."

On page 23 we read,

" *Mechanic Science*[(3)] nor disdains to join
Its wondrous *Aid* to make the *Art Divine:*
From *East* to *West*, hence *Physic* boasts her Sway,
And darts on *All* a more propitious *Ray;*
But fix'd her *Throne* on fair *Britannia's Isle*,
To whom she owes a *Harvey*—and a *Boyle*.[(4)]"

---

(1) " Mr. *Boyle* rescued *Chymistry* from the *Censures* it had long lain under from the *Enthusiasm of Helmont*, &c and has shewn what an infinite *Use* it is to *Philosophy* and *Medicine*, when kept within its proper *Bounds*."

(2) " Mr. *Boyle's Discoveries* of the *Qualities of Bodies*, by the Assistance of *Chymistry*, were so considerable, that the illustrious Sir *Isaac Newton* himself has thought it proper to follow his *Example*—When from the *Effects of Bodies*, he demonstrates their *Laws*—*Actions* and *Powers*—he always brings *Chymical Experiments* for his *Vouchers*."

(3) " Medicine, by late *Improvements* in Philosophy, is become all *Mechanical* and *Corpuscular* . . . "

(4) " He first *discover'd*, or at least brought the *Air-Pump* to *Perfection*, which soon demonstrated the *Absurdity* of that common Notion, that *Nature* abhorr'd a *Vacuum*—Since he has shewn us the true *Origin* of *Qualities* of *Bodies*, Nobody has dar'd to advance the Chimaerical Notion of *Substantial* Forms—By the Help of These, and

other valuable *Discoveries*, many others have been made since his *Death*, and many more probably will be, and his *Reputation* rather increase than diminish in future Ages . . . "

[Notes 1 to 4 are by the original author.]

1749. James CAWTHORN.* *The Vanity of Human Enjoyments*. An Ethic Epistle. To the Right Hon. George Lyttleton, Esq.; One of the Lord's of His Majesty's Treasury, 1749.†

" Can Britain, in her fits of madness pour
One half her Indies on a Roman whore,
And still permit the weeping muse to tell
How poor neglected Desaguliers fell?
How he, who taught two gracious kings to view
All Boyle ennobled, and all Bacon knew,
Died in a cell, without a friend to save,
Without a guinea, and without a grave ? "

1752. Thomas HAYTER.‡ *An Ode to the Author of SIRIS*. Prefixed to George Berkeley, *Farther Thoughts on Tar-Water*. [London], 1752.

" NEWTONUS illic plurima cogitans,
Viroque charus BOYLIUS it comes,
Et SYDENHAMO juncta magni
HIPPOCRATIS spatiatur umbra."

The ode appeared also in the *Gentleman's Magazine*, (1752), **32.** 472 together with an English translation.

" There Newton, studious with extensive aim,
And Boyle, the friend of man, my rev'rence claim:
There great Hippocrates, with Syd'nham join'd,
Shares the sweet friendship of a kindred mind."

---

* James Cawthorn (1719–1761) Poet; headmaster of Tonbridge Grammar School. (*D.N.B.*)

† This poem " was spoken as a school exercise, by Master P. Dalyson, before the Skinner's Company, at their annual visitation of the school " [i.e., Tonbridge] in 1749. (*Gentleman's Magazine*, (1791), **61**, ii. 1082.) The poem was published in James Cawthorn, *Poems*. London, 1771.

‡ Thomas Hayter, (1702–1762) Bishop of Norwich, 1749; translated to London, 1761. (*D.N.B.*).

1755. David GARRICK.* *On Johnson's Dictionary*† in *The Poetical Works of David Garrick Esq.* 2 vols. London, 1785.

" In the deep Mines of Science tho' Frenchman may toil,
Can their Strength be compar'd to Locke, Newton and Boyle?"

1755. Elizabeth TOLLET.‡ *On the Death of Sir Isaac Newton.* Printed in *Poems on Several Occasions.* London, 1755.

" But see the Wonders by thyself achiev'd.
Bacon and Boyle thy Triumphs but fore-run,
As Phosphor rises to precade the Sun."

1768. W. Samson. *The Conciliade.*

This poem contains a reference to Boyle, but I have not seen it.

1791. Erasmus DARWIN.§ *The Botanic Garden: A Poem, in Two Parts.* London, 1791–1789.

In Part I " The Economy of Vegetation ", Canto IV occur the following lines.

" You charm'd, indulgent SYLPHS! their learned toil,
And crown'd with fame your TORRICELL, and BOYLE;
Taught with sweet smiles, responsive to their prayer,
The spring and pressure of the viewless air.
—How up exhausted tubes bright currents flow
Of liquid silver from the lake below,
Weigh the long column of the incumbent skies,
And with the changeful moment fall and rise.
—How, as in brazen pumps the pistons move,
The membrane-valve sustains the weight above;
Stroke follows stroke, the gelid vapour falls,
And misty dew-drops dim the crystal walls;
Rare and more rare expands the fluid thin,
And Silence dwells with Vacancy within.—
So in the mighty Void with grim delight
Primeval Silence reign'd with ancient Night."

The Author supplies appropriate explanatory notes.

* David Garrick (1717–1779). Actor. (*D.N.B.*).

† The poem appeared in *The Public Advertiser*, No. 3681, 22 April 1755. Also in the *Gentleman's Magazine*, (1755), **25**, 190.

‡ Elizabeth Tollet (1694–1754) Poetess.

§ Erasmus Darwin (1731–1802) Physician; amateur in philosophy. (*D.N.B.*)

1797. [ANON]. *The College: A Satire.* London, 1797.

In reference to Science ceasing to be an active pursuit of the Royal College of Physicians in Warwick Lane and being taken over by the Royal Society, the poem continues in Canto II,

" Meanwhile where didst thou go, offended Maid?(1)
On that sad morn, to what kind mansion stray'd
Thy much-insulted form?—The Muse beheld
How conscious dignity thy steps upheld;
She view'd thee with an handmaid's pleasing toil,
Haste to the haunt(2) of Newton and of Boyle."
. . . .
" Then too thy Boyle! with patient voice and slow,
Questioning the four Elements below;
I hear *them*, bound in thy(1) imperial chain,
Disclose the wonders of their jarring reign."

---

(1) Science.
(2) The Royal Society.

1802. [Joseph SYMPSON]. *Science Revived*, or *The Vision of Alfred.* London, 1802.

The likenesses of the most celebrated philosophers are caused to appear by Sylphs, in Book VII, at the command of the Goddess of Science.

" . . . Boyle to view ascends,
By Knowledge class'd among her firmest friends.
Smit with the love of Nature's blooming charms,
He woos the Goddess to his eager arms.
In vain she flies, in vain beneath the ground
Would refuge seek; her dark retreat is found.
Ev'n if she hems herself in brazen tow'rs,
He, master of the strong mechanic pow'rs,
His battering engines plants against the wall,
And soon beholds the shatter'd bulwarks fall:
Or if she hopes to speed her flight through air,
Fir'd by her coyness he pursues her there.
Before his art the vanquish'd air retires,
An art that ARCHIMEDES' shade admires.
Know the receiver fashion'd by his hand,
Transcends whate'er the Syracusan plann'd,
For pistons thence the subtlest air expel
Fast as they heave the water from the well."

1813. [ANON]. *Anecdotes of Eminent Persons.* 2 vols. London, 1813.

In Vol. II, p. 228, the following is found,

" Lismore, long since the Muse's ancient seat,
Of piety and learning in retreat;
Her alma-mater shone as bright as noon
As Oxford, Cambridge, or the great Sorbonne.
Time shifts the scene;—no longer now she boasts
Her churches, colleges, and learned hosts.
Nature, propitious to the favourite soil,
Restor'd her losses with the birth of Boyle:
Centred in him, her ancient splendour shone,
Who made all arts and sciences his own."

1814. *Gentleman's Magazine.* (1814), **84.** 470.

Lines, written by the Rev. T. Maurice*, and recited by Mr. J. L. Edwards, at the Anniversary Dinner of the " Philosophical Society of London," 1814.

" While these a NEWTON'S Heav'n-born fires inflame,
Others aspire to BOYLE'S immortal fame; "

FORGED CORRESPONDENCE

One of the most astounding cases of wholesale forgery is that known as the " affaire Vrain Lucas ", of which a full account has been published.† Over a period of eight years Vrain Lucas sold to Michel Chasles‡ more than 27,000 forged pieces. The only reason for referring to them here is that some of the items are forged letters to or from Robert Boyle. On 8 July 1867 Chasles announced to the Académie des Sciences his possession of a letter from Blaise Pascal to Boyle containing a statement of the laws of attraction.§ On the basis of his " documents " he endeavoured to prove that Pascal, and not Newton, was the first to state the principle of gravitation. Two letters and four notes of Pascal to Boyle he presented to the Académie.||

* Presumably Thomas Maurice (1754–1824) Oriental scholar and historian; vicar of Cudham. (*D.N.B.*).

† H. Bordier & E. Mabille. *Une Fabrique de Faux Autographes ou Récit de l'Affaire Vrain Lucas.* Paris, 1870. It provides the basis of A. Daudet's novel *L'Immortel*, Paris, 1888.

‡ (1793–1880) Geometrician. Membre de l'Académie des Sciences, Paris. Copley Medallist of the Royal Society.

§ *Comptes Rendus* (1867), **65,** 51.

|| *Idem*, (1867), **65,** 89.

Other letters of Pascal and Huygens to Boyle, as well as some written by Boyle himself are mentioned or printed in various issues of the *Comptes Rendus Hebdomadaires des Séances de l'Académie des Sciences*. Needless to say, the authenticity of these items was soon challenged,* The debates thereon occupied the Académie on and off for two years until the time when Chasles was shown to have been the victim of gross deception, and the seller of the documents had been arrested.†

The following is a list of the forged items in so far as they concern Robert Boyle.

Date	Item	Reference
2 Mar. 1648	Letter of Pascal to Boyle.	Mentioned: *C.R.* (1867) **65**, 90.
8 May 1652	Letter of Pascal to Boyle.	Printed: *idem* (1867) **65**, 91 & 92.
2 Sep. N.Y.	Letter of Pascal to Boyle.	
No date	4 notes of Pascal to Boyle.	
6 Jan. [1654]	Letter of Pascal to Boyle.	Printed: *idem* (1867) **65**, 189.
8 Mar. 1654	Letter of Pascal to Boyle.	Mentioned: *idem* (1867) **65**, 90.
2 Jan. 1655	Letter of Pascal to Boyle.	
18 May 1682	Letter of Huygens to Boyle.	Partly printed: *idem* (1867) **65**, 544.
22 Aug 1686	Letter of Huygens to Boyle.	

3 letters of Robert Boyle, one addressed to the king. Mentioned in Bordier & Mabille. *Op. cit.*, p. 53.

76 Letters of Pascal to Boyle. *idem*, p. 63.

Supposititious Works

1. *The Gentleman's Calling.*

This work was published anonymously at London, 1660. Richard Allestree (1619–1681) is generally considered to be the author, though the work has been attributed to John Fell (1625–1686), Lady Dorothy Pakington (?–1679), Richard Sterne (1596?–1683) and others. That Robert Boyle, also, was considered by contemporary opinion to be the probable author is shown by notes made by Lorenzo Magalotti during his first visit to England in 1668. Among the titles of works written by Robert Boyle he includes in one place " *de la Vocation d'un gentilhomme* ", and in another " *Della vocazione di un Gentilhuomo.*"‡

* *Idem* (1867), **65**, 262: *idem* (1868), **66**, 145.

† *Idem* (1869), **68**, 1425; (1869), **69**, 5, 646, 678.

‡ A. M. Crinò. *Fatti e Figure del Seicento Anglo-Tuscano.* Florence, 1957, p. 152, 158.

2. *The Lively Oracles given to us.*

This is another work which is usually attributed to Richard Allestree. It, too, has been ascribed to Boyle.*

3. *Heretickal Philosophie of the Spagirists.*

By way of tail-piece may be added this *imaginary* tract by Boyle published posthumously in 1741, the theft of which formed the basis of a short detective story.†

---

* Madan, No. 3203.

† Charles Irving. "It was Raining." *Evening Standard*, No. 39,587. 17 August 1951.

Illustrations

Plate II. Richard Boyle, 1st Earl of Cork. (1566–1643).
Father of Robert Boyle.

A

Plate III. Catherine Fenton. 1st Countess of Cork. (?–1630).
Mother of Robert Boyle.

Plate V. Effigy of Robert Boyle, *aet.* 4–5.

Plate VI. Eton College, Lower School, viewed from East to West.

Plate VII. Eton College, Lower School, viewed from West to East.

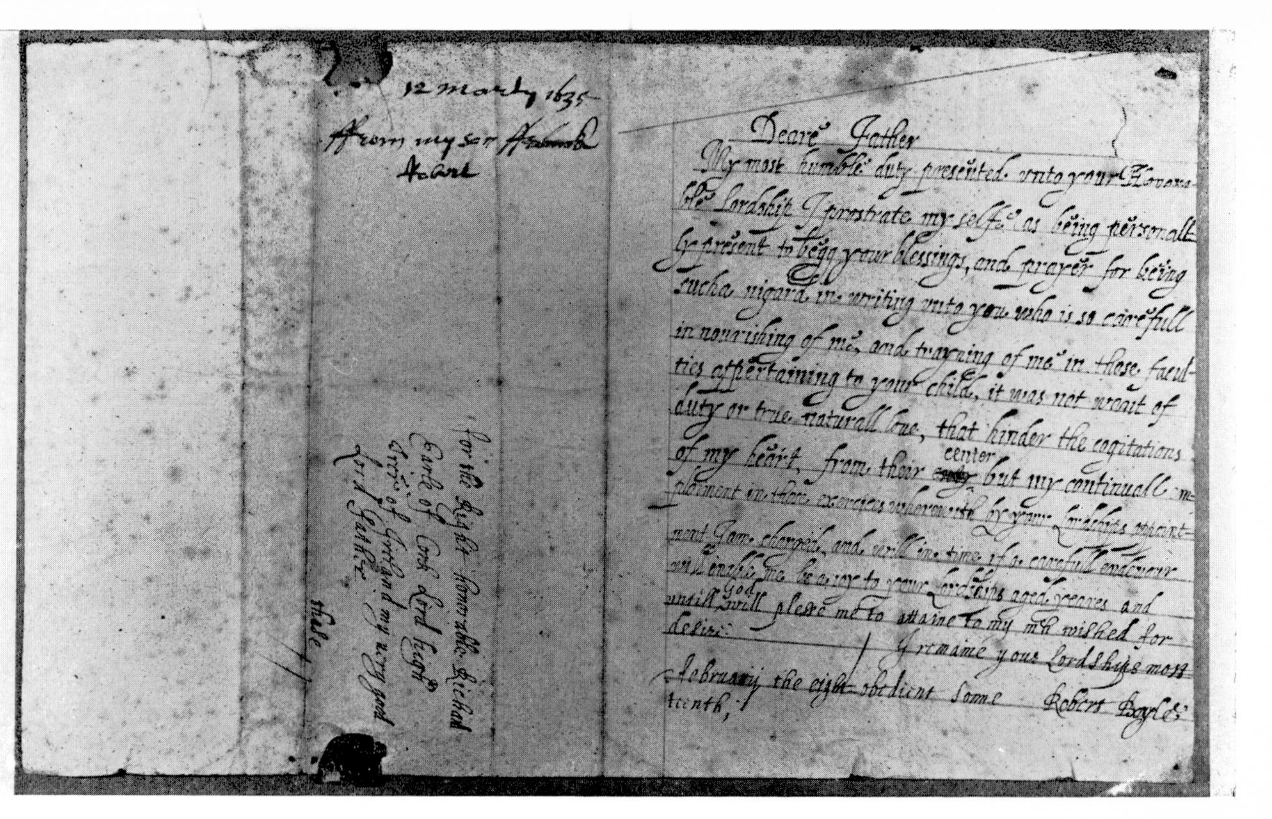

12 march 1635
From my son Robert

Deare Father
My most humble duty presented unto your Honorable Lordship I prostrate my selfe as being personally present to begg your blessings, and prayer for being such a nigard in writing unto you who is so carefull in nourishing of me, and trayning of me in these faculties appertaining to your child, it was not want of duty or true naturall love, that hinder the cogitations of my heart, from their center but my continuall imployment in those exercises wherewith by your Lordships appointment I am charged, and will in time if a carefull endeavour will enable me be a joy to your Lordships aged yeares and untill God will please me to attaine to my much wished for desire.

I remaine your Lordships most obedient sonne
Robert Boyle

February the eighteenth,

for the Right honorable Richard Earle of Cork Lord high Treasurer of Ireland my very good Lord Father
these

Plate VIII. Earliest surviving letter written by Robert Boyle, 18 February $163\frac{5}{6}$.

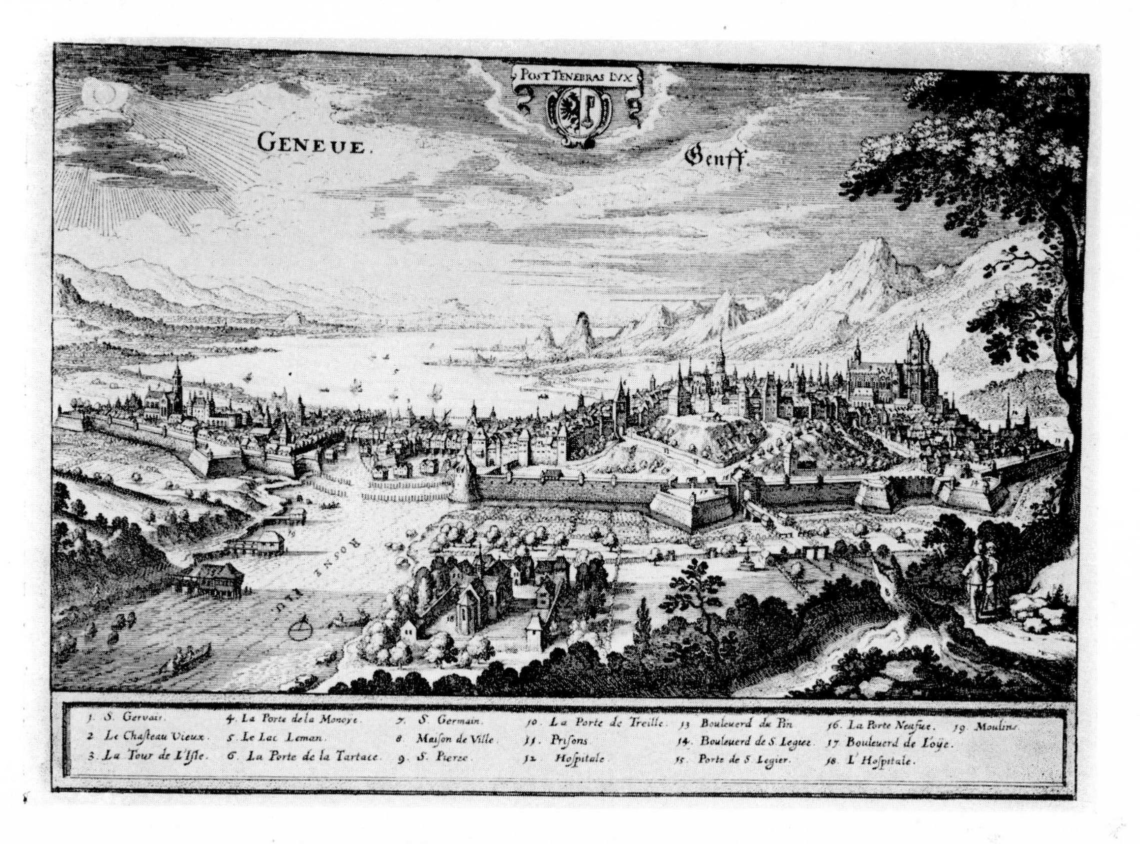

Plate IX. Prospect of Geneva.

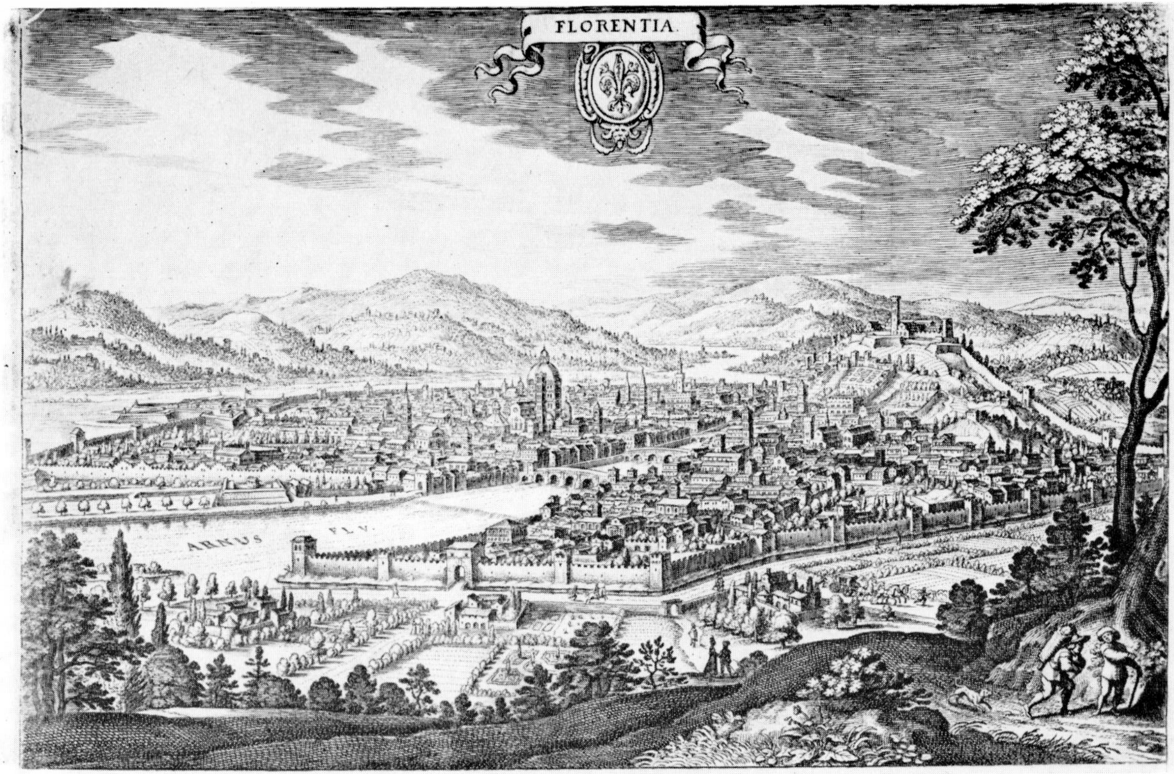

Plate X. Prospect of Florence.

Plate XI. Tomb of Richard Boyle, 1st Earl of Cork, in St. Mary's Church, Youghal.

Plate XII. Stalbridge House, Dorsetshire, *ca.* 1812.

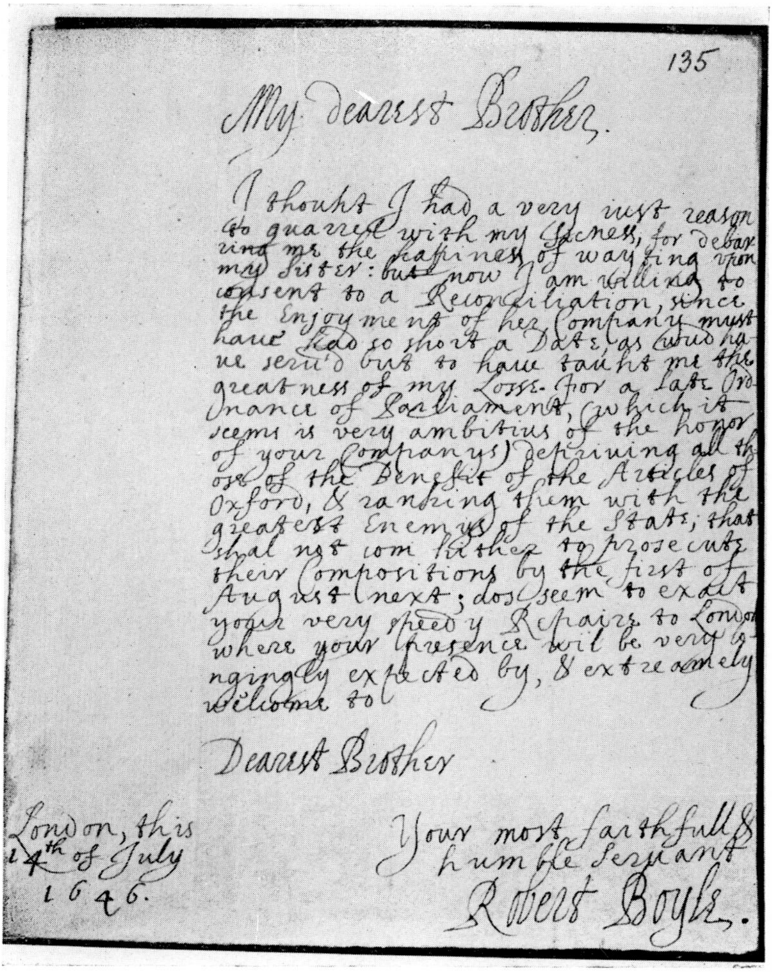

135

My dearest Brother.

I thought I had a very iust reason to quarrel with my sicknes, for debarring me the happiness of wayting upon my Sister: but now I am willing to consent to a Reconciliation, since the Enjoyment of her Company must have had so short a Date, as would have serv'd but to have taught me the greatness of my Losse. For a late Ordinance of Parliament, (which it seems is very ambitious of the honor of your Companys) depriving all those of the Benefit of the Articles of Oxford, & ranking them with the greatest Enemys of the State; that shal not com hither to prosecute their Compositions by the first of August next; dos seem to exact your very speedy Repairs to London where your Presence wil be very longingly expected by, & extreamely welcome to

Dearest Brother

London, this 14th of July 1646.

Your most faithfull & humble Servant
Robert Boyle.

Plate XIII. Robert Boyle's early letter–hand, 1646.

Plate XIV. High Street, Oxford, showing the house (now demolished) next to University College, at one time occupied by Robert Boyle.

Plate XV. House in High Street, Oxford, occupied by Robert Boyle

(Enlargement of part of Plate XIV.)

Plate XVI. Plaque which now marks the site of the house in High Street, Oxford, where Robert Boyle resided.

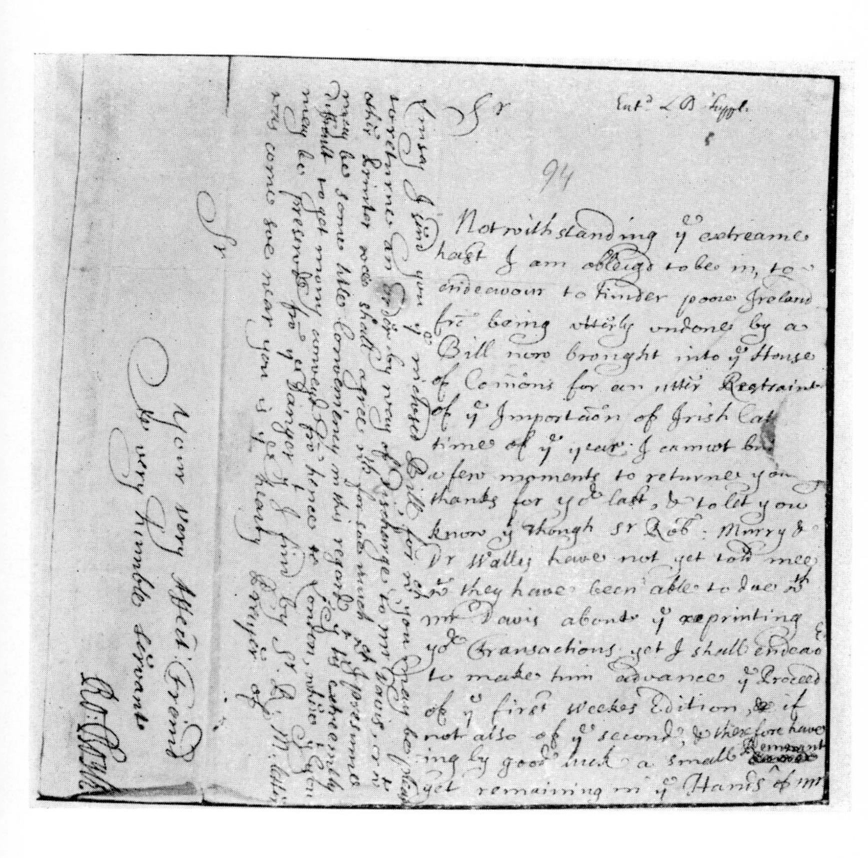
Sr

Entd LB Suppl.

94

Notwithstanding ye extreame
hast I am obleigd to bee in, to
endeavour to hinder poore Ireland
fro being utterly undone by a
Bill now brought into ye House
of Comons for an utter Restraint
of ye Importaon of Irish Cat
times of ye year. I cannot b
a few moments to returne you
thanks for yor last, & to let you
know yt though Sr Rob. Morry &
Dr Wallis have not yet told mee
wt they have been able to doe wth
mr Davis about ye reprinting
yor Transactions yet I shall endeav
to make him advance ye Proceed
of ye first weekes Edition, & if
not also of ye second; & therefore hav
ing by good luck a small Remnant
yet remaining in ye Hands of mr

Sr

Your very Affect. Friend
& very humble Servant

Ro: Boyle

Plate XVII. Robert Boyle's mature letter–hand, [1665].

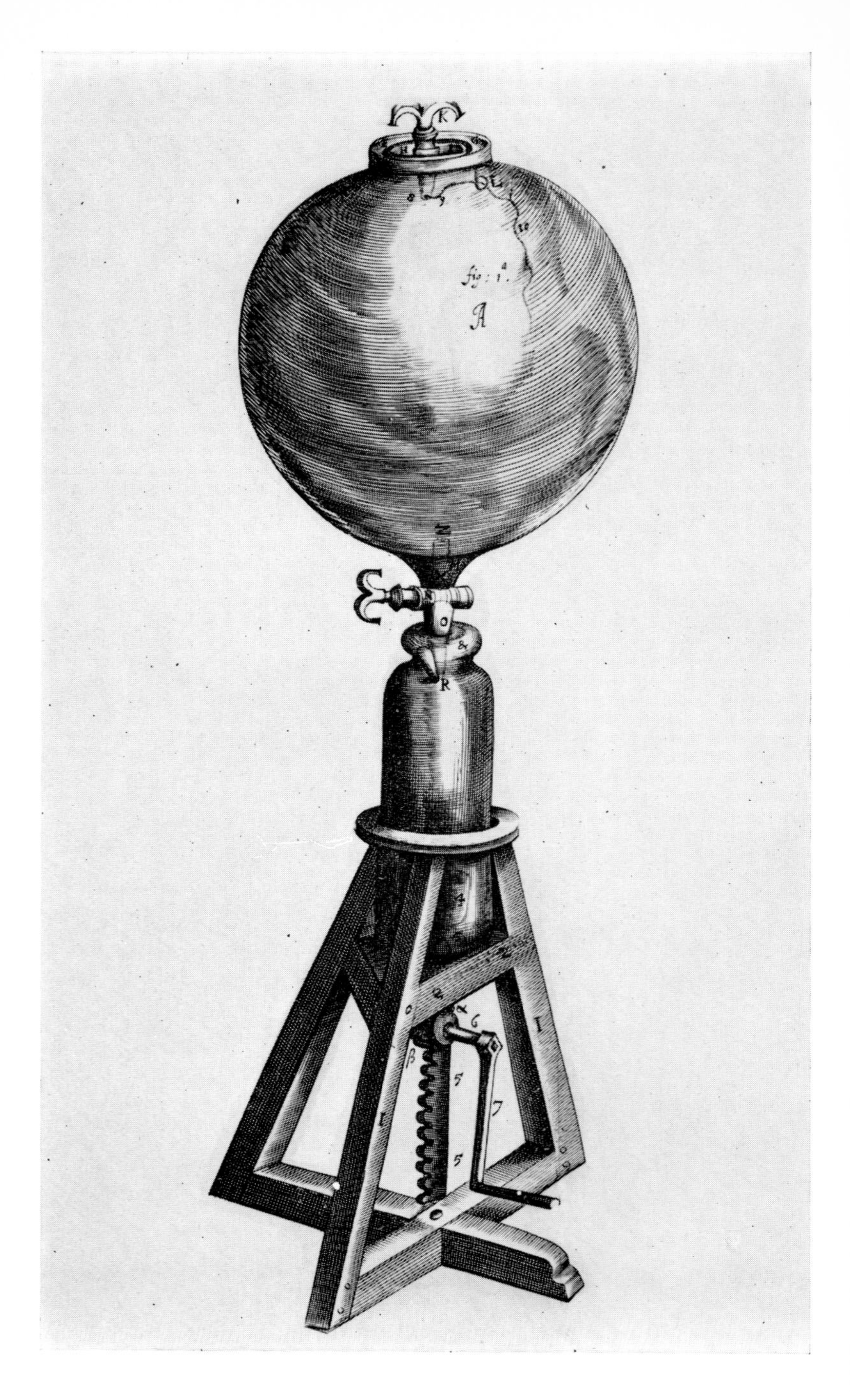

Plate XVIII. Boyle's First Air-Pump.

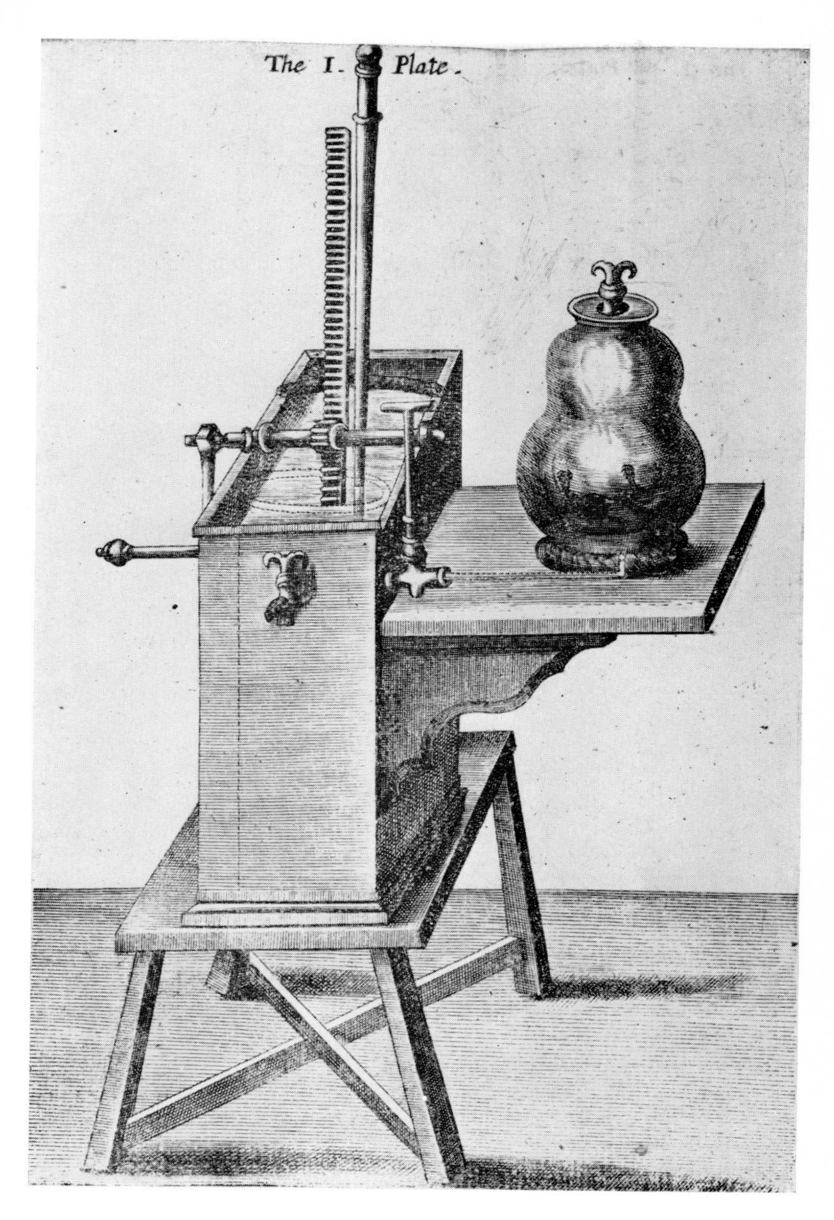

Plate XIX. Boyle's Second Air–Pump.

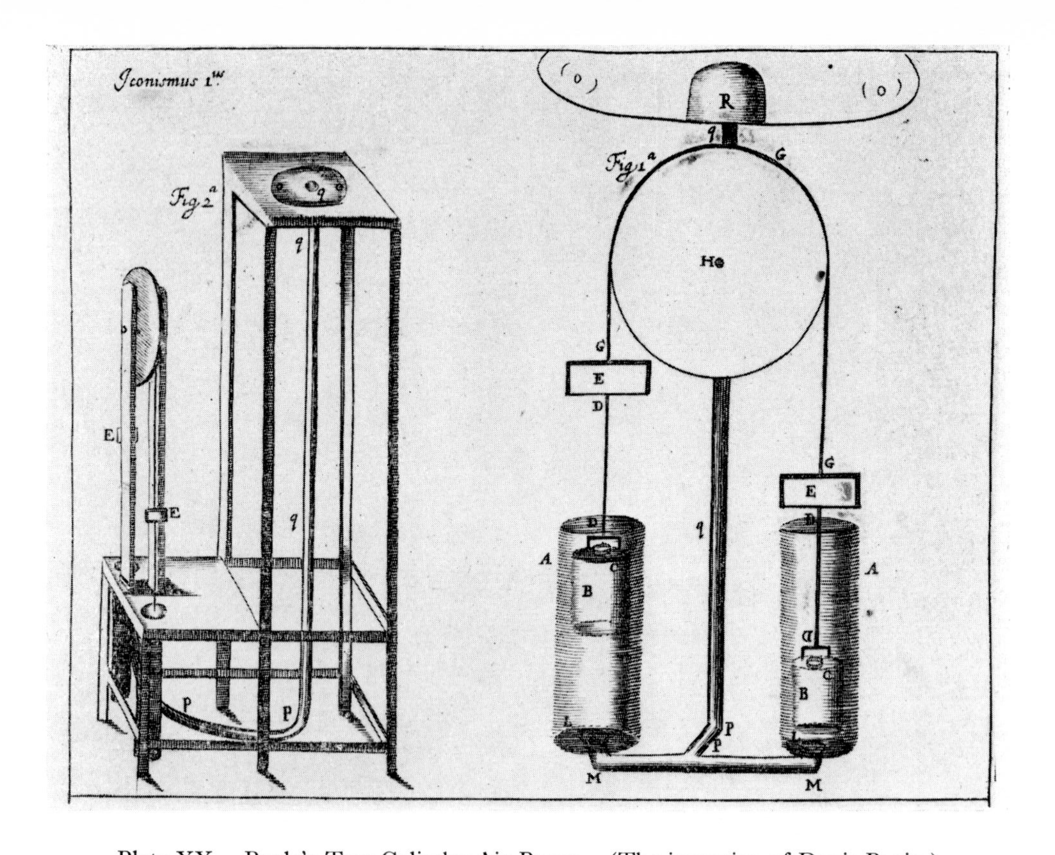

Plate XX. Boyle's Two Cylinder Air-Pump. (The invention of Denis Papin.)

Plate XXI. Robert Boyle, *aet*. 37. *Engraving by William Faithorne.*

The landscape background of the original drawing (see frontispiece) has been replaced at Hooke's suggestion by Boyle's air–pump.

Plate XXII. Prospect of Pall Mall. *Engraving by John Kip. ca*, 1710–20.
St. Jame's Palace in the foreground; the church of St. Martin-in-the-Fields in the middle distance.

B

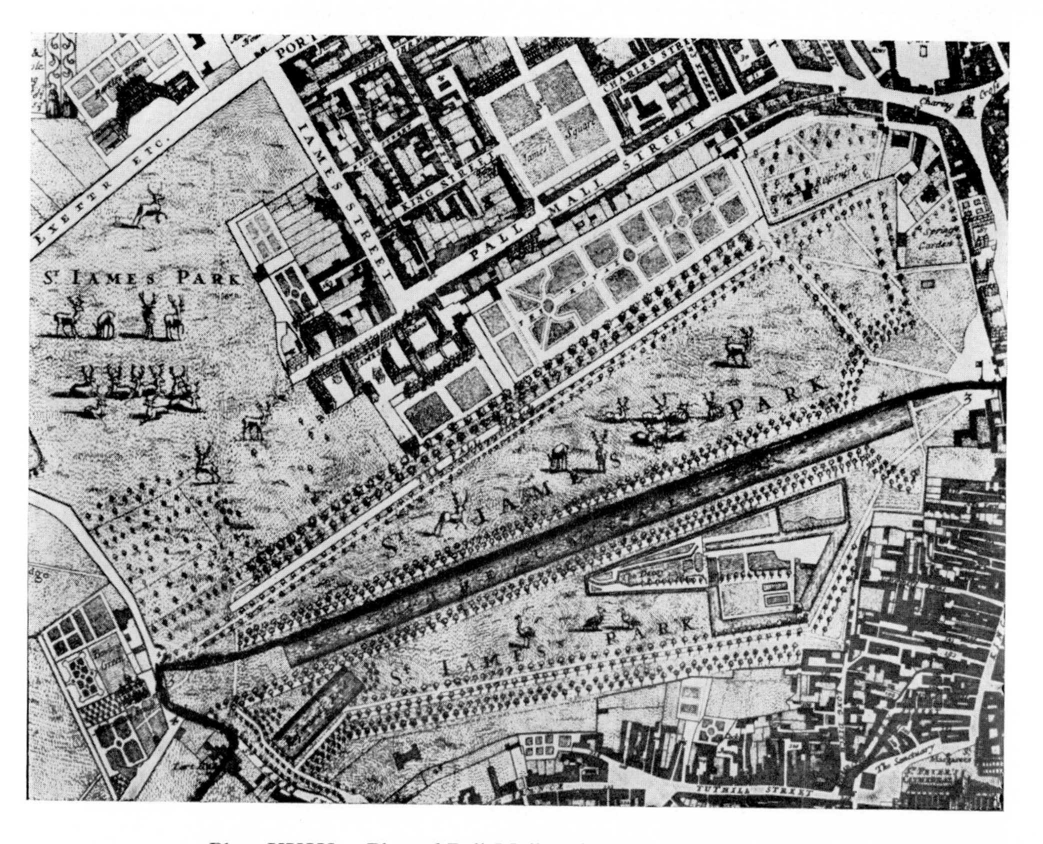

Plate XXIII. Plan of Pall Mall and surroundings, 1682.

Plate XXIV. Katherine Boyle. 2nd Viscountess Ranelagh. (1615–1691).
Sister of Robert Boyle.

Plate XXV. Mary Boyle. 4th Countess of Warwick, *aet.* 53. (1624–1678).
Sister of Robert Boyle.

Engraving by R. White.

TIOMNA NUADH

AN

DTIGHEARNA

Agus ar

SLANUIGHEORA

IOSA CRIOSD

Ar na tarruing go firinneach as GREIGIS go goidheilg.

RE

HUILLIAM O DOMHNUILL.

A LUNNDUIN,

Ar na cur a gcló ré Robert Ebheringtam, an bliadhain daois an Tigherna, 1681.

Plate XXVI. Title page of *Tiomna Nuadh*, London, 1681.

The second edition of the *New Testament* in Irish.

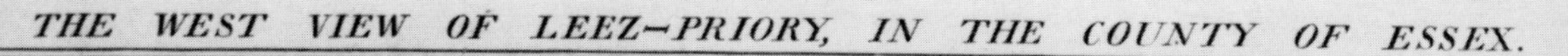

Plate XXVII. West View of Leese-Priory, Essex. *Engraving by S. & N. Buck, 1738.*

A

B

Plate XXVIII. Medals issued to commemorate Robert FitzGerald's invention of an apparatus for making fresh water from sea-water.

Medal A is in pewter.

Medal B is in brass.

Plate XXIX. Ambrosius Gottfried Hanckwitz, F.R.S. (1660–1741).

One of Robert Boyle's operators, and founder of a well-known firm of manufacturing chemists.

Plate XXX. Robert Boyle, *aet.* 62. *Portrait by Johann Kerseboom.*

Plate XXXI. West Prospect of the Church of St. Martin-in-the-Fields.
Robert Boyle and his sister, Katherine, were buried in the chancel of this church.

Engraving by G. Vertue.

Plate XXXII. South Prospect of the Church of St. Martin-in-the-Fields.

Engraving by G. Vertue.

Plate XXXIII. Bolton Abbey Rectory, formerly the Boyle School.

Plate XXXIV a. Boyle School, Yetminster, viewed from the left.

Plate XXXIV b. Boyle School, Yetminster, viewed from the right.

Plate XXXV. View of the Hermitage, Royal Gardens, Richmond.
Engraving by C. du Bosc.

Plate XXXVI. Sectional view of the Hermitage, Royal Gardens, Richmond.

Plate XXXVII a. Plan of the Hermitage, Royal Gardens, Richmond.

The niches of the circular room were occupied by busts of Isaac Newton, John Locke, Samuel Clarke and William Wollaston. The bust of Robert Boyle stood 'on a pedestal in the inmost, and, as it were, the most sacred Recess of the place; behind his head a large Golden Sun, darting his wide spreading Beams all about, and towards the others, to whom his Aspect is directed.'

Plate XXXVII b. Robert Boyle, *aet*. 63.

Obverse of brass cast by Carl Reinhold Berch of Ivory Medallion by Jean Cavalier.

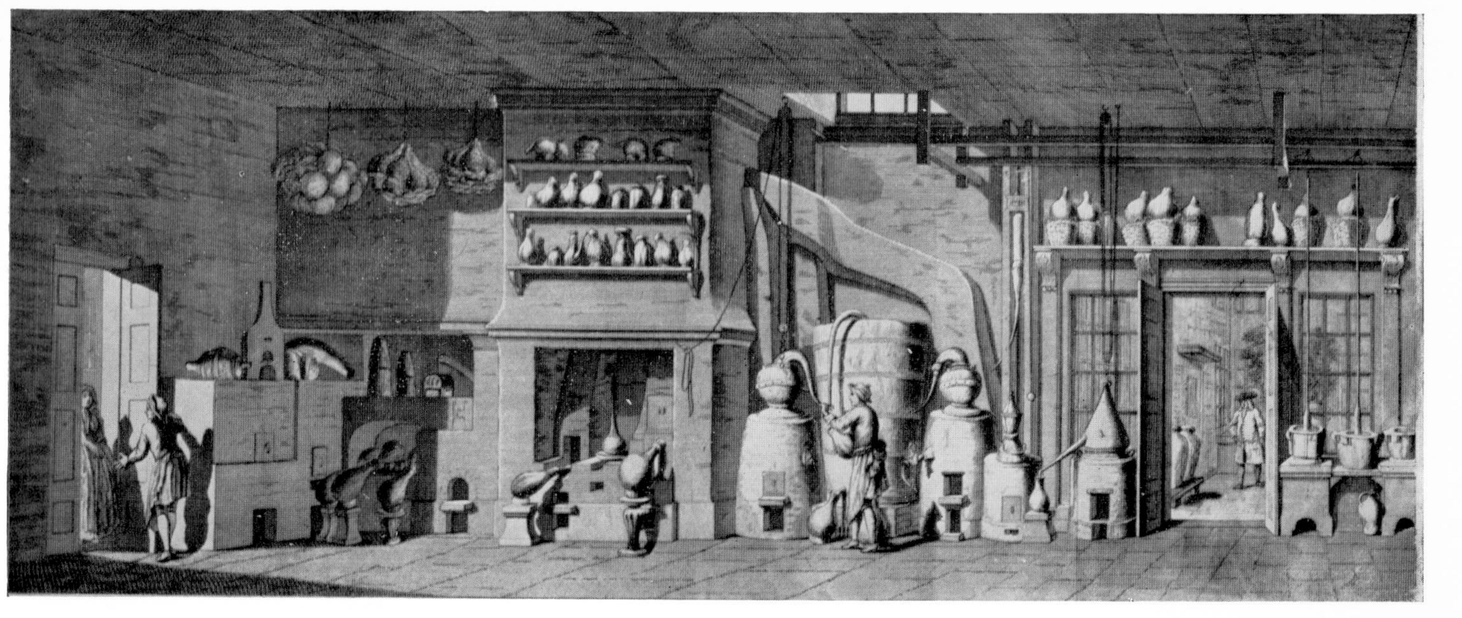

Plate XXXVIII. This plate reproduces one of three engravings which traditionally are considered to represent the interior of the *Golden Phoenix*, the laboratory established in Southampton Street, Covent Garden, London, by Ambrose Godfrey Hanckwitz in 1707. It remained there until 1862–3, when with other properties it was demolished to provide the site for the present Roman Catholic Church of Corpus Christi. The firm of manufacturing chemists established by Hanckwitz, which was well-known in the nineteenth century as Godfrey and Cooke, was acquired by Savory and Moore Ltd, New Bond Street, London in 1916–7. In the absence of any description of Robert Boyle's own laboratory, the above illustration of what was probably the best equipped laboratory of its kind at the beginning of the eighteenth century reveals the prominence of equipment devoted to operations by fire, concerning which Hanckwitz says, " for the space of about forty years I have most frequently busied myself in Operations and Essays relating to Productions and Actions of Fire, and of Heat and Cold in their several Degrees, having made repeated Tryals of the different Actions of both dry and liquid Bodies upon each other, from the slightest intestine Motion, or *Effervescence*, to the most vehement Ebullitions, so as not only to occasion Light and Petillation, but to break out into sudden and violent flames; all which Observations have greatly contributed towards my better perfecting that wonderful Preparation, the *Phosphorus Glacialis*. "

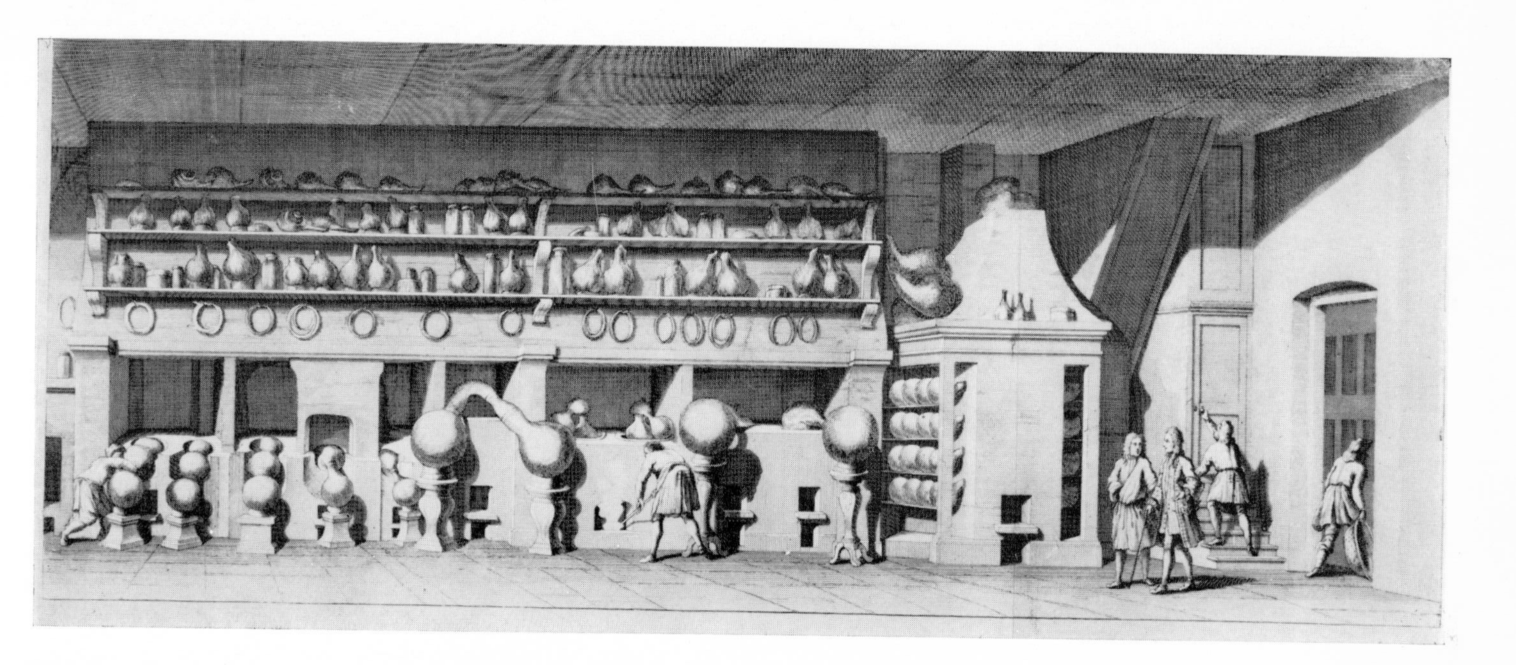

Plate XXXIX. Second view of the interior of the *Golden Phoenix*.

From an engraving in the City of Westminster Public Library, Gardner Collection.

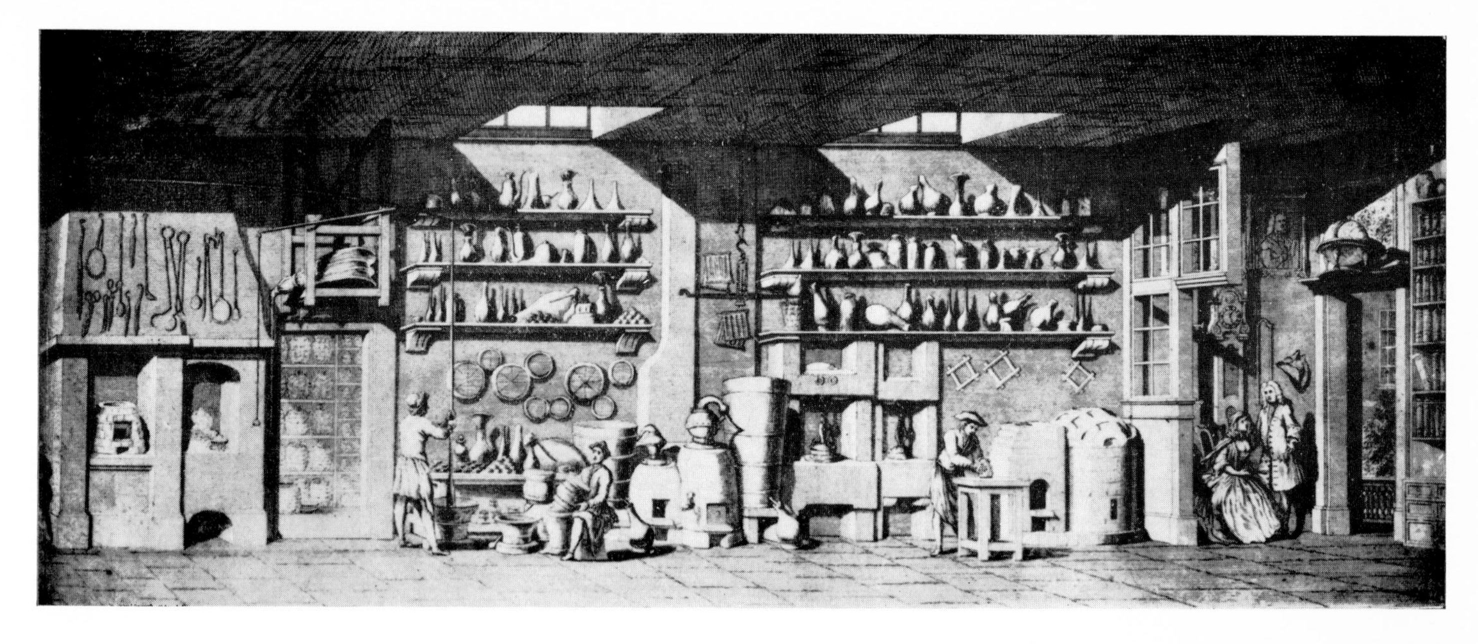

Plate XL. Third view of the interior of the *Golden Phoenix*.

From an engraving in the Wellcome Historical Medical Museum.

APPENDIX I

Robert Boyle's Last Will and Testament

This document is preserved in the Principal Probate Registry,* Somerset House, London. It is written on sixteen large sheets of paper approximately $16\frac{1}{4}'' \times 13''$. With the exception of the tenth sheet they all bear Boyle's signature and/or seal with the Boyle arms: *Party per bend embattled argent and gules.*†

In the Name of God Amen. I Robert Boyle of Stalbridge in the County of Dorsett Esqr youngest Sonn of the late Right Hono^ble^ Richard Earle of Corke deceased being God be praised of good and perfect Memory and taking into due and serious Consideration the certainty of death and the incertainty both of the time and manner of it, Being likewise desireous when I come to die to have nothing to doe but to die Christianly w^th^out being hindred by any voidable distractions from imploying the last houres of my Life in sending upp my desires and meditations before mee to Heaven Doe this Eighteenth day of July in the Third yeare of the Raigne of our Soveraigne Lord and Lady William and Mary by the grace of God King and Queene of England Scotland France and Ireland defenders of the Faith &c And in the yeare of our Lord God One thousand Six hundred Ninety and One Make and Ordaine this my last Will and Testam^t^ in writeing in manner and forme following.

* Prerogative Court of Canterbury, 3 Fane, (1692). Proved 26 January $169\frac{1}{2}$.

† The seal used here shows the Boyle arms without difference. At the time of his father's death RB was the fourth surviving son, the others having died without issue. The cadency mark for the fourth son is a martlet. RB used an armorial seal so differenced in 1652 (see for example B.M., Add. MSS. 32093, fol. 293). RB's coat of arms with a martlet appears on one of the plates in R. Morison's *Historia Plantarum Oxoniensis*, Oxford, 1680, Sect. 2, Tab. 7.

First and chiefely I Com̄end my Soul to Almighty God my Creator wth full Confidence of the pardon of all my Sinns in and through the Merritts and Mediation of my alone Saviour Jesus Christ And my Body I Com̄itt to the Earth to bee decently buried wthin the Citties of London or Westminster in case I die in England wthout Escutcheons or unnecessary pompe and wthout any superfluous Ceremonies* And wthout the Expence of above two hundred & fivety pounds And also wthout being unnecessarily dissected or disembowel'd, and if it shall appeare necessary that my Body be Opened Then my Will is that it bee performed very privately and in the presence of none but the Physitian or Chyrurgion and his Attendants and of my domesticks or very neare Relations And as touching my temporall Estate wherewth God of his goodnesse hath been pleased to endow mee, I dispose thereof in manner and forme following: (That is to say).

I Give and bequeath unto my deare Sister the Lady Katharine Viscountesse Ranelagh a Small Ring usually worne by mee on my left hand having in it Two small diamonds wth an Emerald in the middle wch Ring† being held by mee ever since my youth in great Esteeme and worne for many yeares for a particular reason not Unknowne to my said Sister the Lady Ranelagh I doe earnestly beseech her my said Sister to weare it in remembrance of a Brother that truely honour'd and most dearely Lov'd her.

Item I give to the said Lady Ranelagh all my Manuscripts and Collections of Receipts‡ whether of my owne hand writeing or others Unless I doe otherwise dispose of all or any of them before my decease beseeching her to have a care that they or any of them come not to the hands or perusall of any to whom shee thinkes that if I were alive I should be unwilling to have them comunicated And I doe likewise devise unto her the Sum̄e of Three hundred pounds (and all Interest) which I formerly lent her upon enlargeing the dwelling House in the Pall Mall wch shee now lives in.

Item Whereas by my decease the greatest part of my intailed Lands in Ireland§ will come to my deare and Eldest Brother

---

* RB died in his sister's dwelling house in Pall Mall, Westminster, and was buried in the church of St. Martin-in-the-Fields. His wish that there should be no undue ceremony seems to have been respected. (See p. 185).

† Romantically inclined writers have assumed that this was connected with the Great Earl of Cork's design to marry RB to Anne Howard (see p. 55), and that this was the betrothal ring.

‡ With regard to RB's chemical papers see p. 203.

§ For the complexities of the various Boyle estates see the last will and testament of the Great Earl of Cork. (Townshend. Appendix III.)

Richard now Earle of Burlington and Corke I doe hereby give and bequeath unto him a Sardonixe Seale Ring which I usually weare on my little Finger, beseeching him to accept of it and weare it for my sake as a Testimony of my unfeigned Affection and of my sence of his great kindness and many Favours towards mee, And begg him to beleive That it doth not afflict mee that I have noe Children of my owne to inherit my entailed Lands since they are by that defect to returne to him the truely Honourable Head of our house and Family.

[End of the first sheet. Signed: Ro: Boyle.]

Item I Leave and bequeath unto my deare Brother the Lord Viscount Shannon the best Watch I shall die possest of to putt him in minde of my constant Kindness and Affection wch I endeavour'd to express by my voluntary yearely Expence in keeping upp the Manno^r^ House of Stalbridge w^th^out intending to live in it for his sake.*

Item I give and devise unto my worthy Friend Richard Newman† of Westminster in the County of Middx Esqr a peece of Plate of the value of Tenn pounds sterling as a Testimony of my remembrance of his kindness wch I desire him to continue by affording his Assistance if there be Occasion to my Executo^rs^ hereafter named.

Item I give and bequeath unto my faithfull Friend John Nicholls Gent & Steward of the Courts of my Manno^r^ of Stalbridge a peece of Plate of the Value of Tenn pounds sterling to bee w^th^in the space of One yeare next after my decease deliver'd unto him by my Executo^rs^ as a Testimony of my remembrance and Affection, hereby making him the same earnest request That I have hereby made to Mr Richard Newman.

Item I give and devise unto the Right Reverend Father in God Gilbert Lord Bishop of Salisbury my great Hebrew Bible w^th^

---

* Francis Boyle died in 1699. Stalbridge Manor subsequently passed out of the possession of the Boyle family. At some date not known to me the estate was acquired by the Walter family (Peter Walter died 1735, and his grandson of the same name died 1753 without issue). It then passed to the Bayley family. In 1810 it belonged to Henry William, Earl of Uxbridge, whose estates were sold in 1854 to the Marquess of Westminster, who still owned them in 1869. Stalbridge House was demolished in 1822; there was a farmhouse on the site in 1869. (J. Hutchins. *The History and Antiquities of the County of Dorset*, 3rd edition, 4 vols London, 1861–1873, Vol. **III**, p. 671: *N. & Q.*, **Ser.** ii. Vol. **IV**, 85.)

† Not identified. Perhaps a lawyer.

silver Claspes* as a small Token of my great respect for him and sence of his Favours to mee.

Item I give and devise to Thomas Smith† now my Servant thirty pounds in Case he be wth mee or in my Service att the time of my death But in case of his departure or Removall from mee before that time Then my Will is That he shall have twenty pounds and no more.

Item Whereas my Servant John Warr the younger‡ is indebted unto mee in the Sum̄e of Fifty pounds by Bond or Bill Obligatory my will is That his said debt be Remitted and discharged.

Item Whereas I had Sett apart amongst other things the Sum̄e of foure hundred pounds for certaine Pious uses And whereas his late Ma^ty^ Charles the Second having by his especiall Grace and Favour w^th^out my Seeking or knowledge been pleased to Constitute mee Governo^r^ of the Corporac̄on for propagating of the Gospell amongst the Heathen Natives of New=England and other Parts of America§ hath thereby given mee Opportunity to discerne that Worke to be unquestionably Pious and Charitable

---

* Not identified in the sale catalogue of the bishop's library, 19 March $171\frac{5}{6}$.

† (?–1742). He "lived seventeen years with Mr. Boyle, and was with him at his death." (*Works Fol.*, **I**, Preface: *Works*, **I**, p. ii). He was an apothecary. The records of the Society of Apothecaries provide only one Thomas Smith who can be reasonably identified with this servant of RB. He was bound apprentice 12 September 1667, and made free 5 October 1675. On this reckoning he must immediately have joined RB's household, and would have been a very old man when he died. From 1700, or possibly late 1699, till his death he lived at his dwelling house at the sign of "Boyle's Head." (See p. 222.) His will reveals that he was possessed of considerable property. (Somerset House, P.P.R., P.C.C., 67 Trenley (1742)). He seems to have been a member of Dr. Jabez Earle's presbyterian congregation. This Thomas Smith was granted letters of administration of the estate of John Warr, the younger. (See next note.)

‡ (?–1715). This servant of RB was one of the executors of the will, and charged with the more onerous work in settling the estate. (See p. 206.) Like Thomas Smith, he had been in the service of RB for a good many years, for we find that both of them, together with Hugh Greg, who too is mentioned in this will, witnessed a power of attorney signed by RB on 10 March $168\frac{2}{3}$. John Warr's library was put on sale at fixed prices at Exeter Exchange, Strand, 15 May 1717. (B.M. Sale Catalogue.)

§ The first "Act for the Promoting and Propagating the Gospel of Jesus Christ in New England" was made during the Interregnum, 27 July 1649. (Firth and Rait. Vol. **II**, p. 197). After the Restoration, the Company for Propagation of the Gospel in New England and the parts adjacent in America was granted a Patent of Incorporation (7 February $166\frac{1}{2}$) by which RB was appointed first Governor of the Company "during good behaviour". (*C. S. P. Colonial* (*America*) (*1661–1668*), p. 71: W. Kellaway. *The New England Company 1649–1776*, London, 1961, p. 45 *et seq*). The charter is printed *in extenso* in *Works Fol.*, **I**, *Life*, p. 95: *Works*, **I**, p. cli).

And whereas I have given and paid the Sum̄e of Three hundred pounds towards that Pietie I doe hereby Give and devise the Sum̄e of One hundred pounds more to the said Corporation (Tho' by reason of Sicklyness and Infirmitie I have resign'd the Office of Governo^{r}) to bee sett aside and employ'd as a Stock for the Releife of poore Indian Converts wch I hope will prove of good effect for the Advancem^{t} of the Pious worke for wch they are Constituted and in wch I heartily pray him whose Glory the Worke itselfe tends unto (and I hope the Persons entrusted wth it ayme att) to give them a prosperous Successe.*

Item to the Royall and Learned Society for the Advancement of experimentall Knowledge (wont to meete att Gresham Colledge) I give and bequeath all my raw and unprepared Minneralls, as Ores, Marchasites, Earths, Stones, (excepting Jewells) &c to bee kept amongst their Collections of the like Kind as a Testimony of my great Respect for the Illustrious Society and designe wishing them also a most happy Successe in their laudable Attempts to discover the true nature of the workes of God and praying that they and all other Searchers into Physicall Truths may cordially Referr their attainements to the glory of the Authour of Nature and the benefitt of Mankind.†

Item to my worthy Friend and Physitian Sr Edmund King Knight I give and bequeath a Silver Standish of the value of Thirty pounds sterling as an unfeigned though but slight Testimony of the just Esteeme I have of his Worth and Skill, and the sense I have of his particular Care of mee

[End of the second sheet. Signed: Ro: Boyle.]

and kindnesse for mee.

Item I give to Mr Robert Hooke now Professo^{r} of Mathematicks in Gresham Colledge my best Microscope and my best Load= Stone wch I shall have att the time of my death.

Item I give and bequeath unto Mr John dwight and Mr John Whittacre once my Servants each of them a Ring of Five pounds price.‡

---

* See p. 206.

† With regard to RB's minerals see p. 201.

‡ Of the various servants of RB mentioned in this will, John Dwight, Robert St. Clair and John Milne have not been further identified. John Whittacre is doubtless the same Whitaker who occasionally finds mention in RB's correspondence.

Christopher White is certainly " the skilful and industrious operator " of the Officina Chimica of the Ashmolean Museum, Oxford, where he was already

Item I give and bequeath unto M[r] Christopher White Once my Servant Five pounds Sterling.

Item I give and bequeath unto M[r] John Milne and M[r] Hugh Gregg once my Servants each of them a Ring of Six pounds price.

Item I give to Nicholas Watts my Bayliffe of the Manno[r] of Stalbridge a peece of Plate of the value of Tenn pounds, or Tenn pounds Sterling in money in liew thereof to bee att his owne Choise.

Item I give unto my aforesaid Servant John Warr forty pounds and the one halfe of my wearing Apparrell and Linnen in case he be wth mee att the time of my death, But otherwise to have noe share of my Apparrell and Linnen. And I give to Robert S[t] Clair my Servant the Sum̄e of fifteen pounds and the other halfe of my Apparrell and Linnen in case he continue to be my Servant att the time of my death, If he doth not, Then I give to him a Ring of Six pounds price only. And to dr Frederick Schloer late my Servant a Ring of the price of Eight pounds.

And I Make and Ordaine my said deare Brother Richard Earle of Burlington and Corke and my said deare Sister the Lady Ranelagh* and John Warr the younger Gent my present Servant Sole Executo[rs] of this my Last Will and Testament.

Item I give and devise to my said Executo[rs] All my Estate Right Title and Interest of in and to a certaine Lease bearing date in or

* See the last codicil occasioned by the death of Lady Ranelagh.

installed by 1684. (E. Chamberlayne. *Anglia Notitia*. London, 1684, Part 2, p. 327.) To the chemical library he gave a book, [Joannes Zwelfer] *Pharmacopoeia Augustana Reformata, et eius mantissa*, Goudae, 1653, on the fly-leaf of which is the following inscription:

"*Christopherus White hujus officinae egregius operator sub insigni Boyleo educatus pro suâ in Chymiam benevolentiâ huic Bibliothecülae hunc librum D:D:*" (Bodleian MSS Ashmole 1667).

He is mentioned by V. M. Coronelli in *Viaggi d'Italia in Inghilterra*, Venice, 1697, Vol. **II**, p. 177; and by H. L. Benthem in *Engländischer Kirch- und Schulenstaat*, Lüneberg, 1694, p. 350. When von Uffenbach visited Oxford in 1710, and deplored the neglected state of the Ashmolean laboratory, he remarked on the lack of interest taken in it by the professor of chemistry, Richard Frewin, saying that the operator "Herr White (der ohnedem sehr liederlich seyn soll)" cared even less. (Z. C. von Uffenbach. *Merkwürdige Reisen durch Niedersachsen Holland und Engelland*. 3 vols. Ulm and Memmingen, 1753–4. Vol. **III**, p. 139). If Herr White is Christopher White, then he cannot have died in 1696. (Wood. *Life*. Vol. **III**, p. 55, 199, 227; Vol. **IV**, p. 79; Vol. **V**, Index).

Hugh Gregg wrote *Curiosities in Chymistry*, London, 1691.

Frederick Schloer (=Slare) (1647–1727), physician and chemist; M.D., Oxford; F.R.S. (*D.N.B.*: P. R. Barnett, *Theodore Haak*. 's-Gravenhage, 1962, the index and appendix I of which should be consulted).

about the Month of [Blank] made by the Corporation of the Mynes Royall or their Authority unto William Lord Viscount Brounker S[r] Robert Murray Knight William late Lord Brereton and my selfe of a Myne or Mynes in or neere the parish of Ashbury in Cheshire for One and Forty yeares or thereabouts.*

Item Whereas the Executo[rs] of James Watson late Alderman of the City of dublin and Abraham Riches is late of the same City Merchant or some of them stand indebted unto mee in Nyne hundred and Sixty pounds or thereabouts towards satisfaction whereof I have received the Sum̄e of Fourscore pounds or somewhat more for a House in damaske Street dublin sold by mee to my once worthy Friend S[r] John Temple† Knight deceased formerly Master of the Rolls which House was made over unto mee amongst other things towards satisfaction of the said debt upon a decree which I obteyned in the then called the Court for the Administration of Justice sitting att dublin aforesaid And whereas the said debt of Nine hundred and Sixty pounds or thereabouts saving and except the Fourscore pounds or somewhat more soe as aforesaid before menco̅ned to bee by mee received is still due and owing I give and devise the same wth all Writeings of what kind soever thereunto belonging and the proceeds of Composition for the said debt whensoever it shall happen to be made unto my said Executo[rs]. And I doe hereby Will and declare That all and every of my Bequests aforesaid unto the said Richard Earle of Burlington the Lady Ranelagh and John Warr my Servant joyntly as Executo[rs] and in that Capacitie shall not be for their owne private uses but only in Trust for and towards the paym[t] of my debts and Legacies hereby devised and towards and for performance of this my Last Will and Testam[t] and to and for no other use or uses w[t]soever.

[End of the third sheet. Signed : Ro: Boyle.]

---

* A Mine Royal is one where the ore yields gold and silver exceeding in value the charges of refining and loss of the baser metal it contains. The precious metals were claimed by the sovereign for coinage. The Society of the Mines Royal was established in the tenth year of the reign of Elizabeth I. (J. Pettus. *Fodinae Regales*, London, 1760).

William, Lord Brereton. 3rd Baron, (1631–1680). (G.E.C.)

Ashbury must be an error for Astbury.

† Sir John Temple (1600–1677) (*D.N.B.*).

The deed of sale made to Sir John Temple of RB's house in Damaske Street, Dublin, is mentioned in a " Catalogue of the Writings left in the wooden box with Mrs. Dury." (B.P. **XXXVI**). This list is in RB's hand; its date is about 1656–1657. Mrs. Dury was dead before June 1664. (Turnbull. *H. D. & C.*, p. 297).

And as touching the disposition of all and singular my Castles Manno^{rs} Lordshipps Messuages Lands Tenem^{ts} and Hereditam^{ts} in the Kingdome of Ireland which I have power to dispose of, **This is the Last Will & Testam^{t}** of mee the said Robert Boyle made and declared the day and yeare first above written. (*Viz*^{t}) **I Give & Devise** unto the Right Hono^{ble} Richard Earle of Burlington and Corke, my deare Nephew the Hono^{ble} Henry Boyle Esqr of Castle Martyr in the County of Corke and Sr Robert Southwell of Kingsale in the Kingdome of Ireland All my Estate Right Title Interest and demand whatsoever of in and unto the Castle, Towne, Manno^{r} and Lands (being Nine Plough Lands and halfe or thereabouts) of Buttevant and the Moyetie of the Lands called Buttevant and Rices Lands in the County of Corke severall yeares Since mortgaged to my Father and unto mee for the Sum̄e of Two thousand pounds sterling or thereabouts, and after my Fathers decease (Leased by mee under certaine Covenants and Conditions for One and Thirty yeares unto Lieutenant Colonell Agmondesham Muschampe) for about the Sum̄e of Six score pounds per Annum and now or lately in Lease to denny Muschampe Esqr And also all my Estate Right Title Interest Clayme and demand whatsoever of in and to All and singular the Townes Lands Tenem^{ts} and Hereditam^{ts} of Rathenege and Four pounds Chiefe Rent issuing out of the Lands of Turmore and Ballytrasney by the same of Five Castles all lying in the aforesaid County of Corke And all my Estate Right Title Interest and demand w^{t}soever of in and to the Impropriate Rectory of Adare a̅l̅s̅ Athdare in the County of Lymerick in Ireland. **And also** all my Estate Right Title Interest and demand whatsoever of in and to the Castles Townes and Lands conteyning Two Plough Lands and halfe (be it more or less) of Ballydangan Killcroyne and Bally Brittas heretofore mortgaged to my said Father for the Sum̄e of One Thousand pounds or thereabouts **To have & to hold** all and singular the said Castles Manno^{rs} Lordshipps Messuages Lands Hereditam^{ts} and p^{r}misses of Buttevant Rices Lands Rathenege and the Four pounds a yeare chiefe Rent the Impropriation of Adare the Two Plough Lands and an halfe of Ballydangan Killcroyne and Bally Brittas and all other the p^{r}misses and all Sum̄e and Sum̄es of money therefore or thereout due and payable unto mee wth all my Estate Right Title Interest Clayme and demand whatsoever of in and to the p^{r}misses and every part Member and Appurtenances of them every or any of them And also all deeds Evidences and Writeings w^{t}soever for or concerning the p^{r}misses or any part thereof to them the said

Richard Earle of Burlington the Hono^{ble} Henry Boyle Esqr and Sr Robert Southwell and their heires and Assignes for ever as fully and in as large and ample manner as the p^{r}misses or any part thereof are or is vested in descended Come Given devised or Bequeathed unto Mee the said Robert Boyle upon speciall Trust and Confidence Notwithstanding and to and for the only uses Intents and Purposes in and by this my Last Will and Testam^{t} lymitted and declared and to or for no other use Intent or Purpose whatsoever (That is to say) That they the said Richard Earle of Burlington the Hono^{ble} Henry Boyle Esqr and Sr Robert Southwell and their heires shall after my death by and wth the Advise of my Executo^{rs} Sell the said respective p^{r}misses and pay the Money thereby raised as also the Proffitts untill Sale unto my said Executo^{rs} to and for the uses herein expressed.*

Item I give unto my said Executo^{rs} all and every such Sum̄e and Sum̄es of money as shall bee raised out of or by Sale of my said Lands **And also** all my printed Bookes† Goods Chattells Plate Moneys Jewells and Creddítts whatsoever and herein before bequeathed Lymitted or Appointed unto them to and for the onely uses Intents and Purposes hereafter also in and by these p^{r}sents lymitted

[End of the fourth sheet. Signed: Ro: Boyle.]

expressed and declared and to or for no other use Intent or Purpose whatsoever (That is to say) To the intent and purpose That the said Richard Earle of Burlington Lady Ranelagh and John Warr their heires Executo^{rs} Administrato^{rs} or Assignes shall with what convenient Speede they can after my decease by Sale or Compositions (though upon Termes advantagious to the Buyers) Sell and make money of all the said Castles Lands Mortages Goods and Chattells Cattle Plate Jewells and Creditts by this my Last Will and Testament bequeathed unto them or to

* Henry Boyle. See Gen. Chart [F.g.2].

Agmondesham Muschamp. For his account of some events during the Great Irish Rebellion see *II Lismore*, V, p. 89.

Denny Muschamp is undoubtedly related to the former. For his relationship to the Boyle family see Gen. Chart [F.k.2]. He appears in the Earl of Orrery's rent roll for 1674. (*Calendar of Orrery Papers*).

A Bill of Complaint was lodged by the Attorney General in the Chancery Court on information of Sir Robert Southwell against the executors of RB's will alleging that they had failed in the accounts submitted to the court to account for the income from the Impropriated Rectory of Adare; the amount involved was about £3000. (P.R.O. Chancery Proceedings. Bridges Division. C.5/213/1. 30 May 1700.)

† The fate of RB's library is described above, p. 198 *et seq.*

my said Trustees Richard Earle of Burlington the Hono^{ble} Henry Boyle Esqr and Sr Robert Southwell and shall in some convenient time after my decease out of the Proceeds of the said Sale or Compositions duely and fully Satisfie Pay and discharge as well all the severall debts and Sum̄es of Money (att present not amounting to much) which I now doe or att the time of my decease shall owe to any person or persons whatsoever And the Charges of the Funerall of my Body And also all such legal Engagements as I have undertaken to performe to any person or persons whatsoever especially to my Tenants of my Manno^{r} of Stalbridge **And** from and imediately after my debts and funerall Charges and all such Legacies as I have above by this my Will Given and bequeathed shall bee discharged and paid and my aforesaid Engagements made good, Then my Will is That the rest residue and remainder of the Proceede of the Sale of or Compositions for all the said Lands Mortgages and Chattels hereby devised and Bequeathed unto the said Richard Earle of Burlington Lady Ranelagh and John Warr my said Executo^{rs} shall bee by them Sum̄ed upp And that in the First place they doe distribute among the Poore (having therein a speciall regarde unto the poore of the Parish of Stalbridge and of the parish of Fermoy in the County of Corke in the Kingdome of Ireland and other parishes in that Kingdome where any of my Lands doe lye) the Sum̄e of Three hundred pounds Sterling And next after the distribution thereof the Sum̄e of Two hundred pounds sterling more amongst the much distressed Persons of all Qualities that have been forc'd or frighted out of Ireland and that shall happen to bee in or neere London Bristoll Chester or Leverpoole att the time of my decease which Sum̄es with respect chiefly to some things that Occurr'd in a gracious deliverance that was vouchsafed mee in a great distress* I thinke fitt to Charge my Executo^{rs} with the Payment of after my decease And also in the next place That after the distribution of the Sum̄es of Three hundred pounds and Two hundred pounds as aforesaid and also after the payment of the Sum̄e of Threescore and Six pounds and Four shillings to the above mencōned Sr Robert Southwell to bee distributed by him amongst the Incumbents Residents and Widdowes and Children of such as have been Incumbents and Residents in the severall Parishes of which I have the Impropriations in the Kingdome of Ireland and to such of them and in such proporcōn as the said Sr Robert Southwell shall thinke most needing and deserving of the same That then they

* I am not able to say from what distress RB was delivered.

pay to my Friends and Kindred named in the Schedule to this my Will annexed the respective Sumes therein expressed in the whole amounting to the Sume of fower hundred pounds And after payment thereof I Will and devise all the Residue and Remainder of my Goods and Chattells reall and personall and of the money to bee raised by the Sales aforesaid by this my Last Will above directed to my said deare Sister the Lady Ranelagh with this desire Neverthelesse And my Will is That in case the said Residuary Part of my said Estate shall amount (as I verily beleive it will by farr) to more then eight hundred pounds sterling over and above the Legacies and Bequests above devised and disposed of That then my said deare Sister doe Retaine and Enjoy for her owne use the Sume of Two hundred pounds sterling over and above the Three hundred pounds debt above remitted to her (my Intention being that shee shall bee benefitted in the Totall to the value of Five hundred pounds sterling by this my Last Will and Testament)

[End of the fifth sheet. Signed: Ro: Boyle].

And this Legacy being deducted from the before mencōned Residuary part of my Estate Then my Will is That the Overplus bee laid out by my said deare Sister in case shee survive mee and in case of her death by my Executo[rs] in such manner as by any Codicill or other Writeing under my hand I shall hereafter direct And for want of such direction for charitable and other pious and good Uses att her or their discretion, But I doe cheifely Recomend unto her and them the laying out of the greatest part of the same for the Advance or Propagation of the Christian Religion amongst Infidells* and I hope my Relations and Friends will beleive my Intentions to bee as good towards them now as when I thought I had an Estate more considerable to dispose of then now I have (especially having of late yeares Lost by the breaking of Goldsmiths and others above Fifteene hundred pounds, and since that by the unfaithfullnesse of my Receiver in Ireland above Nine hundred pounds And since that also by the destructive Insurrections and Warr that hath hapned in Ireland the whole Incomb for above Two yeares Last past of my Estate there) and will therefore accept in good part of what Legacies shall bee presented unto them by my Executo[rs] considering that it is the duty of every honest Man to preferr the doeing of Acts of Justice before those of Kindnesse when hee is not sure hee is able to doe both att Once **And** whereas diverse Lands and Tenem[ts] by this

* An account of this charitable bequest is given on p. 206 *et seq.*

my Will above directed to be sold were Originally Mortgaged Estates to the end therefore that the heires of the Mortgagers may have noe just Cause to Complaine nor the Buyers bee mistaken in their Purchases I doe hereby declare direct and Appoint That in case my Cosin [Blank] the heire of Sr Peircye Smith* long since deceased shall desire within the Space of One yeare after notice given him by my Executors to redeeme the Impropriate Rectory of Adare (which was mortgaged by the said Sr Peircye for One thousand pounds) and shall not doe the same That then upon his Releasing of all his Right and Equitie of Redemption, the Sum̄e of Threescore pounds shall bee paid to him out of the proceede and Sale of the said Rectory as a Gift or Legacy from mee to him, But my Will is That the said Sum̄e of Threescore pounds shall not bee paid unto him unlesse he Release as aforesaid within one yeare after the notice aforesaid **And** whereas the Castle Towne and Lands of Ballydangan Killcroyne and Bally Brittas afore mencōned were heretofore transferred by mee to my Honored Cosin Michaell Arch Bishopp of Armagh and Primate of all Ireland† **To have & to hold** to the said Arch Bishopp and his heires from and after the determination of diverse yeares yet to Come and Expire **And whereas** some yeares since when the said Lands were ready to come into my hands I made a promise to Lease the said prmisses to my above named deare Brother the Earle of Burlington for all my Terme and Interest therein under the Rent of Threescore pounds per Annum or thereabout *de claro* I doe hereby Ratifie and Confirme my said promise and doe will and Appoint That hee shall hold and enjoy the same under the Rent of Threescore pounds per Annum during all the residue of my said Terme and Interest which shall bee to expire att the time of my Death. **Lastly** I hereby earnestly desire my said Executors to See this my Will and my Trusts in them reposed faithfully performed as they hope or desire to have their owne Wills faithfully performed **But for as much** as by reason of the Incertainties of the times it may happen (wch God forbidd) That my Estate may meete with some further Misfortunes whereby it shall be soe weakened as not to be sufficient to discharge all the Legacies by this my will bequeathed Then my Will is That after the payment of Three hundred pounds by mee devised to the poore and the Sum̄e

[End of the sixth sheet. Signed : Ro: Boyle.]

* See Gen. Chart [E.d.1. and F.m.]
† See Gen. Chart [E.e.1.].

of Two hundred pounds by mee devised to the much distress'd persons of all Qualities &c as aforesaid mencōned to bee in respect of a deliverance from a distress (which I hereby Will shall bee Observed) That then my other Legatees shall each of them bee respectively abated their Legacies in proportion to what my said Estate shall fall short of my said other Legacies above by mee devised **In witnes** whereof I the said Robert Boyle Revoaking and Anulling all other and former Wills Testaments and Bequests have to every Sheete of this my Last Will and Testament conteyned in Seaven Sheetes of Paper Sett my hand and to the Last of them my Seale the day and yeare first above written.

[Signed] Ro: Boyle. (L. S.)

Signed Sealed Published and Declared by the above named Robert Boyle as his Last Will and Testam[t] the day and yeare first above written in the presence of us

[Signed] Isaac Garnier
John Caddicke
James Oglebey
W[m] Johnson.*

[End of the seventh Sheet.]

A Schedule in my Will referr'd unto for the disposition of Four Hundred Pounds w[ch] I will shall be paid to the Persons herein after Sett downe in such proporcōn to each Person as is herein expressed.† (*viz*[t])

* R. St. Clair, T. Smith and W. Johnson are the only witnesses to RB's will or codicils who have been identified; they were his servants.

† The nephews and nieces named in this schedule are identified as follows:

Earl of Barrymore	Gen. Chart	[F.a.2.]
Earl of Ranelagh		[F.e.4.]
Charles Lord Clifford		[F.d.4.]
Capt. Robert FitzGerald		[F.c.4.]
Capt. Henry Boyle		[F.g.3.]
Countess Dowager of Thanet		[F.d.6.]
Countess Dowager of Clancarty		[F.c.9.]
Viscountess of Powerscourt		[F.g.4.]
Lady Frances Shaen		[F.c.8.]
Lady Katharine FitzGerald		[F.c.7.]
Mrs. Elizabeth Molster		[F.e.2.]

Imprimis To my Honord & Deare Nephews	The Earle of Barrymore Thirty pounds. The Earle of Ranelagh Thirty pounds. Charles Lord Clifford Thirty pounds. Capt Robert Fitz Gerald Thirty pounds. Capt Henry Boyle Thirty pounds.
Item To my Honoured & Dear Neices	The Countess Dowager of Thanet. Thirty pounds. The Countess Dowager of Clancarty Thirty pounds. The Lady Viscountess of Powerscourt Thirty pounds. The Lady Frances Shaen Thirty pounds. The Lady Katharine Fitz Gerald Thirty pounds. And to Mrs Elizabeth Molster, (not to make a Difference betweene her and my other Neices in my Affection, no more then there is in their Relation to mee but) because of her peculiar Circumstances I Give One hundred pounds.

[Signed] Ro: Boyle.

This is the Schedule referr'd to & Mencōned in the Fifth Sheete of this the Last Will & Testam^{t} of the Hono^{ble} Mr Robert Boyle as a part thereof & was by his Appointm^{t} annexed hereunto & (att & before the Ensealing & Publishing this his Will) Signed by him in the presence of us.

[Signed] Isaac Garnier
John Caddicke
James Ogleby
Wm Johnson.

[End of the eighth sheet.]

Whereas by an Act of Parliament in Ireland intituled an Act for the better Execution of his late Matie King Charles the Seconds gracious Declaration for the Settlement of his Kingdome of Ireland and Satisfaction of the severall Interests of Adventurers,

Souldiers and other the Subjects there,* It was Enacted (amongst other things) That I Robert Boyle my Executo[rs] Adm̄strato[rs] and Assignes should and might for and during the Terme of Thirty One yeares Have hold and enjoy all and singular the Impropriations of or belonging to the respective Abbies late dissolv'd Monasteryes, Religious Houses, Priorys or Parishes of Ballytubber in the County of Mayo, Knockmoy, Kilereulta, Oran a̅ls St Maries Athenry and dunmore in the County of Galway and Tyhone in the County of Tipperary or any of them Together with all the Impropriate Tithes and Rectories and Appurtenances of the said Impropriations, Tithes and Rectories or belonging thereunto, which belong unto, or by that Act were vested in his late Ma[ty] to any of the uses therein before mentioned according to the Tenor and effect of such Grant or Grants as had, or then were, or then after should bee passed unto mee thereof by his late Ma[ties] Letters Patents in that behalfe, I or they paying yearely for the same double the Exchequer or Crowne Rent referr'd thereupon in the yeare One thousand Six hundred Forty One **Now** I the said Robert Boyle doe hereby Give and devise to my deare Brother Richard Earle of Burlington and Corke in Ireland the Right Hono[ble] John Lord Viscount Massarine† Clotworthy Skeffington Esq[r] of the County of Antrim in Ireland and S[r] Peter Pett‡ Knight their Executo[rs] Administrato[rs] and Assignes the Fourth part (the whole in four parts to bee

---

* "An Act for the better execution of his Majesties gracious declaration for the settlement of his kingdom in Ireland, and satisfaction of the several interests of adventurers, souldiers, and other his subjects there." 14 & 15 Car. 2. c.2. Section 220 reads,

" Provided nevertheless, and be it enacted by the authority aforesaid; That Robert Boyl, esq; his executors, admins. and assignes, shall and may for and during the term of 31 yrs, have, hold and enjoy all and singular the impropriations of or belonging to the respective abbies, late dissolved monasteries, religious houses, priories or parishes of Ballytabber in the county of Mayo, Knockmoy, Kilcreulta, Oran, *alias* St. Maries, Athenry and Dunmore in the county of Galway, and Tyhone in the county of Tipperary, or any of them, together with all the impropriate tyths and rectories, and appurtenances of the said impropriations, tyths and rectories, or belonging thereunto, which belong unto or by this act are vested in your Majesty to any of the aforesaid uses, according to the tenor and effect of such grant or grants as hath or have been, or hereafter shall be, past unto him thereof by your Majesties letters patent in that behalf; he or they paying yearly for the same double the exchequer or crown rent reserved thereupon in the year one thousand six hundred forty one." (*The Statutes at Large passed in the Parliaments held in Ireland.* 8 vols. Folio, Dublin, 1765, Vol. II, p. 345.)

† For RB's relationship to the Viscounts Massereene, see p. 312.

‡ Sir Peter Pett (1630–1699), lawyer and author, original F.R.S. (*D.N.B.*)

divided) of the said Impropriations and p^{r}misses in and by the said Act granted Limitted or appointed by me to be held and enjoyed during the said Terme of One and Thirty yeares upon Trust to be by them employed and dispos'd for such good and pious uses as I shall from time to time declare or Appoint or in default of such Appointment after the death of mee the said Robert Boyle Then for such pious and good uses as my said Trustees in their best Judgments and Consciences shall thinke to bee most agreeable to my Intent And I devise the one Moyety or Halfendeale of the other Three Parts of the p^{r}misses unto the Executo^{rs} or Assignes of my late deare Sister in Law Margarett Countesse Dowager of Orrery* for all the residue of the Terme that shall be to Come att the time of my death And as to the other Moyetie of the said Three parts I devise the same to the Right Hono^{ble} Elizabeth Countesse Dowager of Anglesey† her Executo^{rs} and Assignes for all the Residue of the terme that shall bee to Come att the Time of my death. And in Case of her dying before mee Then I devise the same to her Executo^{rs} Administrato^{rs} and Assignes In Witness whereof I have hereunto Sett my hand and Seale the Five and Twentieth day of July In the yeare of our Lord God One thousand Six hundred Ninety and One.

[Signed] Ro: Boyle. (L. S.)

Signed Sealed & Published by the above nam'd Robert Boyle as Part of his Last Will in the presence of us.

[Signed] Isaac Garnier
Ro: St-Clair
Wm Johnson
James Oglebey.

[End of the ninth sheet.]

Memorandum That whereas I had thoughts of disposeing to good uses of what Money came to my hands on the Score of my Share of the Grant menc̄oned in the Codicill whereunto this Paper is annex'd to be granted to mee by his late Ma^{ty} King Charles the Second in the Act of Settlement of his Kingdome of Ireland And whereas I myselfe living in England found it difficult to meet wth persons every way qualified to be employ'd

* Margaret, Countess of Orrery, see Gen. Chart [E.b.11.].

† Elizabeth, Countess Dowager of Anglesey (?–1700), daughter of John Manners, 8th Earl of Rutland; married James Annesley, 2nd Earl of Anglesey (1645?–1690). (G.E.C.)

in the laying out of such Money to the uses I design'd, By which intents there remaines yet undispos'd of the Sum̄e of Two hundred pounds, which is now in the hands of the East India Company of Merchants of the City of London, for security whereof I have their Writeing or Bill under their Com̄on Seale dated the Twentieth day of May One thousand Six hundred Eighty and One, Which said Writeing or Bill together wth the benefitt Interest and increase thereof I have by Indenture of Assignement assign'd, transferr'd and Sett over unto Sr Henry Ashurst of London Baronett* by the name of Henry Ashurst of London Esqr in Trust (As by the said Indenture dated the Nynth day of January One thousand Six hundred Eighty and Four relation being thereunto had may Appeare) To and for such uses Intents and Purposes as I should att any time afterwards direct or Appoint And I having hitherto under no direction or Appointm^t^ do hereby (in case I dispose not otherwise of it or any part of it during my Life time) devise Legate and Bequeath the said Sum̄e of money wth the Benefitt Interest and Interest thereof to my Trustees Richard Earle of Burlington and Corke in Ireland The Right Hono^ble^ John Lord Viscount Massarine, Clotworthy Skeffington Esqr and Sr Peter Pett Knight their Executo^rs^ Adm̄strato^rs^ and Assignes to bee by them receiv'd and dispos'd of to the same uses for wch I intrust them wth my Share of the Impropriations &c of Ballytubber Knockmoy &c mentioned in the foregoeing Codicill And in case that w^th^in some reasonable time after my decease my Title to the p^r^misses or any of them, or to the money received by mee on the Score of any of them shall not bee question'd or like to be called in question to the p^r^judice of my Executo^rs^ Then my desire is That the Trustees may distribute the said Sum̄e according as their Consciences and discretion shall direct them among the Preaching Ministers (whether Parsons Recto^rs^ Vicars or Curates) that have bestow'd their paines as Incumbents in the parishes or Churches granted to mee by the aforesaid Act since they came into my quiett possession or to their poore Widdowes or Children The said Trustees being desir'd to have regard in the Distribution to the Pietie and Orthodoxness of the Persons To their paines and Faithfullness in performing the Worke of the Ministry, To the

* Sir Henry Ashurst, Bt. (1645–1711), M.P., Alderman of City of London. Member of the New England Company. (G.E.C. *Baronetage:* A. B. Beaven. *The Aldermen of the City of London*, 2 vols. London, 1908–1913.) He was the son of Henry Ashurst of London, merchant, (1614?–1680) (*D.N.B*). He was succeeded by his only son, another Henry, the 2nd baronet.

smallness of the worth of their Livings, and to the Considerableness of Income they yeild the Impropriato[r] And by this direction I desire and charge my said Trustees to take their Measures in disposeing of the future Income or Proceede of all that I have intrusted them wth Only in case it shall Appeare to them for weighty reasons that somew[t] may be spared for some other eminently good and pious use or uses, They are left at Liberty to use their discretion therein, Provided the Maine part of the above mencōned Money and future Income be employed for the Incouragem[t] of pious Preaching Ministers on the said Impropriations.

[End of the tenth sheet.]

Whereas I have an Intention to Settle in my Life time the Sum̄e of Fifty pounds per Annum for ever or att Least for a Considerable Number of yeares to be for an Annual Salary for some Learned divine or Preaching Minister from time to time to be Elected and Resident within the City of London or Circuite of the Bills of Mortality, who shall be enjoyned to performe the Offices following (*viz*[t])

1 To Preach Eight Sermons* in the yeare for proveing the Christian Religion ag[t] notorious Infidels (*viz*[t]) Atheists, Theists, Pagans, Jews and Mahometans, not descending lower to any Controversies that are among Christians themselves, These Lectures to be on the First Monday of the respective Months of January, February, March, Aprill, May, September, October, November in such Church as my Trustees† herein named shall from time to time appoint.

2 To be Assisting to all Companies and incourageing of them in any Undertakings for Propagating the Christian Religion to Forreigne Parts.

3 To be ready to satisfy such real Scruples as any may have concerning those Matters and to Answer such new Objections or Difficulties as may be started to w[ch] good Answers have not yet been made.

And whereas I have not yet mett wth a convenient Purchase of Lands of Inheritance for accomplishing such my Intention I doe

* This is the Boyle Lecture; see p. 205.

† Sir John Rotherham (1630–1696?) Sergeant-at-Law. (*D.N.B.*)

Thomas Tenison (1636–1715) D.D., Cambridge; bishop of London, 1691; archbishop of Canterbury, 1694. (*D.N.B.*: Edward Carpenter, *Thomas Tenison, Archbishop of Canterbury, His Life and Times*, London, 1948).

John Evelyn (1620–1706), the diarist.

therefore Will and Ordaine (in case it shall please God to take me hence before such Settlement be made) That all that my Messuage or Dwelling House in St Michael Crooked Lane London, wch I hold by Lease for a certaine Number of yeares yet to Come shall stand and be Charged during the Remainder of such Terme as shall be to Come and Expire at the time of my decease wth the paymt of the cleere yearely Rent & Proffitts that shall from time to time be made thereof (Ground Rent, Taxes, and necessary Reparacōns being first to bee deducted) to bee paid to such Learned Divine or Preaching Minister for the time being by Quarterly Paymts (That is to say) Att Midsomer, Michaelmas, Christmas and Lady day, the First paymt to begin att such of the said Feasts as shall first happen next after my decease and shall be made to such Learned Divine or Preaching Minister as shall bee in that Employmt at the time of my death during his Continuance therein. And I will That after my death Sr John Rotherham Serjeant at Law Sr Henry Ashurst of London Knt and Baronett Thomas Tennisson dr in divinity and John Evelyn Senior Esqr and the Survivors or Survivor of them and such person or persons as the Survivor of them shall Appoint to Succeede in the following Trust, shall have the Election and Nomination of such Lecturer And also shall and may Constitute and Appoint him for any Terme not exceeding Three yeares, And att the End of such Terme shall make a new Election and Appointment of the same or any other Learned Minister of the Gospell resideing wthin the City of London or Extent of the Bills of Mortality att their discretions.

And I doe hereby Will and Ordaine this my Codicill to bee a part of my Last Will and Testamt. Witness my hand and Seale this Eight and Twentieth day of July In the yeare of our Lord One thousand Six hundred Ninety and One

[Signed] Ro : Boyle. (L. S.)

Signed Sealed Published & Declared by the honoble Robert Boyle to be a Codicill to and part of his Last Will & Testamt in the presence of us.

[Signed] Isaac Garnier
Ro: St-Clair
Wm Johnson
James Oglebey.

[End of the eleventh sheet.]

Whereas since the Making and Executing of my Last Will and Testament severall Persons are brought to my Remembrance whom I should have made Legatees in my said Will had I thought upon them before my Finishing thereof And being unwilling that they should be wholy deprived of my intended kindnesse to them Wherefore I have thought fitt to insert their severall Names in Legacies in this my Codicill which I hereby declare to bee a part of my said Will. Inprimis I give & bequeath unto Mrs Katharine Molster* daughter of my Neece Mrs Elizabeth Molster the Sum̄e of one hundred pounds to be paid unto her when she shall have Attained to her Age Of One & Twenty yeares or upon the day of her Marriage which shall first happen. Item I give and bequeath unto Sr Robert Southwell the Sum̄e of twenty pounds Item I give & bequeath unto Sr Henry Ashurst Knight and Baronett the Sum̄e of twenty pounds Item I give and bequeath unto Sr William Ashurst† Knight and Alderman of the City of London the Sum̄e of twenty pounds Item I give and bequeath unto Sr John Rotherham Serjeant att Lawe the Sum̄e of twenty pounds Item I give and bequeath unto my Servant William Johnson the Sum̄e of fiveteene pounds Item I give and bequeath unto such a Number of the French Protestants Reffugees inhabitting in and about the Cittys of London and Westminster as my Executors shall thinke fitt to pitch upon and shall judge to be most necessitous the Sum̄e of one hundred pounds to be divided amongst them according to the discretion of my said Executors And whereas Nicholas Courtney of the Inner Temple Esqr is and has been for severall yeares indebted unto mee in the Sum̄e of One hundred and Fifty pounds principall Money besides Interest for the same during the time he has had the said Money Now in Consideracōn that the said Nicholas Courtney hath done some Busienesse for mee in my Life time I doe hereby Remitt and fully discharge him his Executors and Adm̄strators of and from all the Interest Money which now is or shall be due unto mee from him or them att the time of my decease And I doe hereby declare That this Codicill shall be Added to my Last Will and Testament and bee a part thereof In witnes whereof I have hereunto Sett my hand and Seale the Thirtieth

* Katharine Molster, see Gen. Chart [G.b.].

† Sir William Ashurst (1647–1720), Alderman; Lord Mayor of London, 1693–4; Governor of the New England Company, 1696–1720. He was the second son of the Henry Ashurst, merchant, mentioned in the note on p. 273. (John Le Neve. *Monumenta Anglicana*. 5 vols London, 1717–1719. Vol. IV, p. 108.)

day of July In the Third yeare of the Reigne of our Soveraigne Lord and Lady King William and Queene Mary over England &c Annoq[ue] Dn̄i 1691

[Signed] Ro: Boyle. (L. S.)

Signed Sealed Published and Declared by the Hono[ble] Robert Boyle Esq[r] as a Codicill to be added to his Last Will and Testam[t] and to be a part thereof in the p[r]sence of us.

[Signed] Isaac Garnier
Ro: St-Clair.
Jo: Caddick

[End of the twelfth sheet.]

Whereas I have in and by my Last Will and Testament Impowred my most Hono[rd] and Dearest Brother the Right Hono[ble] Richard Earle of Burlington and Corke the Hono[ble] Henry Boyle Esq[r] and S[r] Robert Southwell my Trustees to Grant Bargaine Sell & Dispose off to the best Advantage imediately after my decease All that the Manno[r] and Lands of Buttevant and Rices Lands in the County of Corke w[th]in the Kingdome of Ireland with all and singular their Rights Members and App[ur]tenn̄ces for and during soe long time & Terme as there should be then to Come and unexpired in the said p[r]misses (in case the same should not be redeemed) And have therein declared That the money to be raised by the Sale or Redemption thereof should be by my said Trustees paid unto my Executor[rs] And having for a long time determined to make some reall Expression of the great Respects and Kindness which I have for and beare to my above said Brother and to the Right Hono[ble] Charles Lord Clifford* (Sonn & heire apparent of my said Brother) my Hono[rd] Nephew by making some Guift or Present to the Offspring of my said Brother Now in prosecution of my said Determinac̄on I doe hereby publish and Declare That my Executo[rs] shall make an even and equall Divident of the said Purchase or Redemption Money when paid to them by my said Trustees as aforesaid And One Moyetie or halfe part thereof my Will is That my Executo[rs] shall pay back unto my said Brother the Earle of Burlington and Corke upon speciall Trust and Confidence Neverthelesse and to the Use Intent and Purpose herein after menc̄oned and Declared (That is to say) That he the said Earle shall Apply Lay out Pay

*** See Gen. Chart [F.d.4].**

and Imploy the same Moyetie or halfe part of the said Money within Three months after his Receipt thereof for the Use and best Advantage of such One or more of the Younger Children now Unmarryed of the said Lord Clifford and in such proporc̄on if to more then One as to him the said Earle his Executo^{rs} Adm̄strato^{rs} or Assignes shall seeme fitt, And under this further Trust That in case he the said Earle his Executo^{rs} Adm̄strato^{rs} or Assignes shall fayle to Settle and dispose of the same according to the said Trust within Three Months after the same is paid to him by my Executo^{rs} as aforesaid That then he the said Earle his Executo^{rs} Adm̄strato^{rs} and Assignes shall stand possessed of the said Moyetie or halfe part of the purchase or redemption Money soe paid him aforesaid in Trust for all the Younger Children now Unmarryed as aforesaid of him the said Lord Clifford share and share alike. And as for the other Moyetie or halfe part of the said purchase or Redemption Money which will remaine in the hands of my Executo^{rs} I doe hereby Publish and Declare shall bee Imployed by them to the Use or Uses in my said Last Will declared and Appoynted. And this Codicill I doe hereby declare shall be Added to my Last Will and Testament and bee a part thereof In witnes whereof I have herein to Sett my hand and Seale the First day of August In the Third yeare of the Reigne of our Soveraigne Lord and Lady King William and Queene Mary over England &c Annoq[ue] Dn̄i 1691

[Signed] Ro: Boyle. L. S.

Signed Sealed Published and Declared by the Hono^{ble} Robert Boyle Esqr as a Codicill to be added to and be a part of his Last Will and Testament in the p^{r}sence of us.

[Signed] Isaac Garnier
Ro: St-Clair
Wm Johnson
James Oglebey.

[End of the thirteenth sheet.]

Whereas I have in & by my Last Will & Testament Settled & Appointed the Sum̄e of Two hundred pounds sterling to be paid & distributed by my Executo^{rs} to & amongst such distressed Irish Protestants of all Qualities as have been either forc'd or affrighted out of the Kingdome of Ireland since the late Calamities there & that should happen to bee in or neere London Bristoll Chester or Liverpoole at the time of my decease **Now** in regarde of the present Exigencys & pressing Necessities of severall of those Persons I have thought fitt to Allott & Give out myselfe as a part of that my intended Charity the Sume of Fifty pounds to be divided & distributed amongst such of them as I doe Judge most necessitous, for their present Releife Soe that now my Will is That my Exectors shall be Obliged to pay & distribute to & amongst the said distressed Irish Protestants in my Last Will menconed as aforesaid noe more than the Sume of One hundred & Fifty pounds (being the Residue thereof) for the Reason aforesaid And this my Codicil I doe hereby signifie & Declare shall in Order thereto bee Added to and be a part of my Last Will and Testament In witnes whereof I have hereunto sett my hand & Seale this Fifth day of August In the yeare of our Lord God one thousand Six hundred Ninety and One.

[Signed] Ro: Boyle. (L. S.)

Signed Sealed Published & Declared by the honoble Robert Boyle Esqr for & as a Codicil to be added to & be a part of his Last Will & Testamt in the presence of Us.

[Signed] Ro: St-Clair
Tho: Smith.
Wm Johnson.
James Oglebey.

[End of the fourteenth sheet.]

Whereas I am justly sensible of the great Trouble that I shall give my Dearest and most Honord Brother and Sister the Earle of Burlington and the Lady Viscountess Ranelagh by my having Constituted & Ordained them to bee Two of the Executors in Trust of my Last Will and Testamt, which considering their Age & Qualitie they may bee noe way fitt to undergoe; To the intent therefore that they may bee eased as much as in mee lyes from all the avoidable Trouble of that Office. I doe hereby declare That I have engaged my other Executor in Trust Mr John Warr junior my present Servant to take upon him more imediately the troublesome part of the Execution of my said Last Will and

Testam^t^, Which he is to prosecute by and according to the direction and Appointm^t^ of my other Two Executors and not to absent himselfe from the same during the Terme or Termes herein after mentioned for wch he shall happen to bee employed in the Execution of the said Trust, by Travelling above one dayes Journey from London, unless it shall bee upon an extraordinary Occasion and with the Consent of my other Two Executors. In Consideracōn whereof I doe hereby Give, Devise, Bequeath & settle upon him the said John Warr the Yearely Sum̄e of Forty pounds for his Care and Paines in the execution of the said Trust, Which my Will is shall bee continued to him as an yearely Sallary for the space of Three yeares from the time of the proving of my said Will provided the discharge of my said Trust require soe long time. Otherwise my Will is That this Sallary shall bee continued but for the space of one yeare or at most but in proporcōn for such further time as shall bee necessary for the discharge of the then remaining part of the said Trust. And my further Will is, That besides the Sallary above mencōned my said Executor John Warr shall be allowed all such necessary and reasonable Charges as he shall from time to time expend or bee putt unto in the Performance and Execution of the said Trust. Lastly my Will is, That this Codicill shall be added to and bee a part of my Last Will and Testam^t^ In witness whereof I have hereunto Sett my hand and Seale the Eleaventh day of September In the Third yeare of the Reigne of our Soveraigne Lord and Lady William and Mary by the grace of God of England Scotland France and Ireland King and Queene Defenders of the Faith &c And in the yeare of our Lord God One thousand Six hundred Ninety and One.

[Signed] Ro Boyle. (L. S.)

Sealed Published and Declared by the Hono^ble^
Robert Boyle Esqr for and as a part of his
Last Will & Testam^t^ in the presence of Us.

[Signed] Ro: St-Clair.
Tho: Smith
Jo: Caddicke
Wm Johnson.

[End of the fifteenth sheet.]

Whereas in & by my Last Will & Testam[t] bearing date on or about the Eighteenth day of July last past I did Constitute & Appoint my Dearest Sister Ranelagh One of my Executors in Trust And whereas it has pleased God some few dayes since to take unto himselfe by death my said Deare Sister. I doe therefore in her steede & Place hereby Make Ordaine & Appoint my very Loving Friend S[r] Henry Ashurst of the Parish of S[t] Sepulchers in the County of Middx Kn[t] & Baronett to be one of the Executors in Trust of my said Last Will & Testam[t]. And I doe hereby Desire & Impower my said Executor together wth my other Two surviveing Executors in Trust in my said Last Will named & Appoynted to take speciall Care to discharge the Trusts by mee therein in them reposed & more especially that part of my Will That I had particularly Recom̄ended to the Care of my Dearest Sister Ranelagh deceased And I doe hereby Give & Bequeath unto the said S[r] Henry Ashurst the Sum̄e of Twenty pounds to be payd out by him in a peece of Plate. And I doe hereby Confirme all the Appoyntm[ts] of my said Last Will and Testam[t] (Excepting onely what therein relates to my said deceased Sister Ranelagh which I doe hereby Revoake) And I doe hereby also Confirme all other Codicill or Codicills that have been made by mee since the Date of my above sd̄ Last Will & Testam[t] In witness whereof I have hereunto sett my hand & Seale the Nyne & Twentieth day of December In the Third yeare of the Reigne of our Soveraigne Lord & Lady King William & Queene Mary over England &c Annoq[ue] Dn̄i 1691

[Signed] Ro: Boyle. (L. S.)

Signed Sealed Published & Declared
by the Hono[ble] Robert Boyle Esq[r] as
a Codicill to & a part of his Last Will
& Testm[t] in the presence of Us.

[Signed] Isaac Garnier
Ro: St-Clair
Tho: Smith.
W[m] Johnson.
James Oglebey.

Probatum (cum septem Codicillis annexis) Londini Coram ven̄li viro Dn̄o Thomae Pinfold Milite Legum Doctore surr̄o &c Vicesimo sexto die mensis Januarij Anno Dn̄i (Stylo Angliae) 1691

Juram[tis] Praenobilis et Honorandi Viri Richardi Comitis de Burlington et Corke Dn̄i Henrici Ashurst Bar[ti] et Johannis Warr Jun' Gen' Ex̄torum superstit' Quibus &c De bene &c Jurat' Hon[da] Faemina Catherina nup[er] Comitissa Dotissa de Ranelagh Ex̄torum alia &c ante Testatorem praedemortua

Sed in paro̅a S[ti] Jacobi Westm'

[End of the sixteenth and last sheet.]

APPENDIX II

Rules and Orders Respecting the Charity Left by the Will of The Hon. Robert Boyle

The executors of the Hon. *Robert Boyle* having proved his will, they soon after his death entered into an agreement with Sir *Samuel Gerrard*, Knight, for the purchase of the manor of *Brafferton*, and diverse other lands, tenements and hereditaments in *Brafferton*, in the county of *York*, for the sum of 5400 *l.* the greater half of the said testator's estate; and for settling the said charity, the said executors agreed on the methods following, *viz.* that they the said executors should grant out of the said manor a rent-charge in perpetuity of 90 *l. per ann.* unto the Company for propagating the Gospel in *New England*, and parts adjacent, in *America*, and their successors, to be paid at *Guildhall*, *London*, yearly, at *Michaelmas* and *Lady-day*, by equal portions, clear of all taxes; and that the said Company should apply 45 *l. per ann.* one moiety of the said 90 *l.* for the salary of two ministers, to instruct the natives in or near his Majesty's colonies in *New England* in the Christian knowledge; and the like yearly sum of 45 *l.* the other moiety of the said rent-charge of 90 *l.* the said Company and their successors should transmit unto the president and fellows of *Harvard* College, in *Cambridge*, in *New England*, and their successors, to be by them employed and bestowed for the salary of two other ministers, to teach the said natives in or near the college there the Christian religion; and that the said manor and premises, subject to the said 90 *l. per ann.* should be conveyed to the mayor, commonalty, and citizens of *London*, and their successors, upon trust, that the rents and profits thereof, over and above the said 90 *l. per ann.* (the receiver's salary and other incident charges deducted), should be laid out for the advancement of the Christian religion in, *Virginia*, in such manner, and subject to such methods and rules, as the said Earl of *Burlington*, and the Lord Bishop of *London* for the time being, under their hands and seals, should appoint;

so as such appointment was made on or before *Lady-day*, 1697, and confirmed by a decree of the Court of Chancery.

The said Sir *Samuel Gerrard* dying before the purchase of the said manor of *Brafferton* and premises was compleated, an information was exhibited in Trinity Term, 1695, in the Court of Chancery, by the then attorney-general, and dame *Elizabeth Gerrard*, and *Thomas Owen*, devisees and executors named in the last will and testament of Sir *Samuel Gerrard*, against the said executors of *Robert Boyle*, that they might be obliged to proceed in the said purchase, and settle the same to such pious uses as were agreed on as aforesaid: and the said cause coming on to be heard on 1st *August*, 1695, the court referred it to one of the masters of the said court, to take an account of the personal estate of the said testator come to the hands of the defendants; and to examine and certify whether 5400 *l.* was the major part of the said testator's personal estate, and whether the value of the manor and premises were equivalent to the purchase money agreed to be paid for the same.

And the said cause coming on to be heard after the said master's report, on the 8th of *August*, in the 7th year of King *William*, it was ordered and decreed, that the said defendants should proceed in the said purchase for the said sum of 5400 *l.* and that the rules and methods touching the disposition of the said charity should be ratified and confirmed, with this further addition only, that an account of the said yearly rent of 90 *l. per ann.* after the deaths of the said Earl of *Burlington* and Sir *Henry Ashurst*, should be sent to the president of Trinity College, *Oxford* (of which college the said *Robert Boyle* was a member), for the time being, as well as to the several heirs of the said Earl of *Burlington* and Sir *Henry Ashurst*. And it was further ordered and decreed, that, after the said purchase made, the said defendants should grant the said rent-charge of 90 *l. per ann.* to the said Company and their successors, for the purposes aforesaid; and afterwards should convey the said manor and premises, so charged, unto the mayor and commonalty and citizens of *London*, and their successors, subject to the trusts and purposes in the said rules and agreements made concerning the same.

The said manor and lands were afterwards conveyed to the said Earl of *Burlington*, *Sir Henry Ashurst*, and *John Warr*, and their heirs, in consideration of the said sum of 5400 *l.* and by indenture dated 13th *August*, 1695, in part performance of the trusts of the will of the said *Robert Boyle*, and in obedience to the said decree, they granted a rent-charge in fee of 90 *l. per ann.* out of the said manor and premises unto the said Company and their successors for ever, upon the trusts aforesaid, with powers of distress and entry, for the better securing the said rent-charge. And afterwards, by indentures of lease and release, and bargain and sale inrolled; the lease dated the 30th, and the release and bargain and sale the 31st of *August*, 1695, made between the said Earl of *Burlington*, Sir *Henry Ashurst*, and *John Warr*, of the one part, and the mayor, and commonalty, and citizens of *London*, of the other part; the said Earl of *Burlington*, Sir *Henry Ashurst* and *John Warr*, in full performance and execution of the trusts of the said *Robert Boyle's* will, and in obedience to the said decree, and for and in consideration of the sum of 10 *s.* mentioned to be paid by the said mayor, and commonalty, and citizens of *London*, conveyed the said manor and premises to and to the use of the mayor, and commonalty, and citizens of *London*, and their successors for ever, upon trust nevertheless that the rents, issues, and profits of the said manor and premises, over and above the said yearly rent of 90 *l.* to be paid to the said Company, the receiver's salary, and other necessary and incident charges about the management of the said estate, should be laid out and disposed of for the advancement and propagation of the Christian religion amongst infidels in *Virginia*, in such manner, and subject to such rules and methods, as the Earl of *Burlington*, and the Lord Bishop of *London* for the time being, under their hands and seals, should limit or appoint, so as the same appointment should be made on or before the 25th of *March*, 1697, and so as it should be ratified and confirmed by a decree of the said court within six months after it should be made; and, for default of such appointment, then as the Lord High Chancellor of *England*, or the Lord Keeper of the Great Seal of *England* for the time being, should order or appoint; and in the mean time, and until such appointments made, the residue of the said rents and profits, over and above the said yearly rent of 90 *l.* salary and charges, to be paid into the hands of *Christopher Crofts*, then of the parish of St. *James*, *Westminster*, Gentleman, to be employed by him for the said purpose.

The said Earl of *Burlington*, and *Henry* Lord Bishop of *London*, agreed and appointed certain rules and methods for the settlement of the said charity under their hands and seals, bearing date the 21st day of *December*, 1697, to the purport and effect following, *viz.*

First, That all the yearly rents and profits of the said manor of *Brafferton* and premises aforesaid, as well those incurred due since the purchase thereof, as which should thereafter grow due after a deduction thereout of 90 *l. per ann.* to the Society for propagating the Gospel in *New England*, and other incidental charges, should be by the present and future receiver paid into the hands of *Micajab Perry*, of *London*, merchant, agent in *London* for the president and masters of the said college of *William* and *Mary*, in *Virginia*, and to all future agents for the said College, for the purposes after mentioned; and that such agent's receipts or acquittances should be sufficient discharges to such receivers for what should be so paid.

Secondly, That all sum and sums of money then, or that should thereafter be, received out of the said manor and premises, subject to the deductions aforesaid, should be thereafter remitted to the said president and masters for the time being.

Thirdly, That the said president and masters, and his and their successors, should thereout expend so much as should be necessary for fitting and furnishing lodgings and rooms for such *Indian* children as should be thereafter brought into the said college.

Fourthly, That the said president and masters, and his and their successors, should keep at the said college so many *Indian* children, in sickness and health, in meat, drink, washing, lodging, cloaths, medicines, books, and education, from the first beginning of letters, till they should be ready to receive orders, and be thought sufficient to be sent abroad to preach and convert the *Indians*, at 14 *l. per ann.* for every such child, as the yearly income of the premises, subject to the deductions aforesaid, should amount to.

Fifthly, That the care, instruction, and education of such children, as should be thereafter placed in the said college, should be left to the president and masters thereof for the time being; yet subject therein, as they were in all their trusts, to the visitation and inspection of the rector and governor of the said college for the time being.

Sixthly, That the said president and masters, and their successors, should once in every year transmit to the Earl of *Burlington*, and the Lord Bishop of *London* for the time being, a particular account of what sum or sums of money they should thereafter receive by virtue of the said decree, as also lay out and expend on all or any of the said matters, and the occasions thereof; as also the number and name of the *Indian* children that should be thereafter brought into the said college, together with their progress and proficiency in their studies, and all other matters relating thereto.

Seventhly, That the laying out the monies from time to time thereafter to be remitted, as also the manner and method of educating and instructing such children, and all the matters relating to the said charity and execution of it, should be subject to such other rules and methods as should from time to time thereafter be transmitted to the said president and masters, and his and their successors, by the said Earl of *Burlington*, and the Bishop of *London* for the time being; and, in default thereof, to such rules and methods as the rectors and governors of the said college for the time being should make or appoint; but until such other and further rules were made, the rules and directions thereby given were to take place.

Eighthly, That the name of the benefactor might not be forgotten, the said Earl of *Burlington*, and the Bishop of *London*, did direct and appoint, that the said charity should be called "The Charity of the Honourable *Robert Boyle*, Esquire, of the city of *London*, deceased."

And by an order, bearing date the 9th of *June*, 1698, made by Lord *Somers*, Chancellor of *England*, it was ordered, that the said rules and methods, and all and every the matters and things therein contained, should be established, with the additions and alterations following, *viz.* in the sixth rule, that the yearly account therein appointed to be transmitted to the Earl of *Burlington*, and the Bishop of *London* for the time being, should be from time to time, by the said Earl and Bishop, transmitted into the Court of Chancery, to be filed by the register thereof. And whereas in the seventh rule it was mentioned, that the laying out the money, the manner of educating the children, and all other matters relating to the said charity, or the execution of it, should be subject to such rules and methods as should from time to time thereafter be transmitted to the said president and masters, and his and their successors, by the said Earl of *Burlington*, and the Bishop of *London* for the time being, or, in default thereof, to such rules and methods as the rectors of the said college for the time being should make and appoint; it was ordered, that such other rules and methods, touching the said charity, at any time so made or appointed, should be first confirmed and approved of in the said Court of Chancery. And it was further ordered, that *Micajab Perry*, of *London*, merchant, should be allowed to be the receiver of the rents and profits of the said trust estate, for the purposes in the said rules and methods mentioned, who was from time to time to appoint a receiver under him of the said rents until further order of the said Court, and also from time to time to take and allow such receiver's accounts yearly, and the accounts of the then receiver of the said rents, and of the arrears in his hands, since the said purchase.

The rents and profits of the said manor and lands have, since the said purchase and conveyance, been collected and received by a person appointed from time to time to be receiver thereof, by the agent or principal receiver for the said president, masters, or professors of the said college, residing in *London*; and the surplus of such rents and profits, after payment of the said yearly rent-charge of 90 *l. per ann.* to the said Company, hath from time to time been paid to such agent of the said president, masters, or professors of the said college, residing in *London*, and transmitted to the said president, masters, or professors, according to the directions of the said rules and orders; and the said yearly rent-charge of 90 *l.* hath also from time to time been paid by him to the treasurer of the said Company.

APPENDIX III

Genealogical Chart

The individuals in the following chart are shown in vertical columns lettered [A], [B], etc., so that all in the same column belong to the same generation. When arranged as a normal pedigree, the relationships of the various groups are as follows:

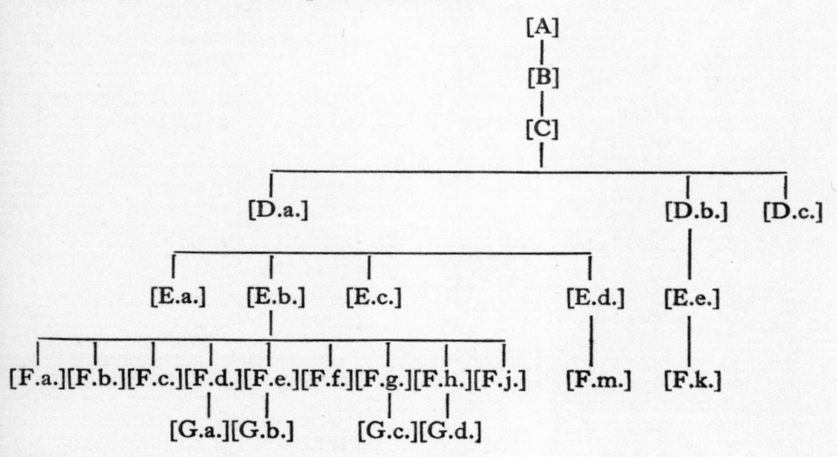

Robert Boyle is [E.b.14.] on the chart.

There is no attempt at genealogical completeness, nor to full information on any person, particularly if remote from our subject. The main purpose is to show those members of the Boyle family, and its ramifications, round about Robert's lifetime.

The information given here is derived mainly from the following sources:

Lismore Papers. 1st series and 2nd series.

G.E.C[ockayne]. *The Complete Peerage.*

G.E.C[ockayne]. *Complete Baronetage.*

J. Lodge. *The Peerage of Ireland.*

Biographia Britannica.

Dictionary of National Biography.

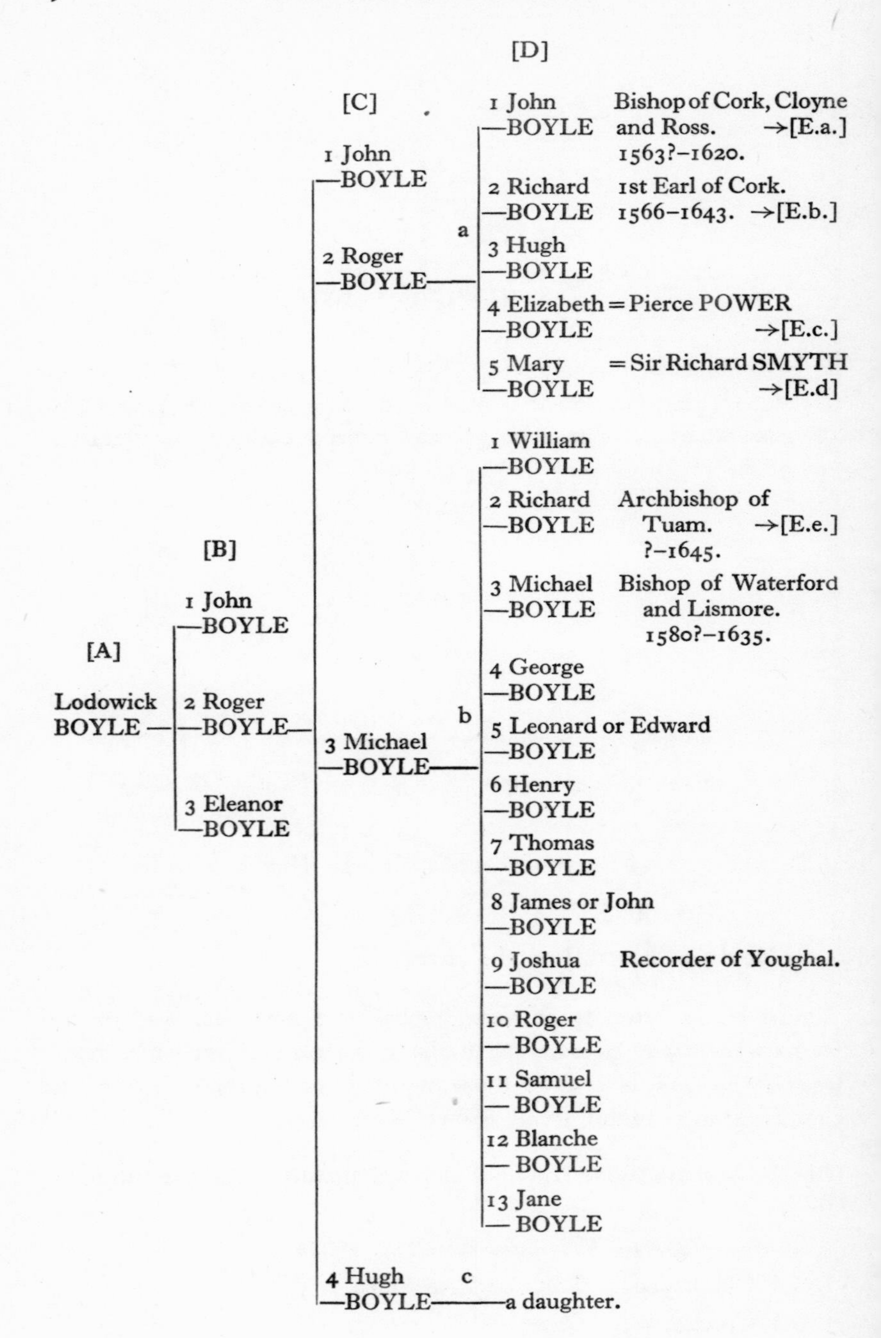
[A]
Lodowick BOYLE
[B]
1 John —BOYLE
2 Roger —BOYLE
3 Eleanor —BOYLE
[C]
1 John —BOYLE
2 Roger —BOYLE
3 Michael —BOYLE
4 Hugh —BOYLE
[D]
a
1 John —BOYLE
Bishop of Cork, Cloyne and Ross. →[E.a.] 1563?–1620.
2 Richard —BOYLE
1st Earl of Cork. 1566–1643. →[E.b.]
3 Hugh —BOYLE
4 Elizabeth —BOYLE = Pierce POWER →[E.c.]
5 Mary —BOYLE = Sir Richard SMYTH →[E.d]
b
1 William —BOYLE
2 Richard —BOYLE
Archbishop of Tuam. →[E.e.] ?–1645.
3 Michael —BOYLE
Bishop of Waterford and Lismore. 1580?–1635.
4 George —BOYLE
5 Leonard or Edward —BOYLE
6 Henry —BOYLE
7 Thomas —BOYLE
8 James or John —BOYLE
9 Joshua —BOYLE
Recorder of Youghal.
10 Roger — BOYLE
11 Samuel — BOYLE
12 Blanche — BOYLE
13 Jane — BOYLE
c
a daughter.

[A]. Lodowick Boyle married Elizabeth, daughter of William Russell. Settled in co. Hereford in reign of Henry VI.

[B.1]. Possessed the Herefordshire estate.

[B.2]. Roger Boyle of co. Hereford married Jane, daughter of Thomas Pattishal of co. Hereford.

[B.3]. Eleanor Boyle married Watkin Russell.

[C.1]. John Boyle of co. Hereford married Alice, daughter of Alexander Hayworth.

[C.2]. Roger Boyle of Preston, near Faversham, co. Kent married Joan, daughter of John Naylor of Canterbury 16 October 1566. Roger died 24 March 157$\frac{6}{7}$. She was born 15 October 1529, and died at Faversham 20 March 158$\frac{6}{7}$. Both were buried in the parish church of Preston, where a monument to their memory was erected in 1629 by their son Richard (*I Lismore*, **II**, p. 294). An illustration of this is given in *Orrery Papers*, **I**, p. 310. The epitaph is printed *ibid*, **II**, p. 320, and in John Lewis, *The History and Antiquities of the Abbey and Church of Favresham in Kent*, 1727, **I**, p. 26.

[C.3]. Michael Boyle, a merchant in London, married Jane, daughter of William Peacock, mercer of London.

[C.4]. Hugh Boyle of London, mercer.

[D.a.] See page 292.

[D.b.] See page 293.

[D.a.1]. John Boyle; born 1563?; entered King's School, Canterbury, 1578; Prebendary of Litchfield, 1611; Bishop of Cork, Cloyne and Ross, 1618; died 10 July 1620; buried at Youghal. He married Elizabeth, daughter of Matthew Lucy; she married (2) Sir William Hull, 10 January $162\frac{1}{2}$. *Athenae*, **II**, 860: Townshend p. xii: Venn, *sub voce: I Lismore*, **II**, p. 101: *ibid*, **I**, p. 253: *ibid*, **II**, p. 34. Monumental inscription in Lodge, **I**, p. 150.

[D.a.2]. Richard Boyle; born 13 October 1566 at Canterbury; entered King's School, Canterbury, 1580; admitted to Corpus Christi College, Cambridge, 1583; student of Middle Temple, London. Knighted, 1603; P.C.(Ireland), 1613; created Baron of Youghal, 1616; Viscount of Dungarvon, and Earl of Corke, 1620; one of the Lords Justices (Ireland), 1629; Lord High Treasurer (Ireland), 1631; P.C. (England), 1640; died 15 September 1643 at Youghal; buried in St. Mary's Abbey there. His tomb is illustrated on Plate XI, and its monumental inscription is given in *II Lismore*, **V**, p. 288, Appendix B. He married (1) Joan, daughter of Capt. William Apsley of Limerick 6 November 1595; she died 14 December 1599, and was buried in the church of Buttevant, co. Cork. He married (2) Katherine, daughter of Sir Geoffrey Fenton, Principal Secretary of State (Ireland) 25 July 1603; she died at Dublin 16 February $16\frac{29}{30}$, and was buried in St. Patrick's Cathedral there. The magnificent monument erected by the Earl to the memory of Katherine, her parents and grandparents is described and illustrated in *Annals of Science* (1959, published 1962), **15**, 194 (R. E. W. Maddison, "The Portraiture of the Honourable Robert Boyle, F.R.S.").

[D.a.3]. Hugh Boyle; baptized 4 May 1576 at Preston, nr. Faversham, co. Kent; killed abroad in the wars. *Orrery Papers*, **II**, p. 321: *Lismore*, **II**, p. 102.

[D.a.4]. Elizabeth Boyle married Pierce Power of Lisfinnon Castle; buried at Youghal. *II Lismore*, **V**, p. 228, Appendix B.

[D.a.5]. Mary Boyle married Sir Richard Smyth of Ballynetra; died 17 January $163\frac{2}{3}$, and was buried at Youghal. *I Lismore*, **II**, p. 394: *ibid*, **III**, p. 175: *II Lismore*, **V**, p. 228, Appendix B.

[D.b.2]. Richard Boyle. Dean of Waterford, 1604; Archdeacon of Limerick, 1605; Bishop of Cork, 1620; Archbishop of Tuam, 1638. He died 19 March $164\frac{4}{5}$, buried in Cathedral of St. Finbar. The Earl of Cork (D.a.2) describes him as " my faythles & vnthanckfull kinseman, whome I haue raised from being a poor schoolmaster . . . and hath been by my onely favor advanced to all that he is now come vnto . . . " *I. Lismore*, **IV**, p. 185: Venn. He married Martha, daughter of Richard (or John) Wright of Catherine Hill, co. Surrey.

[D.b.3]. Michael Boyle; entered Merchant Taylor's School, London, 1587; D.D., Oxon., 1611; Archdeacon of Cork, 1614; Dean of Lismore, 1614; Bishop of Waterford, 1619. He died at Waterford 27 (?29) December 1635; buried in the cathedral there. *I Lismore*, **IV**, p. 146: *Athenae*, **II**, 887. He married (1) Dorothy Fish: (2) Christian Bellot.

[D.b.4]. George Boyle assisted the Earl of Cork in his ironworks: Knighted, 30 August 1624. He married Una Bourck. Evidently dead before 1628. *I Lismore*, **I**, p. 2: *ibid*, **II**, p. 137.

[D.b.9]. Joshua Boyle. Recorder of Youghal 1641–1651. R. Caulfield. *The Council Book of the Corporation of Youghal*, Guildford, 1878, p. 624.

[E]

a

1 John —BOYLE
2 George —BOYLE
3 Edward —BOYLE
4 Barbara —BOYLE = Sir John BROWNE
5 Mary —BOYLE = Stephen CROW
Katherine —BOYLE = William TYNT

b

1 Roger —BOYLE
2 Alice —BOYLE = (1) David BARRY, 1st Earl of Barrymore. → [F.a.] = (2) John BARRY. 1605?–1642.
3 Sarah —BOYLE = (1) Sir Thomas MOORE. = (2) Robert DIGBY, 1st Baron Digby of Geashill. → [F.b.] ?–1642.
4 Lettice —BOYLE = George GORING 1629–1657.
5 Joan —BOYLE = George FITZGERALD, 16th Earl of Kildare. → [F.c.] 1612–1660.
6 Richard —BOYLE 2nd Earl of Cork, 1st Earl of Burlington. → [F.d.] 1612–1698.
7 Katherine —BOYLE = Arthur JONES, 2nd Viscount Ranelagh. → [F.e.] ?–1670.
8 Geoffrey —BOYLE
9 Dorothy —BOYLE = (1) Sir Arthur LOFTUS. → [F.f.] = (2) Mr. TALBOT.
10 Lewis —BOYLE Viscount Boyle of Kinalmeaky. 1619–1642.
11 Roger —BOYLE 1st Earl of Orrery. → [F.g.] 1621–1679.
12 Francis —BOYLE 1st Viscount Shannon. → [F.h.] 1623–1699.
13 Mary —BOYLE = Charles RICH, 4th Earl of Warwick. → [F.j.] 1616–1673.
14 Robert —BOYLE The subject of the present account.
15 Margaret —BOYLE

[E.a.1]. John Boyle attended Trinity College, Dublin. He married Anne —. *I Lismore*, **II**, p. 29, 237: *ibid*, **V**, p. 54.

[E.a.2]. George Boyle. Mentioned, *I Lismore*, **II**, pp. 29, 238.

[E.a.3]. Edward Boyle attended Trinity College, Dublin. He married Mary, daughter of Sir William Hull, *I Lismore*, **II**, pp. 29, 178.

[E.a.4]. Barbara Boyle; christened 3 February 161$\frac{3}{4}$. She married Sir John Browne of Hospital, co. Limerick, *I Lismore*, **I**, p. 37: *ibid*, **II**, p. 101.

[E.a.5]. Mary Boyle married Stephen Crow. *I Lismore*, **III**, p. 134.

[E.a.6]. Katherine Boyle; christened 7 October 1619. She married William Tynt, son of Sir Robert Tynt, 1 August 1639. *I Lismore*, **I**, p. 232: *ibid*, **V**, pp. 98, 100.

[E.b.1-15]. are the Earl of Cork's children by his second wife.

[E.b.1]. Roger Boyle; born at College House, Youghal, 1 August 1606; sent to school at Deptford; died at Sayes Court, the residence of his mother's uncle, Sir Richard Browne, 10 October 1615; buried in the parish church of Deptford. *I Lismore*, **I**, p. 37, 84: *ibid*, **II**, p. 101. The monumental inscription is given in *Works Fol.*, **I**, *Life*, p. 4: *Works*, **I**, p. x.

[E.b.2]. Alice Boyle; born at College House, Youghal, 20 March 160$\frac{6}{7}$; died 23 March 166$\frac{6}{7}$; buried in St. Patrick's Cathedral, Dublin. She married (1) David Barry, 1st Earl of Barrymore, 29 July 1621. He was born about 10 March 160$\frac{4}{5}$; was mortally wounded at the battle of Liscarrol; died 29 September 1642; buried in the Chapel of St. Mary, Youghal. She married (2) John Barry of Liscarrol. *I Lismore*, **I**, p. 268: *ibid*, **II**, p. 21.

[E.b.3]. Sarah Boyle; born at Dublin, 29 March 1609; died 14 July 1633; buried in St. Patrick's Cathedral, Dublin. She married (1) Sir Thomas Moore, second son of Gerald or Garret Moore, Viscount Moore of Drogheda, 29 July 1621; he died 1 December 1623. She married (2) Robert Digby, son of Sir Robert Digby and Lettice, Baroness Offaley, granddaughter and heir general of Gerald, Earl of Kildare, 25 December 1626. He married a second time; he died 7 June 1642, and was buried in St. Patrick's Cathedral, Dublin. *I Lismore*, **I**, p. 256; *I Lismore*, **II**, pp. 21, 112, 203.

[E.b.4]. Lettice Boyle; born at Dublin 25 April 1610; died [11?] April 1643. She married George Goring, son and heir of George Goring, 1st Earl of Norwich, 25 July 1629. He was born 14 July 1608; died 25 July 1657 (N.S.) at Madrid; buried in St George's Chapel there. No surviving issue. *I Lismore*, **II**, p. 109: *II Lismore*, **V**, p. 124: *The Memoirs of Ann, Lady Fanshawe*, London, 1907, pp. 414–415.

[E.b.5]. Joan Boyle; born at Lismore 14 June 1611; died 11 March 165$\frac{5}{6}$; buried in St. Patrick's Cathedral, Dublin. She married George FitzGerald, 16th Earl of Kildare, 15 August 1630. He was baptized 23 January 161$\frac{1}{2}$; died early in 1660; buried at Kildare. *I Lismore*, **II**, pp. 109, 115: *ibid*, **III**, p.48.

[E.b.6]. Richard Boyle; born at College House, Youghal 20 October 1612; Knighted, 1624; travelled abroad, 1632–4; defeated the Irish near Liscarrol, 1642; Governor of Youghal, 1641; M.P., for Appleby, 1640–3; succeeded his father as 2nd Earl of Cork, 1643; succeeded to the estates of the Clifford family on death of his wife's father, the Earl of Cumberland, 1643; created Baron Clifford of Lanesborough, 1644; Lord Treasurer, 1660–95; P.C. (Ireland), 1660; created Earl of Burlington, 1664; Recorder of York, 1685–8; died at Burlington House, London, 15 January $169\frac{7}{8}$; buried at Londesborough. He married Elizabeth, Baroness Clifford, daughter and heiress of Henry Clifford, 5th Earl of Cumberland and 1st Lord Clifford, 3 July 1634 at Skipton Castle, Yorkshire. She was born 18 September 1613; buried 20 January $169\frac{0}{1}$. *I Lismore*, **III,** pp. 144, 156: *ibid*, **IV,** p. 46: M. F. Keeler. "The Long Parliament, 1640–1641." *Memoirs of the American Philosophical Society*, Philadelphia, (1954), **36,**

[E.b.7]. Katherine Boyle; born at College House, Youghal, 22 March $161\frac{4}{5}$; died 23 December 1691; buried in the church of St. Martin-in-the-Fields, London. She married Arthur Jones, eldest son and heir of Roger Jones, 1st Viscount Ranelagh, at Dublin 4 April 1630. He died 7 January $16\frac{69}{70}$; buried in St. Patrick's Cathedral, Dublin. *I Lismore*, **II,** p. 110; *I Lismore*, **III,** p. 24, 25.

[E.b.8]. Geoffrey Boyle; born at Youghal 10 April 1616; died there 20 January $161\frac{6}{7}$; buried there. *I Lismore*, **I,** p. 106, 133, 140; *II Lismore*, **V,** Appendix B.

[E.b.9]. Dorothy Boyle; born in the Chantry House, Youghal, 31 December 1617; died 26 March 1668 according to *G.E.C.*, but before 21 March 1668 according to Lady Warwick's Diary. She married (1) Sir Arthur Loftus 13 February $163\frac{1}{2}$, who died young; (2) a Mr Talbot (?).

[E.b.10]. Lewis Boyle; born 23 May 1619; Knighted, and created Baron of Bandon Bridge and Viscount Boyle of Kinalmeaky, 1628; admitted to Trinity College, Dublin, 1630; admitted to Gray's Inn, 1636; killed at the Battle of Liscarrol 3 September 1642; buried at Lismore. He married at the Chapel Royal, Whitehall, Elizabeth, daughter of William Feilding, 1st Earl of Denbigh, 26 December 1639. She was created Countess of Guildford 14 July 1660; and died at Colombes, near Paris, about September 1667. No issue.

[E.b.11]. Roger Boyle; born at Lismore 25 April 1621; Knighted, and created Baron of Broghill, 1628; admitted to Trinity College, Dublin, 1630; admitted to Gray's Inn, 1630; Master of the Ordnance (Ireland), 1648; M.P., for Cork at Westminster, 1654: for Edinburgh 1656; President of the Council in Scotland, 1655; Commissioner of Government for Ireland, 1660; created Earl of Orrery, 1660; M.P., for Arundel, 1660; P.C. (Ireland), 1660; Governor of co. Clare, 1661; Constable of Limerick Castle, 1661; P.C. (England), 1665; impeached for high treason, 1669; died at Castlemartyr, co. Cork, 16 October 1679; buried at Youghal. His epitaph is given in Morrice, p. 51. He married Margaret, daughter of Theophilus Howard, 2nd Earl of Suffolk, 27 January $164\frac{0}{1}$. She was baptized at St. Martin-in-the-Fields, London, 11 February $162\frac{2}{3}$; and buried at Isleworth, Middlesex, 24 August 1689.

[E.b.12]. Francis Boyle; born at Lismore 25 June 1623; admitted to Eton College, 1635; distinguished himself at the Battle of Liscarrol, 1642; created Viscount Shannon, 1660; P.C. (Ireland); Governor of city and county of Cork, 1672; buried at Youghal 19 April 1669. He married at the Chapel Royal, Whitehall, Elizabeth, daughter of Sir Robert Killigrew, 24 October 1638. She was baptized at St. Margaret's, Lothbury, London, 16 May 1622; and buried in Westminster Abbey, 4 January $168\frac{0}{1}$. The letters patent conferring the title of Viscount Shannon, Paris, 25 March 1654, were sold by Sotheby & Co., 21 December 1965; lot 661. The grant was confirmed at the Restoration.

[E.b.13]. Mary Boyle; born at Youghal 11 November 1624. Mary herself always kept her birthday on 8 November, but the precision with which her father has recorded the date as 11 November (*I Lismore,* **II,** p. 111), invites the conclusion that he is right. She died at Leigh, co. Essex, 12 April 1678; buried at Felsted, co. Essex. Robert and Katherine Boyle were executors of her will. She left £300 to the former and £600 to the latter. She married Charles Rich, 2nd son of Robert, 3rd Earl of Warwick, at Shepperton, co. Middlesex, 21 July 1641. He was born 1616; died 24 August 1673; and buried at Felsted, co. Essex. A. Walker. *Eurika, Eurika. The Virtuous Woman found . . . her loss . . . bewailed . . . in a Sermon . . . at the Funeral of . . . Mary, Countess Dowager of Warwick.* London, 1678: J. Foster. *London Marriage Licences* 1531–1869. London, 1887.

[E.b.14]. Robert Boyle; born at Lismore 25 January 162^{6}_{7}; died at London 31 December 1691.

[E.b.15]. Margaret Boyle; born in house of Viscount Grandison in Channel Row, Westminster, London, 30 April 1629; died in Sir Randell Clayton's house near Cork 28 June 1637; buried in the church of St. Mary, Youghal. (*I Lismore,* **II,** pp. 112, 317; *ibid,* **III,** p. 31; *ibid,* **V,** p. 17 and Appendix B.)

[E]

c
1 Roger —POWER
2 Richard —POWER

d

1 Percy —SMYTH	Knight; Governor of Youghal. (*I Lismore*, **II**, p. 102). =(1) Mary MEAD 1621. (*I Lismore*, **II**, p. 8). =(2) Isabella USHER 1635. (*I Lismore*, **IV**, p. 70) → [F.m.]
2 Boyle —SMYTH	1616–1637. Died at Genoa; buried at sea. (*I Lismore*, **I**, p. 131; B.M., Add. MSS. 19832. f. 42).
3 Katherine —SMYTH	=William SUPPLE 1622. (*I Lismore*, **I**, p. 124).
4 Dorothy —SMYTH	=(1) Rev. Thomas BURT, 1633. =(2) Arthur FREKE
5 Alice —SMYTH	=(1) W. WISEMAN 1635. Son-in-law of Edmund Spenser see p. 310. =(2) Redmond ROCHE.

e

1 Michael —BOYLE	Archbishop of Armagh. → [F.k.] 1609?–1702.
2 Richard —BOYLE	1623–1649.
3 Elizabeth —BOYLE	=Sir Robert TRAVERS 1647.
4 Alice —BOYLE	=Henry DELAUNE.
5 Anne —BOYLE	=John DAVANT; both died 1641.
6 Catherine —BOYLE	died young.
7 Jane —BOYLE	=William HULL.
8 Catherine —BOYLE	=John FITZGERALD, 1641.
9 Dorothy —BOYLE	=(1) Sir Hewet HALSH. =(2) Henry TURNER. =(3) Dr. Thomas ROBERTS.
10 Martha — BOYLE	=(1) Col. OSBALDESTON, 1641. =(2) Lt. Col. John NELSON. =(3) Sir Matthew DEANE.
11 daughter? —	

[E.e.1]. Michael Boyle; born about 1609; died 1702. M.A., Trinity College, Dublin, and Oxford; D.D., Oxford. Privy councillor (Ireland), 1660; Chancellor of Ireland, 1665. Bishop of Cork, Cloyne and Ross, 1660; bishop of Dublin, 1663; archbishop of Armagh, 1675. He married (1) Margaret Synge; died 1641. (2) Mary O'Brien.

[F]

a

1 Katherine —BARRY	1628–?	= Edward DENNY.
2 Richard —BARRY	2nd Earl of Barrymore. 1630–1694	= (1) Susan KILLIGREW. = (2) Martha LAWRENCE. = (3) Dorothy FERRAR.
3 James —BARRY	?–1664.	
4 Ellen —BARRY	1631–?	= Sir Arthur DENNY.

b

1 Kildare —DIGBY	2nd Baron Digby. 1631?–1661.	
2 Catherine —DIGBY	died young.	
3 Mary or Mabel —DIGBY	died young.	
4 Lettice —DIGBY	?–1656.	= William DILKE.
5 Catherine —DIGBY	?–1661.	
6 a daughter	born & died 23 May 1643.	

c

1 Elizabeth —FITZGERALD	1631–1632. (*I Lismore*, **III**, pp. 111, 116).	
2 Richard —FITZGERALD	born 1633; died young. (*I Lismore*, **III**, pp. 175, 179).	
3 Wentworth —FITZGERALD	17th Earl of Kildare. 1634–1664.	
4 Robert —FITZGERALD	1637–? (*I Lismore*, **V**, p. 26).	
5 Jane —FITZGERALD		
6 Eleanor —FITZGERALD		
7 Catherine —FITZGERALD	?–1714.	
8 Frances —FITZGERALD	1639–?	= Sir James SHAEN, Bt ?–1695.
9 Elizabeth —FITZGERALD	?–1698.	= (1) Callaghan MACCARTY, 3rd Earl of Clancarty. ?–1676. = (2) Sir William DAVIS. ?–1687.

[F.a.2]. Richard Barry; baptized 4 November 1630; died November 1694. He married (1) Susan Killigrew in 1649. If she is the daughter of Sir William Killigrew (1606–1695), then she was baptized 1 April 1629. He married (2) Martha, daughter of Henry Lawrence, President of Cromwell's Council, 2 November 1656; died 1664; and (3) Dorothy, daughter of John Ferrar, 3 February $166\frac{5}{6}$.

[F.b.1]. Kildare Digby; born *ca.* 1631; died 11 July 1661 at Dublin; buried in St. Patrick's Cathedral there. He married *ca.* 1652 Mary Gardiner, who died 23 December 1692.

[F.c.3]. Wentworth FitzGerald; born 1634; died 5 March $166\frac{3}{4}$; buried in Christ Church Cathedral, Dublin. He married *ca.* 1655 Elizabeth, daughter of 2nd Earl of Clare. She died 30 June 1666 and was buried with her husband.

[F.c.8]. Frances FitzGerald; born at Dublin 10 August 1639; married James Shaen 28 July 1656. He was created baronet, 1663; original F.R.S. died 1695. (*I Lismore*, V, p. 101).

[F.c.9]. Elizabeth FitzGerald; died in Dublin, and was buried in the church of St. Martin-in-the-Fields, London, 15 February $169\frac{7}{8}$. She married (1) Callaghan Maccarty, 3rd Earl of Clancarty, who died 21 November 1676; and (2) Sir William Davis, Chief Justice of the King's Bench, Ireland, 1682. He died 24 September 1687.

[F]

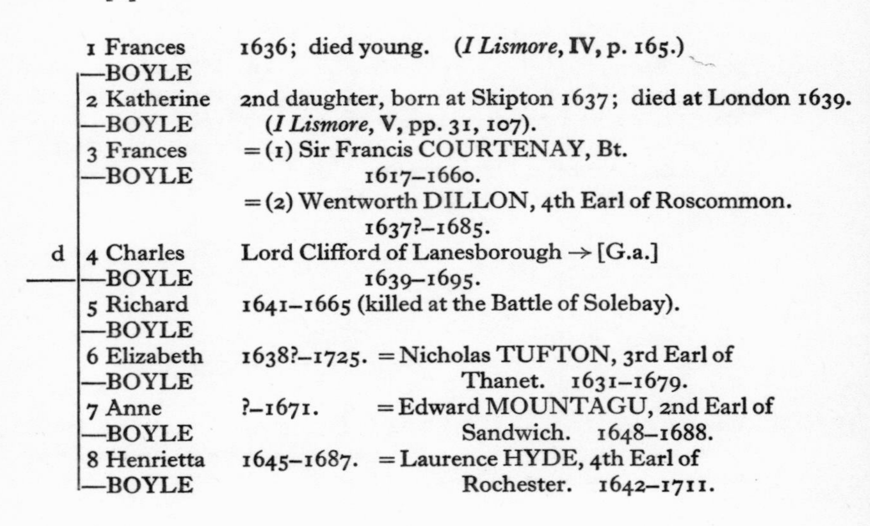

d

1 Frances —BOYLE 1636; died young. (*I Lismore*, IV, p. 165.)

2 Katherine —BOYLE 2nd daughter, born at Skipton 1637; died at London 1639. (*I Lismore*, V, pp. 31, 107).

3 Frances —BOYLE =(1) Sir Francis COURTENAY, Bt. 1617–1660.
=(2) Wentworth DILLON, 4th Earl of Roscommon. 1637?–1685.

4 Charles —BOYLE Lord Clifford of Lanesborough → [G.a.] 1639–1695.

5 Richard —BOYLE 1641–1665 (killed at the Battle of Solebay).

6 Elizabeth —BOYLE 1638?–1725. =Nicholas TUFTON, 3rd Earl of Thanet. 1631–1679.

7 Anne —BOYLE ?–1671. =Edward MOUNTAGU, 2nd Earl of Sandwich. 1648–1688.

8 Henrietta —BOYLE 1645–1687. =Laurence HYDE, 4th Earl of Rochester. 1642–1711.

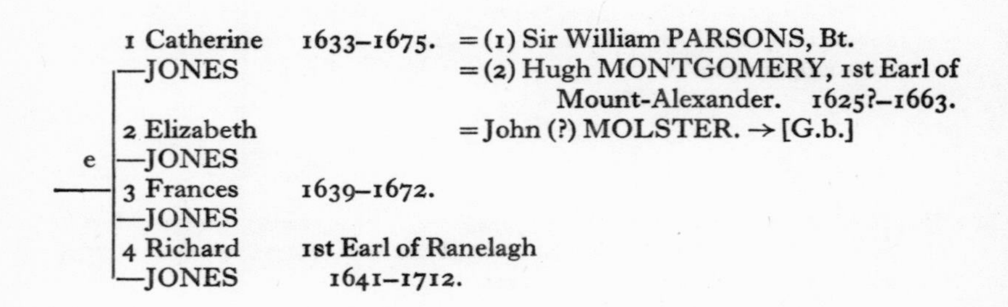

e

1 Catherine —JONES 1633–1675. =(1) Sir William PARSONS, Bt.
=(2) Hugh MONTGOMERY, 1st Earl of Mount-Alexander. 1625?–1663.

2 Elizabeth —JONES =John (?) MOLSTER. → [G.b.]

3 Frances —JONES 1639–1672.

4 Richard —JONES 1st Earl of Ranelagh 1641–1712.

[F.d.3]. Frances Boyle; alive December 1672. She married (1) Sir Francis Courtenay, Bt; born 1617; died 2 March 16$\frac{59}{60}$; and (2) Wentworth Dillon, 4th Earl of Roscommon, April 1662. He was born *ca.* 1637; died 17 January 168$\frac{4}{5}$; buried in Westminster Abbey. He married his second wife, Isabella Boynton, 9 November 1674.

[F.d.4]. Charles Boyle; born in the house of Sir Thomas Stafford in the Savoy, London, 17 November 1639; M.P., F.R.S.; died at Londesborough, Yorkshire, 12 October 1694; buried there. He married (1) Jane, daughter of William Seymour, Duke of Somerset, 7 May 1661. She was born 7 July 1637; died 23 November 1679; buried in Westminster Abbey. He married (2) Arethusa, daughter of George Berkeley, 1st Earl Berkeley, before 12 May 1688. She died 11 February $174\frac{2}{3}$, and was buried at Cranford, Middlesex. (*I Lismore*, **V**, p. 115).

[F.d.5]. Richard Boyle; M.P., for co. Cork; original F.R.S. (T. S. Patterson· " Richard Boyle, Esq., M.P., F.R.S., and—incidentally—some of his Relatives." *Annals of Science*, (1936), **I**, 62: R.C.H.M. 11th Rept. Part VI, p. 85: *Memoirs of Ann, Lady Fanshawe*, London, p. 530: P. du Moulin (the younger). *παρεργα Poematum libelli tres*, Cambridge, 1670, which is dedicated to RB, contains an epitaph on Richard Boyle.

[F.d.6]. Elizabeth Boyle; died 1 September 1725, aged 88 years. She married Nicolas Tufton, 3rd Earl of Thanet, 11 April 1664. He was born 7 August 1631; died 24 November 1679.

[F.d.7]. Anne Boyle; died 14 September 1671. She married Edward Mountagu, 2nd Earl of Sandwich. He was born 3 January $164\frac{7}{8}$; died November 1688.

[F.d.8]. Henrietta Boyle; died 12 April 1687, aged 42 years; buried in Westminster Abbey. She married Laurence Hyde, 4th Earl of Rochester, 1665. He was baptized 15 March $164\frac{1}{2}$; died 2 May 1711; buried with his wife.

[F.e.1]. Catherine Jones; born at Athlone Castle 23 December 1633; died at Dublin 8 October 1675. She married (1) Sir William Parsons, Bt.; he died 31 December 1658; and (2) Hugh Montgomery, 1st Earl of Mount-Alexander, 20 October 1660. He was born *ca.* 1625; died 15 September 1663. (*I Lismore*, **III**, p. 224).

[F.e.2]. Elizabeth Jones. Her marriage (sometime about April 1677) to Molster, said to be a footman, was a *mésalliance*. (*Diary of Mary, Countess of Warwick*, B.M., Add. MSS. 27355.) RB's niece by her marriage to a person of inferior social station disappeared from history.

[F.e.3]. Frances Jones; born at Stalbridge 17 August 1639; died 28 March 1672; buried in the church of St. Martin-in-the-Fields, Westminster. Monumental inscription now in the crypt. (*I Lismore*, **V**, p. 102: [E. Hatton.] *A New View of London*, 1708, Vol. I, pp. 342–3.) RB refers to the illness of his niece in one of his papers. (*Phil. Trans.* (1672), Vol. 7, No. 89, pp. 5108–16: *Works Fol.*, **III**, pp. 305–6; *Works*, **III**, p. 654.)

[F.e.4]. Richard Jones; born in the house of his uncle, Viscount Dungarvon, in Long Acre, London, 8 February $164\frac{0}{1}$; pupil of John Milton; travelled abroad with Henry Oldenburg; original F.R.S.; created Earl of Ranelagh, 1677; built Ranelagh House at Chelsea; died 5 January $171\frac{1}{2}$; buried in Westminster Abbey. He married (1) Elizabeth, daughter of Francis Willoughby, 5th Baron Willoughby of Parham, 28 October 1662; she died 1 August 1695; buried in Westminster Abbey; and (2) Margaret, dowager Baroness Stawell, daughter of James Cecil, 3rd Earl of Salisbury, 9 January $169\frac{5}{6}$. (*I Lismore*, **V**, p. 169).

[F]

f

1 Adam —LOFTUS		
2 Adam —LOFTUS	1st Viscount Lisburne. ?–1691.	
3 Richard —LOFTUS		
4 Robert —LOFTUS		
5 Letitia —LOFTUS		
6 Eleanor —LOFTUS		
7 Jane —LOFTUS		

g

1 —a son	?–1642. (*II Lismore*, V, p. 115.)	
2 Roger —BOYLE	2nd Earl of Orrery. → [G.c.] 1646–1682	
3 Henry —BOYLE	?–1693.	
4 Elizabeth —BOYLE	?–1709.	= Folliott WINGFIELD, 2nd Viscount Powerscourt. 1642 1718.
5 Anne —BOYLE	died young.	
6 Margaret —BOYLE	?–1683.	= William O'BRIEN, 2nd Earl of Inchiquin *ca.* 1640–1692.
7 Katherine —BOYLE	?–1681.	= Richard BRETT.
8 Barbara —BOYLE	?–before 1682.	= Arthur CHICHESTER, 3rd Earl of Donegall 1666–1706.

h

1 Richard —BOYLE	alive in 1679. → [G.d.]	
2 Charles —BOYLE		
3 Elizabeth —BOYLE		= John JEPHSON.

j

1 Elizabeth —RICH	1642; died young.
2 Charles —RICH	1643–1664.

[F.f.2]. Adam Loftus; killed at the siege of Limerick; buried in St. Patrick's Cathedral, Dublin. He married Lucy, daughter of George Brydges, 6th Baron Chandos; she died before 12 July 1689.

[F.g.2]. Roger Boyle; baptized 24 August 1646; Vice-president of Munster, 1669; died 29 March 1682; buried at Youghal. He married Mary, daughter of Richard Sackville, 5th Earl of Dorset, 6 February $166\frac{4}{5}$. She was born 4 February $164\frac{7}{8}$; died 6 November 1710.

[F.g.3]. Henry Boyle; died 1693 in Flanders. He married Mary, daughter of Murrough O'Brien, 1st Earl of Inchiquin.

[F.g.4]. Elizabeth Boyle; died 17 October 1709; buried in St. Patrick's Cathedral, Dublin. She married, before 29 September 1660, Folliott Wingfield, 2nd Viscount Powerscourt, whilst he was in the wardship of Roger, 1st Earl of Orrery. He was baptized 2 November 1642; buried 17 February $171\frac{7}{8}$ in St. Patrick's Cathedral, Dublin.

[F.g.6]. Margaret Boyle; buried in the church of St. Martin-in-the-Fields, London, 27 December 1683. She married William O'Brien, 2nd Earl of Inchiquin. He was born about 1640; died January $169\frac{1}{2}$ in Jamaica, and was buried there.

[F.g.7]. Catherine Boyle; died 3 September 1681 aged 28; buried at Richmond, Surrey. Monumental inscription in Lodge, Vol. **I**, p. 193. She married Richard Brett of Somersetshire.

[F.g.8]. Barbara Boyle; died 20 November 1682; buried in St. Patrick's Cathedral, Dublin. She married before 1676 Arthur Chichester, 3rd Earl of Donegall. He was born 1666; killed before the fort of Monjuich, Spain, 10 April 1706; buried at Barcelona.

[F.h.1]. Richard Boyle; alive in September 1679, but died in lifetime of his parents. He married Elizabeth, daughter of Sir John Ponsonby, 7 September 1668.

[F.h.2]. Charles Boyle. If this be the nephew mentioned by G. Pierre in his letter dated 13 May 1678 (N.S.), then Charles died before 22 April 1678 (O.S.). (See p. 171).

[F.h.3]. Elizabeth Boyle; married John Jephson, of Moyallow.

[F.j.1]. Elizabeth Rich; born at Warwick House, 1642; died young.

[F.j.2]. Charles Rich; born 28 September 1643; died 16 May 1664 at Leigh; buried at Felsted. He married Anne, daughter of William Cavendish, 3rd Earl of Devonshire, 2 September 1662. She died 18 June 1703.

[F]

k. 1 Mary ?–1641.
——BOYLE
1st.

2 Elizabeth = Denny MUSCHAMP.
—BOYLE
3 Martha = Sir William DAVIS.
—BOYLE ?–1687.
4 Eleanor = William HILL.
—BOYLE
5 Honour = (1) Thomas CROMWELL, Earl of Ardglass.
—BOYLE = (2) Francis CUFFE.
k. 6 Mary died young.
——BOYLE
2nd. 7 Margaret ?–1710. = Rev. Dr. Samuel SYNGE.
—BOYLE ?–1708.
8 Murragh ?–1718. = (1) Mary PARKER.
—BOYLE (2) Anne COOKE.
9 John died young.
—BOYLE
10 Richard died young.
— BOYLE

1 ? Born 14 May 1637. (*I Lismore*, **IV**, p. 7).
m —SMYTH
2 ? Born 15 December 1642. (*I Lismore*, **IV**, p. 217).
—SMYTH One of these is named Boyle Smyth. (Townshend p. 497).

[G]

a — 1st.	1 Richard BOYLE	1666–1675.	
	2 Charles BOYLE	3rd Earl of Cork, 2nd Earl of Burlington. Before 1674–1704.	
	3 Henry BOYLE		
	4 William BOYLE	died young	
	5 Frances BOYLE	died young	
	6 Elizabeth BOYLE	1662–1703.	= James BARRY, 4th Earl of Barrymore. 1667–1748.
	7 Jane BOYLE	died young	
	8 Mary BOYLE	died 1709 aged 39.	= James DOUGLAS, 2nd Duke of Queensbury. 1662–1711.
	9 Arabella BOYLE	?–1750.	= Henry PETTY, Earl of Shelburne. 1675–1751.
	10 Frances BOYLE	died young.	
a — 2nd.	Arethusa BOYLE	= James VERNON.	

[G]

b — Katherine MOLSTER.

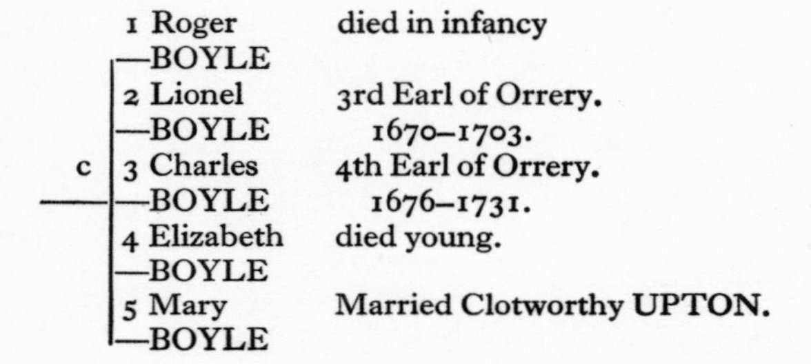

d
- 1 Richard BOYLE — 2nd Viscount Shannon. 1675?–1740.
- 2 Francis BOYLE
- 3 John BOYLE
- 4 Elizabeth BOYLE

[G.b.] Katherine Molster; mentioned only in RB's will. He gave her £100.

[G.c.2]. Lionel Boyle; born 1670; died 24 August 1703, married Mary, illegitimate daughter of Charles Sackville, 6th Earl of Dorset. (See [G.d.1].)

[G.d.1]. Richard Boyle; born 1675?; died 20 December 1740. Married (1) Mary, widow of his cousin, Lionel Boyle, 3rd Earl of Orrery. (See [G.c.2]. She died 26 June 1714.) He married (2) Grace, daughter of John Senhouse.

Relationship of Robert Boyle to Edmund Spenser, the poet

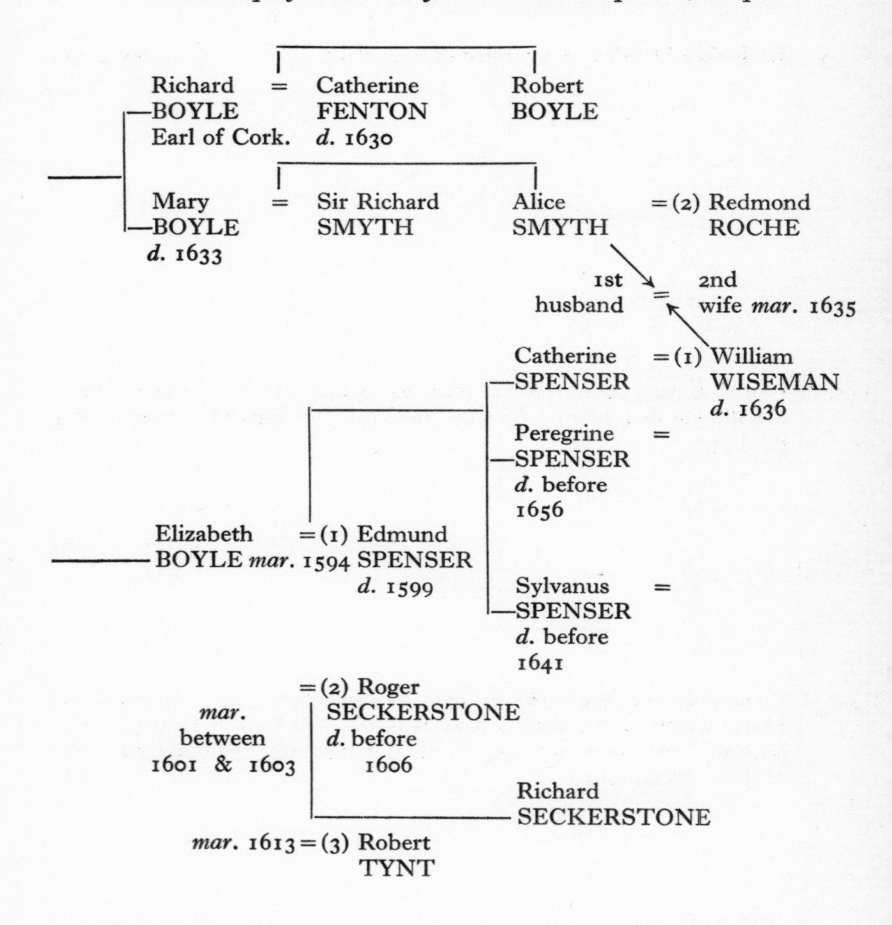

Elizabeth Boyle was probably a cousin of the Earl of Cork.

A. B. Grosart. *The Complete Works . . . of Edmund Spenser*, 9 vols, 1882–1884. Vol. I: *I Lismore*, I, p. xiv *et seq.*

Relationship of Robert Boyle to John Evelyn, the diarist

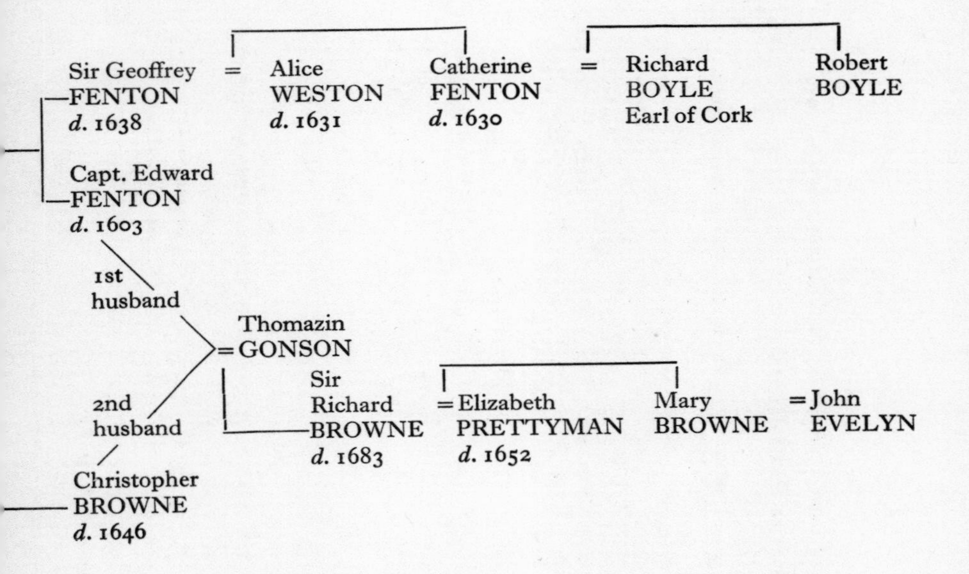

C. G. S. Foljambe. *Evelyn Pedigrees and Memoranda.* London, 1893: *Evelyn.* (de Beer.) Vol. **I,** Tables 7 and 8; Vol. **IV,** p. 304, n. 2.

Relationship of Robert Boyle to the Viscounts Massereene

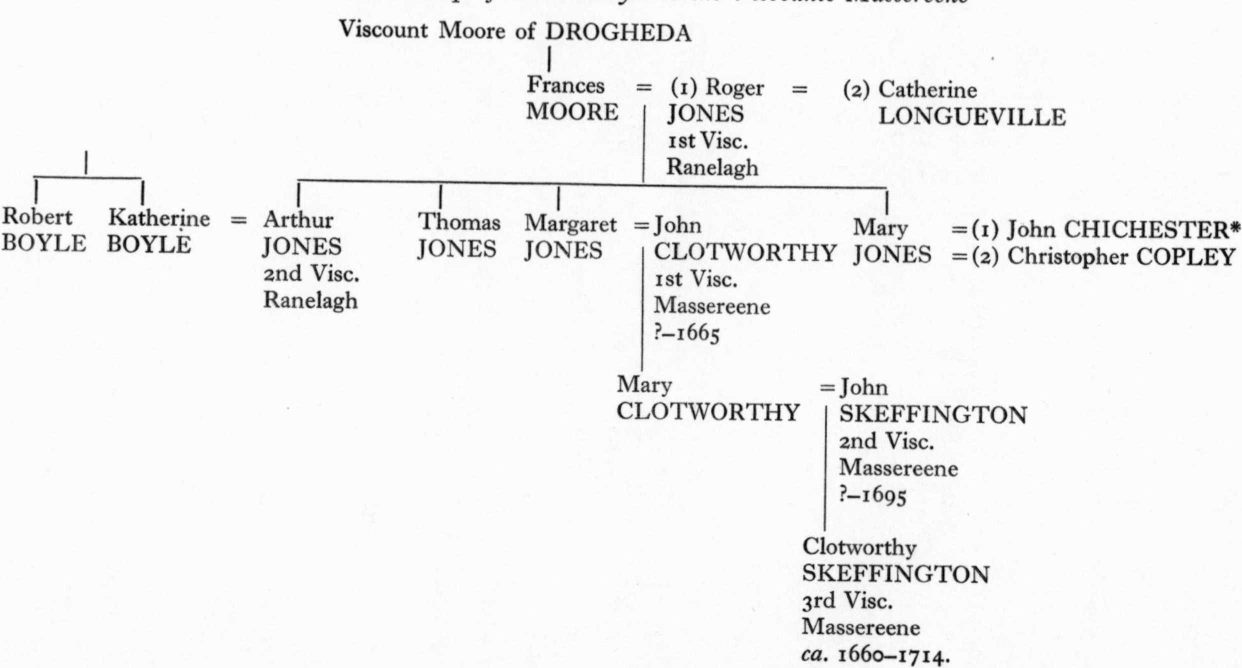

* See next pedigree for continuation. See also note ¶, p. 62.

Relationship of Robert Boyle to Jane Itchingham

(This is an extension of the previous pedigree.)

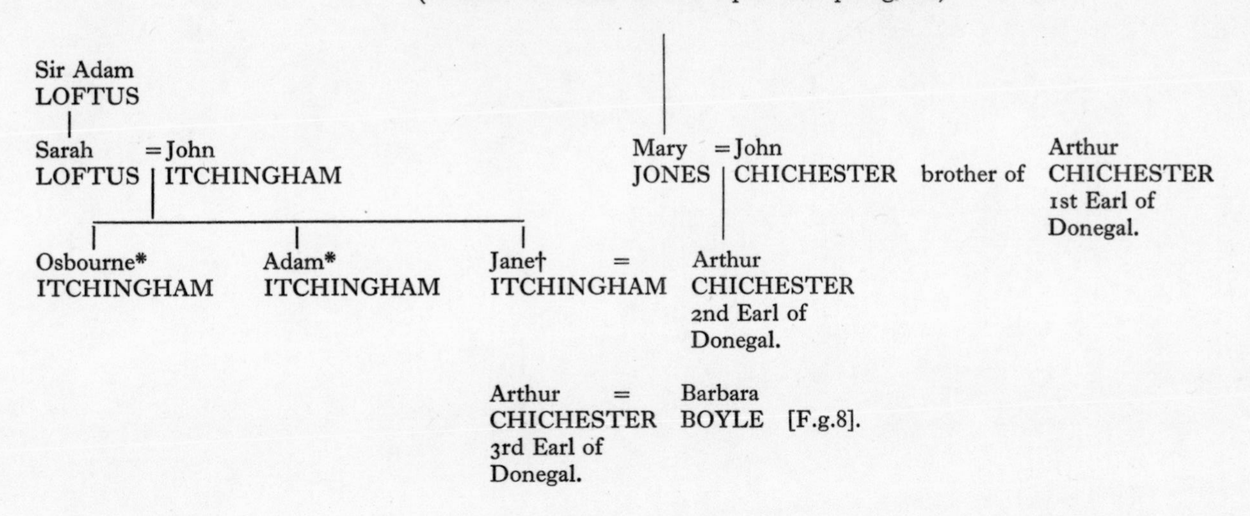

* Died unmarried.

† Married a second time; died before 22 May 1712.

Addenda

p. 3, line 23. & in MS is followed by *da* deleted.

p. 4, line 25. *annually* in MS is followed by *co* deleted.

line 33. *Sport* in MS is followed by /Blank/.

p. 6, line 24. /Blank/ in MS is followed by *aff* deleted.

p. 7, line 11. *Dayes* in MS is followed by *st* deleted.

p. 9, line 11. *seduce* in MS has a final *s* deleted.

p. 15, line 10. *to* in MS is followed by *dr* deleted.

p. 16, line 2. *happy* in MS is followed by *an D* deleted.

line 17. *vomit* in MS is followed by *fr* deleted.

p. 17, note g. Add *fan* after *taught his.*

p. 18, line 8. *auersion* in MS has a final *s* deleted.

p. 20, line 4. *Home* in MS is followed by *w* deleted.

line 25. *Copys* in MS is altered from *Copis.*

p. 21, line 1. *Fate* in MS is followed by *a* deleted.

p. 195. note *. Add. This work is a translation of the spurious *Institutiones et Experimenta Chemiae* (Paris, 1724). In the second and third editions (London, 1741 and 1753 respectively), which are translations of Boerhaave's authentic *Elementa Chemiae* (Leyden, 1732), the passage quoted above on pages 193–195 appears as a footnote. Seeing that the passage in question does not occur in the original authentic Latin version, nor in T. Dallowe's translation thereof (London, 1735), it seems fairly certain that the authorship of this glowing tribute to RB must be ascribed to the translators Shaw and Chambers, the former being the more likely author of it. Nevertheless, in his inaugural oration after being appointed Professor of

Chemistry at the University of Leyden, Boerhaave spoke as follows:

'Boyle, who passed his days in experimenting and spent much money on the promotion of science, courageously took upon himself the task of executing what the great mind of Francis Bacon had conceived. For his work he received a well-deserved reward, and he became famous especially for his work in chemistry. It was thanks to his chemistry that he was useful to the philosophers and of assistance to the physicians, and so he was loved and honoured by both.' (*Sermo academicus de chemia suos errores expurgante . . . XXI Septembris* 1718. Quoted from G. A. Lindeboom, *Herman Boerhaave: The Man and his Work.* London, 1968, pp. 115–116.)

p. 221. note †. Add. *In The Experimental History of Colours* (*Works Fol.* **II,** p. 8; *Works* **I,** p. 673) RB records the case of a lady, who, as a result of a fall, received injuries which caused her to see dazzling, rainbow colours for a space of several weeks. These cases which came to RB's notice recall the similar one of an artist whose vision was disturbed by vividly coloured circles, ascribed by Tyndall to diffraction of light by small particles in the humours of the eye. (J. Tyndall *Six Lectures on Light*, London, 1873, p. 92.)

p. 284. There is no evidence that RB was a member of Trinity College, Oxford.

Index

by M. A. Hennings

Index

PMC COLLEGES
LIBRARY
CHESTER, PENNSYLVANIA

Date Due